T0184502

Communications
in Computer and Information Science   788

Commenced Publication in 2007
Founding and Former Series Editors:
Alfredo Cuzzocrea, Xiaoyong Du, Orhun Kara, Ting Liu, Dominik Ślęzak,
and Xiaokang Yang

More information about this series at http://www.springer.com/series/7899

Azlinah Mohamed · Michael W. Berry
Bee Wah Yap (Eds.)

Soft Computing in Data Science

Third International Conference, SCDS 2017
Yogyakarta, Indonesia, November 27–28, 2017
Proceedings

 Springer

Editors
Azlinah Mohamed
Universiti Teknologi MARA
Shah Alam, Selangor
Malaysia

Bee Wah Yap
Universiti Teknologi MARA
Shah Alam, Selangor
Malaysia

Michael W. Berry
University of Tennessee at Knoxville
Knoxville, TN
USA

ISSN 1865-0929   ISSN 1865-0937   (electronic)
Communications in Computer and Information Science
ISBN 978-981-10-7241-3   ISBN 978-981-10-7242-0   (eBook)
https://doi.org/10.1007/978-981-10-7242-0

Library of Congress Control Number: 2017959596

Printed on acid-free paper

This Springer imprint is published by Springer Nature
The registered company is Springer Nature Singapore Pte Ltd.
The registered company address is: 152 Beach Road, #21-01/04 Gateway East, Singapore 189721, Singapore

Preface

We are pleased to present the proceedings of the Third International Conference on Soft Computing in Data Science 2017 (SCDS 2017). SCDS 2017 was held in the Royal Ambarrukmo Hotel in Yogyakarta, Indonesia, during November 27–28, 2017. The theme of the conference was "Science in Analytics: Harnessing Data and Simplifying Solutions." Data science can improve corporate decision-making and performance, personalize medicine and health-care services, and improve the efficiency and performance of organizations. Data science and analytics play an important role in various disciplines including business, medical and health informatics, social sciences, manufacturing, economics, accounting, and finance.

SCDS 2017 provided a platform for discussions on leading-edge methods and also addressed challenges, problems, and issues in machine learning in data science and analytics. The role of machine learning in data science and analytics is significantly increasing in every field from engineering to life sciences and with advanced computer algorithms, solutions for complex real problems can be simplified. For the advancement of society in the twenty-first century, there is a need to transfer knowledge and technology to industrial applications to solve real-world problems. Research collaborations between academia and industry can lead to the advancement of useful analytics and computing applications to facilitate real-time insights and solutions.

We were delighted this year to collaborate with Universitas Gadjah Mada, and this has increased the submissions from a diverse group of national and international researchers. We received 68 submissions, among which 26 were accepted. SCDS 2017 utilized a double-blind review procedure. All accepted submissions were assigned to at least three independent reviewers (at least one international reviewers) in order to have a rigorous and convincing evaluation process. A total of 49 international and 65 local reviewers were involved in the review process. The conference proceeding volume editors and Springer's CCIS Editorial Board made the final decisions on acceptance, with 26 of the 68 submissions (38%) being published in the conference proceedings.

We would like to thank the authors who submitted manuscripts to SCDS 2017. We thank the reviewers for voluntarily spending time to review the papers. We thank all conference committee members for their tremendous time, ideas, and efforts in ensuring the success of SCDS 2017. We also wish to thank the Springer CCIS Editorial Board, organizations, and sponsors for their continuous support.

We sincerely hope that SCDS 2017 provided a venue for knowledge sharing, publication of good research findings, and new research collaborations. Last but not least, we hope everyone benefited from the keynote, special, and parallel sessions, and had an enjoyable and memorable experience at SCDS 2017 and in Yogyakarta, Indonesia.

November 2017

Azlinah Mohamed
Michael W. Berry
Bee Wah Yap

Organization

Patron

Hassan Said   Universiti Teknologi MARA, Malaysia

Honorary Chairs

Azlinah Mohamed	Universiti Teknologi MARA, Malaysia
Agus Harjoko	Universitas Gadjah Mada, Indonesia
Michael W. Berry	University of Tennessee, USA
Yasmin Mahmood	Malaysia Digital Economy Corporation, Malaysia
Lotfi A. Zadeh (posthumous)	University of California, Berkeley, USA
Fazel Famili	University of Ottawa, Canada

Conference Chairs

Bee Wah Yap	Universiti Teknologi MARA, Malaysia
Sri Hartati	Universitas Gadjah Mada, Indonesia

Program Committee

Sharifah Aliman	Universiti Teknologi MARA, Malaysia
Shuzlina Abdul Rahman	Universiti Teknologi MARA, Malaysia
Mohamad Asyraf Abdul Latif	Universiti Teknologi MARA, Malaysia
Nur Atiqah Sia Abdullah	Universiti Teknologi MARA, Malaysia
Isna Alfi Bustoni	Universitas Gadjah Mada, Indonesia
Andi Dharmawan	Universitas Gadjah Mada, Indonesia
Aufaclav Z. K. Frisky	Universitas Gadjah Mada, Indonesia
Nurzeatul Hamimah Abdul Hamid	Universiti Teknologi MARA, Malaysia
Guntur Budi Herwanto	Universitas Gadjah Mada, Indonesia
Roghib M. Hujja	Universitas Gadjah Mada, Indonesia
Zainura Idrus	Universiti Teknologi MARA, Malaysia
Venoruton Budi Irawan	Universitas Gadjah Mada, Indonesia
Faizal Makhruz	Universitas Gadjah Mada, Indonesia
Norizan Mat Diah	Universiti Teknologi MARA, Malaysia
Nur Huda Nabihan Md Shahri	Universiti Teknologi MARA, Malaysia
Nur Aziean Mohd Idris	Universiti Teknologi MARA, Malaysia
Siti Shaliza Mohd Khairy	Universiti Teknologi MARA, Malaysia

Aina Musdholifah   Universitas Gadjah Mada, Indonesia
Sofianita Mutalib   Universiti Teknologi MARA, Malaysia
Nuru'l-'Izzah Othman   Universiti Teknologi MARA, Malaysia
Paholo Iman Prakoso   Universitas Gadjah Mada, Indonesia
Sugeng Rahardjaa   Universitas Gadjah Mada, Indonesia
Mardhani Riasetiawan   Universitas Gadjah Mada, Indonesia
Marshima Mohd Rosli   Universiti Teknologi MARA, Malaysia
Azhari S. N.   Universitas Gadjah Mada, Indonesia
Faisal Saeed   Universiti Teknologi Malaysia, Malaysia
Anny Kartika Sari   Universitas Gadjah Mada, Indonesia
Nazrul Azha   Universiti Teknologi MARA, Malaysia
  Mohamed Shaari
Shuhaida Shuhidan   Universiti Teknologi MARA, Malaysia
Suhartono   Institut Teknologi Sepuluh Nopember, Indonesia
Shanti Sukmajaya   Universitas Gadjah Mada, Indonesia
Triyogatama Wahyu   Universitas Gadjah Mada, Indonesia
  Widodo'
Ezzatul Akmal   Universiti Teknologi MARA, Malaysia
  Kamaru Zaman

International Scientific Committee

Adel Al-Jumaily   University of Technology Sydney, Australia
Azlinah Mohamed   Universiti Teknologi MARA, Malaysia
Siti Zaleha Zainal Abidin   Universiti Teknologi MARA, Malaysia   '
Afiahayati   Universitas Gadjah Mada, Indonesia
Tahir Ahmad   Universiti Teknologi Malaysia, Malaysia
Dhiya Al-Jumeily   Liverpool John Moores University, UK
Khairul Anam   University of Jember, Indonesia
Nordin Abu Bakar   Universiti Teknologi MARA, Malaysia
Mohammed Bennamoun   University of Western Australia, Australia
Min Chen   Oxford University, UK
Simon Fong   University of Macau, Macau, SAR China
Sumanta Guha   Asian University of Technology, Thailand
Sri Hartati   Universitas Gadjah Mada, Indonesia
Dariusz Krol   University of Wroclaw, Poland
Siddhivinayak Kulkarni   Universiti of Ballarat, Australia
Mazani Manaf   Universiti Teknologi MARA, Malaysia
Yasue Mitsukura   Keio University, Japan
Daud Mohamed   Universiti Teknologi MARA, Malaysia
Yusuke Nojima   Osaka Perfecture University, Japan
Ashraf Osman   Alzaiem Alazhari University, Sudan
Jose Maria Pena   Technical University of Madrid, Spain
Sigit Priyanta   Universitas Gadjah Mada, Indonesia
MHD. Reza M. I. Pulungan   Universitas Gadjah Mada, Indonesia
Agus Sihabuddin   Universitas Gadjah Mada, Indonesia

Retantyo Wardoyo	Universitas Gadjah Mada, Indonesia
Richard Weber	University of Chile, Santiago, Chile
Wahyu Wibowo	Insititut Teknologi Sepuluh Nopember, Indonesia
Moh. Edi Wibowo	Universitas Gadjah Mada, Indonesia
Edi Winarko	Universitas Gadjah Mada, Indonesia
Jasni Mohamad Zain	Universiti Teknologi MARA, Malaysia

International Reviewers

Moulay Akhloufi	University of Moncton, Canada
Anas AL-Badareen	Aqaba University of Technology, Jordan
Hanaa Ali	Zagazig University, Egypt
Samaher Al-Janabi	Babylon University, Iraq
Alaa Aljanaby	Aspire2 International, New Zealand
Dhiya Al-Jumeily	Liverpool John Moores University, UK
Karim Al-Saedi	University of Mustansiriyah, Iraq
Khairul Anam	University of Jember, Indonesia
Mohammadali Behrang	FARAB Co., Iran
Michael Berry	University of Tennessee, USA
Neelanjan Bhowmik	University Paris-Est, France
Rodrigo Campos Bortoletto	São Paulo Federal Institute of Education, S&T, Brazil
Ranjan Dash	College of Engineering and Technology, India
Diego De Abreu	Universidade Federal do Pará, Brazil
Ashutosh Dubey	Trinity Inst. Technology and Research Bhopal, India
Noriko Etani	Kyoto University, Japan
Ensar Gul	Istanbul Sehir University, Turkey
Rohit Gupta	Thapar University, India
Albert Guvenis	Bogazici University, Turkey
Ali Hakam	UAE University, United Arab Emirates
Sri Hartati	Gadjah Mada University, Indonesia
Nikisha Jariwala	VNSGU, India
Harihar Kalia	Seemanta Engineering College, India
S. Kannadhasan	Tamilnadu Polytechnic College, India
Shadi Khalifa	Queen's University, Canada
Harish Kumar	King Khalid University, India
Raghvendra Kumar	LNCT Group of College, India
Bidyut Mahato	ISM Dhanbad, India
Amitkumar Manekar	KL University, India
Norian Marranghello	São Paulo State University – UNESP, Brazil
Somya Mohanty	University of North Carolina at Greensboro, USA
Ayan Mondal	Indian Institute of Technology, Kharagpur, India
Arif Muntasa	Trunojoyo University, Indonesia
Gowrishankar Nath	Ambedkar Institute of Technology, India
Tousif Khan Nizami	Indian Institute of Technology Guwahati, India
Tri Priyambodo	Universitas Gadjah Mada, Indonesia
Sigit Priyanta	Universitas Gadjah Mada, Indonesia

Pakawan Pugsee	Chulalongkorn University, Thailand
Ali Qusay Al-Faris	University of the People, USA
Pramod K. B. Rangaiah	Jain University, India
Vinay Sardana	Abu Dhabi Transmission and Despatch Company, UAE
Krung Sinapiromsaran	Chulalongkorn University, Thailand
Siripurapu Sridhar	LENDI Institute of Engineering and Technology, India
Suhartono Suhartono	Institut Teknologi Sepuluh Nopember, Indonesia
Deepti Theng	G.H. Raisoni College of Engineering, India
J. Vimala Jayakumar	Alagappa University, India
Yashwanth Sai Reddy Vyza	TU Delft, The Netherlands
Wahyu Wibowo	Institut Teknologi Sepuluh Nopember, Indonesia
Ke Xu	AT&T Labs, USA

Local Reviewers

Rafikha Aliana A. Raof	Universiti Malaysia Perlis, Malaysia
Ahmad Shahrizan Abdul Ghani	Universiti Malaysia Pahang, Malaysia
Nur Atiqah Sia Abdullah	Universiti Teknologi MARA, Malaysia
Shuzlina Abdul-Rahman	Universiti Teknologi MARA, Malaysia
Asmala Ahmad	Universiti Teknikal Malaysia Melaka, Malaysia
Azlin Ahmad	Universiti Teknologi MARA, Malaysia
Suzana Ahmad	Universiti Teknologi MARA, Malaysia
Tahir Ahmad	Universiti Teknologi Malaysia, Malaysia
Nor Azizah Ali	Universiti Teknologi Malaysia, Malaysia
Sharifah Aliman	Universiti Teknologi MARA, Malaysia
Muthukkaruppan Annamalai	Universiti Teknologi MARA, Malaysia
Zalilah Abd Aziz	Universiti Teknologi MARA, Malaysia
Azilawati Azizan	Universiti Teknologi MARA, Malaysia
Nordin Abu Bakar	Universiti Teknologi MARA, Malaysia
Sumarni Abu Bakar	Universiti Teknologi MARA, Malaysia
Bee Wah Yap	Universiti Teknologi MARA, Malaysia
Muhammad Hafidz Fazli Bin Md Fauadi	Universiti Teknikal Malaysia Melaka, Malaysia
Bong Chih How	Universiti Malaysia Sarawak, Malaysia
Yun Huoy Choo	Universiti Teknikal Malaysia Melaka, Malaysia
Bhagwan Das	Universiti Tun Hussein Onn Malaysia, Malaysia
Norizan Mat Diah	Universiti Teknologi MARA, Malaysia
Soo Fen Fam	Universiti Teknikal Malaysia Melaka, Malaysia
Nurzeatul Hamimah Abdul Hamid	Universiti Teknologi MARA, Malaysia
Mashitoh Hashim	Universiti Pendidikan Sultan Idris, Malaysia
Zaidah Ibrahim	Universiti Teknologi MARA, Malaysia
Zainura Idrus	Universiti Teknologi MARA, Malaysia

Organizers

Organized by

Hosted by

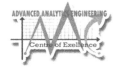

In Co-operation with

Sponsors

Contents

Data Visualization

Fuzzy Logic

Prediction Models and E-Learning

Text and Sentiment Analytics

Deep Learning and Real-Time Classification

Evaluation of Randomized Variable Translation Wavelet Neural Networks

Khairul Anam[1]([⊠]) and Adel Al-Jumaily[2]

[1] University of Jember, Jember 68121, Indonesia
khairul@unej.ac.id
[2] University of Technology - Sydney, Sydney, NSW 2007, Australia
Adel.Al-Jumaily@uts.edu.au

Abstract. A variable translation wavelet neural network (VT-WNN) is a type of wavelet neural network that is able to adapt to the changes in the input. Different learning algorithms have been proposed such as backpropagation and hybrid wavelet-particle swarm optimization. However, most of them are time costly. This paper proposed a new learning mechanism for VT-WNN using random weights. To validate the performance of randomized VT-WNN, several experiments using benchmark data form UCI machine learning datasets were conducted. The experimental results show that RVT-WNN can work on a broad range of applications from the small size up to the large size with comparable performance to other well-known classifiers.

Keywords: Wavelet · Neural network · Random weight

1 Introduction

A wavelet neural networks (WNNs) is a type of artificial neural networks that incorporate wavelet theory in the networks as the activation function [19]. WNN has been applied to many applications [6], such as groundwater lever forecasting [1], electricity price forecasting [13], nonlinear time-series modelling [11], monthly rainfall prediction [14], transportation control system [7], breast cancer recognition [16] and myoelectric pattern recognition [17]. Although involving wavelet theory, WNN still utilizes same learning algorithm as an ordinary WNN i.e. backpropagation [19].

To improve the performance of WNN, Ling et al. [12] proposed variable translation wavelet neural networks (VTWNNs). VTWNM is a type of WNN in which the translation parameter of the wavelet function is varied according to the variation of input. To train the weight of VTWNN, Ling et al. [12] implemented a hybrid particle swarm optimization wavelet mutation (HPSOWM). The experimental results showed that VTWNN performed better than WNN and feed-forward neural network. However, the training procedure of the VTWNN using the extension of PSO (HPSOWM) is complex and time-consuming.

To overcome time-consuming process, the idea of random weights has emerged since decades. The earlier work that considers a constant random weight in the hidden layer was proposed by [15]. They examined the performance of single layer feed-forward neural networks (SLFN) whose hidden weights were determined

© Springer Nature Singapore Pte Ltd. 2017
A. Mohamed et al. (Eds.): SCDS 2017, CCIS 788, pp. 3–12, 2017.
https://doi.org/10.1007/978-981-10-7242-0_1

randomly and kept fixed. Meanwhile, the weights of the output layers were optimized and calculated numerically using a fisher method. The main idea in the fisher method is a numeric calculation of the invers of the hidden output multiplied by the vector of the target. They found the output weights were significantly more importance than the hidden layer weights. In other words, the fixed random hidden layer weight does not influence the performance of the system as long the output weight optimized properly.

Similar to Schmidt et al. [15], Huang et al. [9] optimized the calculation the output weight by putting constraints. This learning mechanism was called as extreme learning machine (ELM). ELM has been tested in various implementation such as classification [3], regression [10] and clustering [8]. Another algorithm for random weight is random vector functional link (RVFL) [4]. More survey on the randomized neural network can be found in [18].

Based on this fact, this paper extends the theory VTWNM using random weight. Instead of using backpropagation or HPSOWM, the VTWNM selects random hidden layer weights and calculate the output weights using least-square optimization. The efficacy of the proposed system will be tested on various benchmark data collected from UCL machine learning repository [5]. The idea to implement random weight in VTWNN was conducted by [2]. However, the model was tested on the case of myo-electric patter recognition only. The general evaluation is necessary to verify its efficacy for the random VTWNN for general application.

2   Variable Translation Wavelet Neural Networks (VTWNNs)

Figure 1 shows the reconstruction of VTWNN model proposed by [12]. The output function of the model for arbitrary samples $(x_k, t_k) \in R^n \times R^o$ with M hidden nodes is

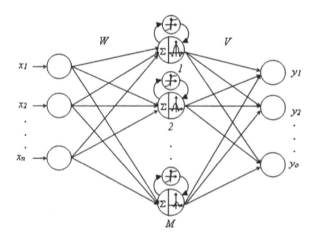

Fig. 1. The variable translation WNN (VTWNN)

$$f_i^k(\mathbf{x}) = \sum_{j=1}^{M} V_{ij}\psi_{a_jbj}(w_j, c_j, \mathbf{x}_k) \qquad i = 1, 2, \ldots, O \tag{1}$$

where

$$\psi_{a_jbj}(x) = \frac{1}{\sqrt{a_j}}\psi\left(\frac{x - b_j}{a_j}\right), \qquad j = 1, 2, \ldots, M \tag{2}$$

In the Eq. (2), a_j and b_j are dilatation and translation parameters of the wavelet, respectively. The input of the hidden layer H_j when M is the number of hidden node and N is the number of input is given by:

$$H_j(x) = \sum_{i=1}^{N} x_i.w_{ji} + c_j \qquad j = 1, 2, \ldots, M \tag{3}$$

where x_i are the input variables and w_{ji} are the weights of the connection between ith input and jth hidden nodes. Meanwhile c_j denotes the bias of jth hidden layer. From Eq. (3), the output of the hidden node can be calculated using:

$$\psi_{a_jb_j}(H_j(x)) = \psi\left(\frac{H_j(x) - b_j}{a_j}\right), \qquad j = 1, 2, \ldots, M \tag{4}$$

In the VTWNN model proposed by Ling et al. [12], the dilatation parameter a_j is equal to j, so:

$$\psi_{a_jb_j}(H_j(x)) = \psi\left(\frac{H_j(x) - b_j}{j}\right), \qquad j = 1, 2, \ldots, M \tag{5}$$

In VTWNN or WNN in general, the mother wavelet $\psi_{a_ib_i}$ can be in different model such as Mexican Hat function [12], as described in Fig. 2.

Following Ling et al. [12], this paper selects this function as the mother wavelet $\psi_{a_ib_i}$. It is defined as

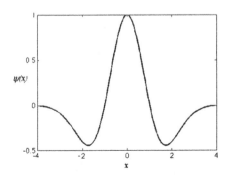

Fig. 2. The mother wavelet of the Mexican hat

$$\psi(x) = e^{-x^2/2}(1-x^2) \tag{6}$$

Finally, the output of the activation function of VTWNN is:

$$\psi_{a_jb_j}(H_j) = e^{-0.5\left(\frac{H_j-b_j}{j}\right)^2}\left(1-\left(\frac{H_j-b_j}{j}\right)^2\right) \tag{7}$$

According to its name, VTWNN varies the value of the translation parameters b_j according to the input information. It is driven by a nonlinear function as shown in Fig. 3 and defined by:

$$b_j = f(P_j) = \frac{2}{1+e^{-P_j}} - 1 \tag{8}$$

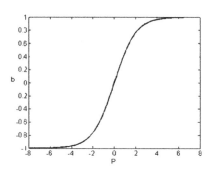

Fig. 3. A sigmoid function to calculate b_j

3   Randomized VTWNN

Different from VTWNN developed in [12], this paper applies random weight for the hidden layer and then calculates the output weight. Because the weights of the hidden layer are set random, the output of the hidden layer can be considered as a constant. Therefore Eq. (1) can be can be written as a linear system:

$$\mathbf{F=VQ} \tag{9}$$

where V is the output weight and Q is the output if hidden layer as described below.

$$\mathbf{Q} = \begin{bmatrix} \psi_{a_1b_1}(H_1(\mathbf{x}_1)) & \cdots & \psi_{a_Mb_M}(H_M(\mathbf{x}_1)) \\ \vdots & \vdots & \vdots \\ \psi_{a_1b_1}(H_1(\mathbf{x}_L)) & \vdots & \psi_{a_Mb_M}(H_M(\mathbf{x}_L)) \end{bmatrix}_{LxM} \tag{10}$$

$$V = \begin{pmatrix} v_1^T & v_2^T & \cdots & v_O^T \end{pmatrix}_{MxO}^T \tag{11}$$

Assume, desired output is defined as:

$$\mathbf{Z} = (z_1^T \quad z_2^T \quad \cdots \quad z_L^T)_{LxO} \tag{12}$$

Then, the learning mechanism can be simplified by:

$$\mathbf{VQ} = \mathbf{Z} \tag{13}$$

The aim of learning is to calculate the output weight as follow:

$$\hat{\mathbf{V}} = \mathbf{Q}^\dagger \mathbf{Z} \tag{14}$$

where $\mathbf{Q}^\dagger$ is a pseudoinverse of $\mathbf{Q}$. Finally, the training algorithm of the randomized VTWNN is presented in (Fig. 4).

```
Begin
   Load input x, target z
   Randomize hidden weights w and bias c
   Calculate input hidden layer H          // Eq. (3)
   Calculate translation parameter b        // Eq. (8)
   Calculate output hidden layer Q          // Eq. (10)
   Calculate the output weights V           // Eq. (14)
end
```

Fig. 4. Pseudo code for randomized VTWNN

4 Performance Evaluation

The evaluation of the randomized VTW performances was conducted by using benchmark datasets that are available online from the UCI machine learning website [5]. The selected datasets are presented in Table 1. The experiment was conducted based on the size of the data. For small and medium size data, the experiments will be performed using 5-fold cross validation for small and medium size data. Meanwhile, a 3-fold cross validation will be applied to the large size data. Special for large size data,

Table 1. Data specification for benchmarking

Dataset	Group	# data	# features	#classes
Iris	Small size	150	4	3
Glass		214	9	6
Vehicle	Medium size	846	18	4
Vowel		990	10	11
Satimage	Large size	6435	36	6
Letter		20000	16	26
Shuttle		58000	9	7

there is no randomization on the data. Instead, the data was just divided into three groups and then validated using cross-validation technique.

This experiment considered seven different classifiers: randomized WNN (RWNN), randomized VT-WNN (RVT-WNN), randomized single layer feedforward neural networks (RSLFNs), radial basis extreme learning machine (RBF-ELM), support vector machine from libsim (LIBSVM), linear discriminant analysis (LDA) and k-nearest neighbour (kNN).

Before conducting the experiment, the optimal parameters of the classifiers were found. For instance, LIBSVM with radial basis kernel relies on parameter regulation C and gamma. A grid-search method is used to select the optimum number of nodes in the node based ELM. As for C and gamma in RBF-ELM and LIBSVM, the optimal parameters were taken from [10] due to the similarity in data and classifiers. Table 2 provides all parameters used in the experiment (Table 3).

Table 2. The optimal parameters used by each classifier in the UCI dataset experiments

Dataset	# Nodes			RBF-ELM		LIBSVM		kNN
	RWNN	RVT-WNN	RSLFN	C	gamma	C	gamma	k
Iris	30	30	20	1	1	1	0.25	10
Glass	30	30	20	32	1	1	0.25	10
Vehicle	190	170	210	64	8	2^{14}	4	10
Vowel	290	350	440	32	0.5	1	1	10
Letter	970	980	920	8	0.25	2^{10}	2^{-4}	10
Satimage	740	640	770	16	0.25	1	1	10
Shuttle	900	500	970	2^{20}	2^{-10}	2^{10}	0.25	10

Table 3. The accuracy of seven classifiers on various data using 5-fold cross validation for small and medium size data and 3-fold cross validation for large size data

Dataset	Accuracy (%)						
	RWNN	RVT-WNN	RSLFN	RBF-ELM	LIBSVM	kNN	RBF-ELM
Iris	96.00	96.67	96.67	96.67	96.67	98.00	96.00
Glass	65.03	65.38	66.29	69.23	63.48	57.50	63.98
Vehicle	82.50	81.68	80.61	84.17	71.51	78.47	70.21
Vowel	94.44	93.54	93.84	94.65	91.21	60.81	84.75
Satimage	87.35	87.26	87.99	90.57	89.91	82.70	88.66
Letter	62.78	62.65	61.99	69.96	46.56	33.01	67.31
Shuttle	99.74	99.57	99.72	99.90	98.59	85.30	99.81

In addition, Table 4 shows that RVT-WNN could perform moderately across seven different datasets. The comparison of RVT-WNN and RSLFN shows that both classifiers are comparable. One-way ANOVA test shown in Table 5 explains that the accuracy gap between them is not significant ($p > 0.05$). It means that they are

Table 4. The accuracy of seven classifiers on various data using 5-fold cross validation for small and medium size data and 3-fold cross validation for large size data

Dataset	Accuracy (%)						
	RWNN	RVT-WNN	RSLFN	RBF-ELM	LIBSVM	kNN	RBF-ELM
Iris	96.00	96.67	96.67	96.67	96.67	98.00	96.00
Glass	65.03	65.38	66.29	69.23	63.48	57.50	63.98
Vehicle	82.50	81.68	80.61	84.17	71.51	78.47	70.21
Vowel	94.44	93.54	93.84	94.65	91.21	60.81	84.75
Satimage	87.35	87.26	87.99	90.57	89.91	82.70	88.66
Letter	62.78	62.65	61.99	69.96	46.56	33.01	67.31
Shuttle	99.74	99.57	99.72	99.90	98.59	85.30	99.81

comparable. As for RSLFN, the accuracy of RVT-WNN is significantly better RSLFN only in "Letter" dataset while the rest of them are not significantly different ($p < 0.05$).

Furthermore, Tables 4 and 5 indicate that RBF-ELM is the most accurate classifier across seven datasets except the "Iris" dataset. Nevertheless, the accuracy of RVT-WNN is comparable to RBF-ELM in all datasets ($p > 0.05$) except in "Letter" dataset ($p < 0.05$). Moreover, RVT-WNN has the same performance as LIBSVM except on Vehicle and Letter datasets in which RVT-WNN is better. For the rest of the classifiers, Tables 4 and 5 show that RVT-WNN is significantly better than LDA in two datasets: Vowel and Letter. Moreover, RVT-WNN is better than kNN in Vowel dataset while kNN is better than RVT-WNN in Letter datasets ($p < 0.05$). Overall, RVT-WNN is comparable to RBF-ELM and LIBSVM and slightly better than LDA and kNN in most datasets.

Table 5. One way ANOVA test results on the comparison of RVT-WNN and other classifiers (the black box shows $p < 0.05$)

RVT-WNN -->	RWNN	RSLFN	RBF-ELM	LIBSVM	LDA	kNN
				p-value		
Iris	0.771	1.000	1.000	1.000	0.524	0.809
Glass	0.942	0.835	0.377	0.682	0.075	0.755
Vehicle	0.585	0.424	0.080	0.000	0.121	0.000
Vowel	0.432	0.807	0.354	0.064	0.000	0.005
Satimage	0.999	0.980	0.922	0.934	0.885	0.966
Letter	0.765	0.043	0.000	0.000	0.000	0.000
Shuttle	0.996	0.997	0.993	0.980	0.679	0.995

In addition to the classification performance, the processing time of classifiers is one of the discussion objectives. Table 6 presents the training time while Table 7 provides the testing time. Table 6 shows that the training time of RVT-WNN is one of the slowest classifiers, compared to other classifiers over all datasets. It becomes worse when RVT-WNN works on big data like "Letter" dataset. The RVT-WNN is the slowest classifier taking around 40 s to learn Letter datasets. However, an anomaly occurred when RVT-WNN worked on Shuttle datasets. RVT-WNN took around 33 s,

Table 6. The training time of seven classifiers on various data using 5-fold cross validation for small and medium size data and 3-fold cross validation for large size data

Dataset	Training time (ms)						
	RWNN	RVT-WNN	RSLFN	RBF-ELM	LIBSVM	LDA	kNN
Iris	45.20	48.33	41.33	3.51	0.00	15.61	38.53
Glass	69.53	68.40	31.93	1.81	2.00	1.83	9.62
Vehicle	403.13	353.67	438.13	16.03	56.00	2.76	40.11
Vowel	776.00	1,014.80	1,334.67	20.95	30.00	4.82	28.72
Satimage	6,412.22	6,236.67	5,436.33	562.04	723.33	49.07	799.55
Letter	29,202.20	40,664.33	17,541.87	11,442.31	20,140.00	72.39	4,618.73
Shuttle	62,114.33	33,069.22	47,162.44	123,161.44	10,820.00	49.54	1,472.41

Table 7. The testing time of seven classifiers on various data using 5-fold cross validation for small and medium size data and 3-fold cross validation for large size data

Dataset	Testing time (ms)						
	RWNN	RVT-WNN	RSLFN	RBF-ELM	LIBSVM	LDA	kNN
Iris	10.00	16.27	6.47	0.99	0.00	1.74	6.28
Glass	10.73	17.20	6.60	0.46	2.00	1.47	5.80
Vehicle	38.07	46.00	28.07	2.47	56.00	1.88	9.88
Vowel	51.40	72.67	34.47	3.69	30.00	2.93	12.43
Satimage	709.00	1,082.00	134.67	112.45	723.33	14.85	375.33
Letter	2,391.53	4,751.87	276.87	873.45	20,140.00	28.72	1,177.10
Shuttle	11,241.89	8,536.11	1,001.22	10,101.54	10,820.00	36.02	786.39

faster than RWNN, RSLFN even much more quickly than RBF-ELM that took 123 s. Similar results happen on the testing time. This happened because RVT-WNN used a lower number of nodes than RSLFN or RVT-WNN when working on the Shuttle dataset.

One thing that is not normal on LIBSVM; the training time and the processing time of LIBSVM is the same while the other classifiers took a shorter testing time than training time. The fastest classifier is LDA, which took only around 72 ms on Letter dataset and about 49 ms on Shuttle dataset. Overall, the training time of RVT-WNN is slow, but it can be compensated by using less number of nodes with comparable performance to other classifiers.

5 Discussion and Conclusion

RVT-WNN has been implemented in wide range classification problems using UCI machine learning datasets. The experimental results show that RVT-WNN could work on a wide range of datasets from small size to large size data. RVT-WNN is comparable to RWNN in all datasets, and it is better than RSLFN on the Letter dataset.

Moreover, RVT-WNN attained better accuracy than LDA in Vowel and Letter datasets while it shows the same performance as LDA with the other datasets. Comparison with RBF-ELM indicates that RVT-WNN is comparable to it except on the Letter dataset. On this dataset, RBF-ELM is better than RVT-WNN.

Other results show that RVT-WNN has the same performance as LIBSVM except on Vehicle and Letter datasets in which RVT-WNN is better. Finally, the comparison with kNN indicates that RVT-WNN is better than kNN on Vowel dataset while kNN is better than RVT-WNN on Letter dataset. Overall, RVT-WNN is a promising classifier for many classification applications.

References

1. Adamowski, J., Chan, H.F.: A wavelet neural network conjunction model for groundwater level forecasting. J. Hydrol. **407**, 28–40 (2011)
2. Anam, K., Al-Jumaily, A.: Adaptive wavelet extreme learning machine (AW-ELM) for index finger recognition using two-channel electromyography. In: Loo, C.K., Yap, K.S., Wong, K.W., Teoh, A., Huang, K. (eds.) ICONIP 2014. LNCS, vol. 8834, pp. 471–478. Springer, Cham (2014). https://doi.org/10.1007/978-3-319-12637-1_59
3. Anam, K., Al-Jumaily, A.: Evaluation of extreme learning machine for classification of individual and combined finger movements using electromyography on amputees and non-amputees. Neural Netw. **85**, 51–68 (2017)
4. Antuvan, C.W., Bisio, F., Marini, F., et al.: Role of muscle synergies in real-time classification of upper limb motions using extreme learning machines. J. Neuroeng. Rehabil. **13**, 76 (2016)
5. Asuncion, A., Newman, D.: The UCI Machine Learning Repository (2007)
6. Cao, J., Lin, Z., Huang, G.-B.: Composite function wavelet neural networks with extreme learning machine. Neurocomputing **73**, 1405–1416 (2010)
7. Chen, C.-H.: Intelligent transportation control system design using wavelet neural network and PID-type learning algorithms. Expert Syst. Appl. **38**, 6926–6939 (2011)
8. Huang, G., Song, S., Gupta, J.N., et al.: Semi-supervised and unsupervised extreme learning machines. IEEE Trans. Cybern. **44**, 2405–2417 (2014)
9. Huang, G.-B., Zhu, Q.-Y., Siew, C.-K.: Extreme learning machine: theory and applications. Neurocomputing **70**, 489–501 (2006)
10. Huang, G.B., Zhou, H., Ding, X., et al.: Extreme learning machine for regression and multiclass classification. IEEE Trans. Syst. Man Cybern. Part B (Cybern.) **42**, 513–529 (2012)
11. Inoussa, G., Peng, H., Wu, J.: Nonlinear time series modeling and prediction using functional weights wavelet neural network-based state-dependent AR model. Neurocomputing **86**, 59–74 (2012)
12. Ling, S.H., Iu, H., Leung, F.H.-F., et al.: Improved hybrid particle swarm optimized wavelet neural network for modeling the development of fluid dispensing for electronic packaging. IEEE Trans. Industr. Electron. **55**, 3447–3460 (2008)
13. Pindoriya, N.M., Singh, S.N., Singh, S.K.: An adaptive wavelet neural network-based energy price forecasting in electricity markets. IEEE Trans. Power Syst. **23**, 1423–1432 (2008)
14. Ramana, R.V., Krishna, B., Kumar, S., et al.: Monthly rainfall prediction using wavelet neural network analysis. Water Resources Manag. **27**, 3697–3711 (2013)

15. Schmidt, W.F., Kraaijveld, M.A., Duin, R.P.: Feedforward neural networks with random weights. In: Proceedings of the 11th IAPR International Conference on Pattern Recognition, pp. 1–4. IEEE (1992)
16. Senapati, M.R., Mohanty, A.K., Dash, S., et al.: Local linear wavelet neural network for breast cancer recognition. Neural Comput. Appl. **22**, 125–131 (2013)
17. Subasi, A., Yilmaz, M., Ozcalik, H.R.: Classification of EMG signals using wavelet neural network. J. Neurosci. Methods **156**, 360–367 (2006)
18. Zhang, L., Suganthan, P.N.: A survey of randomized algorithms for training neural networks. Inf. Sci. **364**, 146–155 (2016). %@ 0020-0255
19. Zhou, B., Shi, A., Cai, F., Zhang, Y.: Wavelet neural networks for nonlinear time series analysis. In: Yin, F.-L., Wang, J., Guo, C. (eds.) ISNN 2004. LNCS, vol. 3174, pp. 430–435. Springer, Heidelberg (2004). https://doi.org/10.1007/978-3-540-28648-6_68

Road Lane Segmentation Using Deconvolutional Neural Network

Dwi Prasetyo Adi Nugroho and Mardhani Riasetiawan(✉)

Universitas Gadjah Mada, Yogyakarta, Indonesia
dwi.prasetyo.a@mail.ugm.ac.id, mardhani@ugm.ac.id

Abstract. Lane departure warning (LDW) system attached to modern vehicles is responsible for lowering car accident caused by inappropriate lane changing behaviour. However the success of LDW system depends on how well it define and segment the drivable ego lane. As the development of deep learning methods, the expensive light detection and ranging (LIDAR) guided system is now replaced by analysis of digital images captured by low-cost camera. Numerous method has been applied to address this problem. However, most approach only focusing on achieving segmentation accuracy, while in the real implementation of LDW, computational time is also an importance metric. This research focuses on utilizing deconvolutional neural network to generate accurate road lane segmentation in a realtime fashion. Feature maps from the input image is learned to form a representation. The use of convolution and pooling layer to build the feature map resulting in spatially smaller feature map. Deconvolution and unpooling layer then applied to the feature map to reconstruct it back to its input size. The method used in this research resulting a 98.38% pixel level accuracy and able to predict a single input frame in 28 ms, enabling realtime prediction which is essential for a LDW system.

Keywords: Convolutional Neural Network · Semantic segmentation
Ego lane segmentation · Lane departure warning

1 Introduction

With the development of urban transportation system, safety become one of the most important aspect when developing advanced driving system. According to WHO (2009), road incident is the highest cause of death for people in age 15–29, higher than suicide or HIV/AIDS [1]. This road accident according to IFRCRCS is up to 80–90% caused by bad driving behavior, such as improper lane changing [2].

However, road accident caused by the improper lane changing behavior can be reduced by 11–23% with the implementation of Lane Departure Warning (LDW) system [3]. LDW works by segmenting the drivable lane of the road and guide the driver to follow this lane. Most of the segmentation approach are using LIDAR sensor. However, this sensor is relatively expensive and unable to give precise information of color and texture of the road. On the other side, segmentation can also be done by analyzing image captured by digital camera that is relatively cheaper compared to light detection and ranging (LIDAR) sensor and able to capture road detail in high resolution images.

© Springer Nature Singapore Pte Ltd. 2017
A. Mohamed et al. (Eds.): SCDS 2017, CCIS 788, pp. 13–22, 2017.
https://doi.org/10.1007/978-981-10-7242-0_2

While there are many approach to segment drivable lane using digital image, most of them only focus on getting high accuracy. In the real implementation, computational time is also an important metric. An LDW system should be able to warns the driver immediately when the driver leave their drivable lane improperly. The use of fully convolutional network (FCN) that is first introduced in [4] opens possibility for real time image segmentation. However, the simple upsampling method employed in FCN produce inaccurate reconstruction of segmented image.

2 Related Works

Road and lane segmentation is an active research for past few years. Methods, both using traditional handcrafted features as well as deep learning based are employed to solve the segmentation problem.

A work by Aly [15] make use of Inverse Perspective Mapping (IPM) transformation followed by Random Sample Consensus (RANSAC) to extract road lane candidates from input images. RANSAC spline fitting method then applied to the candidates to retrieve appropriate road lane. Similar work to extract road lane features has been done by Niu et al. [14]. Features extracted using curve fitting that produce small lane segment followed by clustering it. The final segmentation obtained by using particular features such as distance between dashed line and its vanishing point.

He et al. [12] used the dual-view convolutional network to solve the segmentation problem. The first view shows road from the driver's perspective. The second view is the result of IPM transformation, resulting a bird eye view of the road. Each view feeded into two different Convolutional Neural Network (CNN) that are then joined to a single fully connected layer at the end.

An attempt to reduce the dependency on human labeling on solving road and lane segmentation was done in [13]. OpenStreetMap data was used to reconstruct the surrounding environment, resulting additional training data. Fully Convolutional Network (FCN) is then trained on the combined data.

Contextual Blocks was introduced in [11] as a way to provide contextual information to the classifier. These blocks are classification, contextual blocks and road blocks. In order to determine the class of a classification blocks, it uses the feature extracted from its own block as well as the corresponding contextual and road blocks. The feature is then concatenated to a single vector an inputted to a standard MLP.

Mohan in [5] proposed the use of deconvolutional neural network to address road segmentation problem. The deconvolutional layer (also known as transposed convolution) is used as learnable upsampling part of a deep neural network. This research also introduce multipatch training method, in which the input image is separated to several patch and the same number of network is trained on each patch independently. Although this method produce outputs with high F-score, it takes 2 s per image on its inference time. Other methods with similar F-score are able to segment an input in less than 0.2 s.

A convolutional patch network for road and lane segmentation is proposed in [6]. On this method, a standard CNN is trained to predict each pixel on a input image by looking at the value of its surrounding pixel. The output of the network is binary decision stating whether the current pixel is part of the road/lane or not. The inference process should then be done to each pixel on input image. This means that the inference process must be iterated up to the total number of pixel in the image. Thus, resulting a 30 s inference time of a single input image.

Fully Convolutional Network (FCN) has been employed to address slow inference time problem [7]. This network remove the fully convolutional layers on a standard Convolutional Neural Network. By removing the fully connected layers, the number of weight to be multiplied on a forward phase reduced significantly. This research also employ Network in Network architecture that is claimed to reduce inference time. However, the FCN architecture is only used when inferencing a new image. The network is trained in patch based and the network transformation to become FCN requires a complex prerequisite.

3   Method

3.1   Network Architecture

The network architecture is consist of two part. The first part is the encoder, which is used to generate segmentation map from input image. The second part is the decoder, which is used to reconstruct the segmentation map back to its input size.

The final network is structured by stacking several encoders followed by a same the number of decoders as shown in Fig. 1.

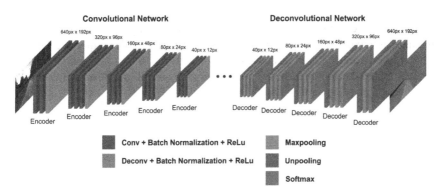

Fig. 1. Encoder-decoder architecture

3.2   Encoder Architecture

The encoder is used to encode input image to a segmentation map. It consist of a series of convolution and pooling operation. The encoder segment the image by classifying each pixel on its input image to either part of ego lane or not. By classifying all pixel on the input image, all pixel will be labeled. Thus, giving different color code to each pixel label will output a segmentation map of the input image.

Convolution. The first operation in the encoder part is the convolution operation. In our final architecture, we use 64 filters with the size 5 × 5. These filter is initialized with random number using normal distribution. The filter convolved through its input with the stride 1.

Maxpooling. The maxpooling layer is used to add translation invariant property to the network. This layer also enlarge the receptive field on the following encoder, enabling the network to give coarse prediction of its input image. However, the use of maxpooling in the encoder resulting a feature map that is spatially smaller than its input size.

3.3   Decoder Architecture

The decoder is used to reconstruct the segmentation map created by the encoder part. This can be done by using deconvolution and unpooling operation.

The output of the last decoder has the same spatial size with the input image and the ground truth. We then add a cross-entropy loss function layer to calculate how far the difference of the segmentation result generated by the network with its respective ground truth. The comparison is done pixel wise, that is by comparing each pixel on the segmentation result with pixel on the same position on the ground truth image. This error value is then back propagated through the network to correct its weights and produce better segmentation.

Unpooling. Reconstructing from low resolution to high resolution image is not trivial because there are some information that is lost during the maxpooling step in the encoder part. To overcome this problem, [9] propose the use of unpooling layer to reconstruct high resolution image from low resolution image. The position of maximum value in maxpooling step is stored in a switch variable. The unpooling layer then use this information to place current pixel value to the position of its maximum value when maxpooled that is stored in the switch variable and fill the other position with 0. The maxpooling operation is shown in Fig. 2.

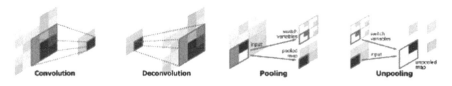

Fig. 2. Unpooling and deconvolution operation

Deconvolution. The deconvolution operation can be thought as inverse of convolution operation. It convolves to each single pixel on its input image with a M × N kernel, then multiply the current convolved pixel with the kernel. The multiplication result a M × N output matrix. A certain activation function then applied to the multiplication result. The M × N activated matrix then placed in the output space of the deconvolution layer. The deconvolution layer also works with certain stride value S. As the layer move one step in the input space, it moves S steps in the output space. When two output overlap each other, those overlapping value is added. The deconvolution operation can be used to convert sparse image generated by unpooling layer to dense image. This operation is shown in Fig. 2.

As for recap, the unpooling layer is used to increase the spatial size of its input image. The sparse output generated by unpooling layer is then filled using deconvolutional layer using its learnable filter. By stacking several decoders, the segmentation map size can be reconstructed back to its input image size (Fig. 3).

Fig. 3. Sample input dataset and the segmentation ground truth

3.4   Training Deconvolution Neural Network

The encoder-decoder architecture is trained on 95 image and ground truth pair of The KITTI Vision Benchmark Database. This dataset is a collection 2D images of road lane photograph taken from driver perspective. The ground truth consist of binary images with the same spatial dimension as the dataset images with each pixel is labeled as ego-lane or not ego-lane.

The input images is roughly 1242 × 375 pixels by size. However, the exact size across images are not equal. In addition, the original size of the input images is too big for available computational power. Therefore, we reduce the spatial size of all input images and its corresponding ground truth to 640 * 192.

We use Adam algorithm [10] with starting learning rate of 0.0001 to train our final architecture. The forward phase of the training algorithm will compute the predicted value of a every pixel. A cross-entropy loss function then computed using predicted value and its ground truth. The loss/error then backpropagated through the network's architecture to determine the update needed to be done to each of network's parameters (weights).

Since we introduce deep neural network with many parameters, we add dropout layer to each of our encoder and decoder to reduce overfitting. We also add batch normalization layer to each encoder and decoder to speed up the training process.

We employ 7-fold cross validation on the training process. The data were initially partitioned to 7 part. On each fold, six parts of the data is used for training and the remaining one is used for validation. The network is trained for 27 epochs within each fold.

4 Experimental and Evaluation Result

4.1 Overall Result

In this section, we present result of our method in twelve different architecture setups. On this experiment, we vary the architecture depth, that is the number of encoder and decoder used in the architecture, and the filter/kernel size. We varying the architecture depth with the number of encoder-decoder pair of 3, 5, 6, and 7 and varying the kernel size to be 3×3, 5×5, and 7×7. Other hyperparameters are set to constant. Our experiments result is shown in Table 1.

Table 1 shows that with more encoder-decoder used, the network achieve a higher accuracy. Meanwhile the inference time is highly depends on the filter size used. The larger filter size gives longer inference time.

Table 1. Architecture experiments

Model name	Architecture	Kernel size	Loss	Accuracy	Inference time
Model3-K3	3 Encoder - 3 Decoder	3×3	0.1447	0.9233	0.024 s
Model3-K5	3 Encoder - 3 Decoder	5×5	0.1330	0.9214	0.029 s
Model3-K7	3 Encoder - 3 Decoder	7×7	0.1126	0.9396	0.035 s
Model4-K3	4 Encoder - 4 Decoder	3×3	0.1193	0.9333	0.024 s
Model4-K5	4 Encoder - 4 Decoder	5×5	0.0859	0.9697	0.028 s
Model4-K7	4 Encoder - 4 Decoder	7×7	0.0709	0.9772	0.035 s
Model5-K3	5 Encoder - 5 Decoder	3×3	0.0904	0.9673	0.024 s
Model5-K5	5 Encoder - 5 Decoder	5×5	0.0522	0.9838	0.028 s
Model5-K7	5 Encoder - 5 Decoder	7×7	0.0558	0.9834	0.035 s
Model6-K3	6 Encoder - 6 Decoder	3×3	0.1216	0.9560	0.024 s
Model6-K5	6 Encoder - 6 Decoder	5×5	0.1949	0.9559	0.029 s
Model6-K7	6 Encoder - 6 Decoder	7×7	0.1241	0.9576	0.035 s

The correlation between filter size and inference time is related to the number of parameter/weight of a filter. Filters on the convolution and deconvolution layer contains weight that will be multiplied with its input on inference process to produce segmentation. On a 3×3 filter, there are 9 weights on a filter. Meanwhile, on 5×5 there are 25 weights on a filter. Increasing the filter size to 7×7 produce 49 weights on a filter.

The different on that three configuration is not significant when we only sees it as a single filter. However, in our architecture, we use 64 filter for each encoder or decoder. Each encoder is always paired to a decoder. Thus, we use 128 filter for each encoder-decoder pair. Moreover, we use 3 up to 6 encoder-decoder pair in our experiment. This high number of filter that is used introduce significant different on number of weight that is used on different filter size. Table 2 shows comparison of number of weight that is used on different filter size. The difference of number of weight in a complete architecture shown on Table 3.

Table 2. Number of weight in an encoder-decoder pair

Kernel size	Number of weights on each Filter	Number of weights for each encoder-decoder pair
3 × 3	9	1152
5 × 5	25	3200
7 × 7	49	6272

Table 3. Number of weight in 5 encoder-decoder architecture

Architecture	Total number of weight (using 5 encoder-decoder pair)
3 × 3	5760
5 × 5	16000
7 × 7	31360

Although that the number of weights grow significantly as small increase done on the filter size, the networks depth (number of encoder-decoder pair used) shows only a little change on the number of weight that is used when it is varied. The reason is that by adding a new encoder-decoder pair on a network, we will only create 128 x (filter size) new weight. Table 4 shows the number of weight that is used on a four different architecture with the same filter size.

Table 4. Number of weights on different architectures

Architecture	Total number of weight (using 3 × 3 filter size)
3 Encoder - 3 Decoder	3456
4 Encoder - 4 Decoder	4608
5 Encoder - 5 Decoder	5760
6 Encoder - 6 Decoder	6912

4.2 Qualitative Result

In this section we shows qualitative result on four different architecture depths, as shown in Fig. 4. This comparison suggest that as the number of encoder-decoder pair increased, the network produce better segmentation. The better segmentation can be identified by smaller false positive and false negative.

The better segmentation result is produced by employing more encoder-decoder pair. Initial layer on the encoder part tend to learn primitive feature such as edge and curve. Meanwhile, the later layer tend to learn more complex feature such as geometric shapes and other lane feature.

However, each encoder layer reduces the its input spatial size to into half of its original size. In our experiment, we only use up to 6 encoder-decoder pair.

Further addition of encoder-decoder pair will produce segmentation map that is too small and difficult to upsampled. This limitation is showed in Table 5.

Fig. 4. Qualitative result over architectures

Table 5. Output size shrinking

Encoders layer	Output size
Encoder 1	310 × 96
Encoder 2	160 × 48
Encoder 3	80 × 24
Encoder 4	40 × 12
Encoder 5	20 × 6
Encoder 6	10 × 3

Not all images in the testing set is well segmented. Figure 4 shows several image that is difficult to segment. The three images share similar characteristic for having shadow present in the image. We found that the training set does not contain many image with such characteristic. This makes the network not able to give a good generalization for those image, and produce inferior segmentation result.

4.3   Inference Time Evaluation

In order to get a better understanding on how this architecture could process the segmentation task in short time frame, we conduct a comparative experiment with the convolutional patch based method as proposed in [6].

In this method, input images are broken down into patches. Each patch is 16 × 16 pixels. The 16 × 16 pixel will be used as context to predict 4 × 4 pixel on its center. Patches are created starting from top left of an image to its bottom right. The stride of 4 pixel is used when creating patches. Thus, from an input image with size of 640 * 192, there are 7680 patches created. Summing up all 95 input images resulting 729600 image patches.

Table 6 shows the comparison of the two architecture in term of pixel accuracy and its inference time.

Table 6. Output size shrinking

Parameter	Convolutional patch	Deconvolutional neural network
Accuracy	47.50%	98.38%
Average inference time	7.936 s	0.028 s
Average training time on each epoch	320 s	15 s

On the convolutional patch method, classification is done on each 16×16 path. As the size of each patch is small, the inference time is only 2 ms. However, since there are 7680 different patches on a single input image that need to be evaluated, the total inference time of an image is 15 s.

On the other hand, as shown in Table 1 the average inference time for our architecture is 29 ms for each input image. On our architecture, the inference process only done once on each input image. This makes the inference time significantly faster.

5   Conclusion

We employ encoder-decoder architecture to solve road lane segmentation problem. The use of fully convolutional network coupled with gradual upsampling approach produce segmentation result that is not only accurate but also computationally feasible for real-time application. Our best model scores 98.38% accuracy on 7 fold cross validation and able to inference a single image frame in 28 ms.

The segmentation accuracy is related to the network depths, while the inference time is highly correlated to the filter size used in convolution and deconvolution operation. Deeper architecture produce better segmentation. However, the larger the filter size, the longer the time needed to inference a single image.

However, several optimization could be done to achieve better result. The small number of dataset used for training resulting bad segmentation on several test images. Our experiment reduce original size of the dataset to its half due to computational limitation. A better hardware could be used so that the training can be done in its original size. A skip connection, as introduced in [4] can also be employed so that the upsampling is not only takes account of its last segmentation map but also the prior segmentation maps.

References

1. World Health Organization: Global status report on road safety: time for action (2009). http://apps.who.int/iris/bitstream/10665/44122/1/9789241563840_eng.pdf. Accessed 2 Sept 2016

2. IFRCRCS: Practical Guide on Road Safety: A Toolkit for Red Cross and Red Crescent Societies (2007). http://www.grsproadsafety.org/themes/default/pdfs/GRSPRed%20Cross%20Toolkit.pdf. Accessed 7 Sept 2016
3. Kusano, K.D., dan Gabler, H.C.: Comparison of expected crash and injury reduction from production forward collision and lane departure warning systems. Traffic Injury prevent. **16** (sup2), S109–S114 (2015)
4. Jonathan, L., Shelhamer, E., dan Darrell, T.: Learning deconvolution network for semantic segmentation. In: Proceedings of the IEEE Conference on Computer Vision and Pattern Recognition (2015)
5. Mohan, R.: Deep deconvolutional networks for scene parsing (2014). https://arxiv.org/pdf/1411.4101.pdf. 15 November 2014. Accessed 22 Oct (2016)
6. Brust, C.A., et al.: Efficient convolutional patch networks for scene understanding. In: CVPR Scene Understanding Workshop (2015)
7. Mendes, C.C.T., Fremont, V., dan Wolf, D.F.: Exploiting fully convolutional neural networks for fast road detection. In: IEEE International Conference on Robotics and Automation (ICRA), pp. 3174–3179 (2016)
8. Noh, H., Seunghoon H., dan Bohyung H.: Learning deconvolution network for semantic segmentation. In: Proceedings of the IEEE International Conference on Computer Vision 2015, pp. 1520–1528 (2015)
9. Zeiler, M.D., Fergus, R.: Visualizing and understanding convolutional networks. In: Fleet, D., Pajdla, T., Schiele, B., Tuytelaars, T. (eds.) ECCV 2014. LNCS, vol. 8689, pp. 818–833. Springer, Cham (2014). https://doi.org/10.1007/978-3-319-10590-1_53
10. Kingma, D., Jimmy B.: Adam: a method for stochastic optimization (2014). https://arxiv.org/pdf/1412.6980.pdf. 30 Januari 2017. Accessed 31 Mar 2017
11. Mendes, C., Frémont, V., Wolf, D.: Vision-Based Road Detection using Contextual Blocks (2015). https://arxiv.org/pdf/1412.6980.pdf. 30 Januari 2017. Accessed 10 Mar 2017
12. He, B., Ai, R., Yan, Y., dan Lang, X.: Accurate and robust lane detection based on dual-view convolutional neural network. In: Intelligent Vehicles Symposium (IV), pp. 1041–1046 (2016)
13. Laddha, A., Kocamaz, M.K., Navarro-Serment, L.E., dan Hebert, M.: Map-supervised road detection. In: Intelligent Vehicles Symposium (IV), pp. 118–123 (2016)
14. Niu, J., Lu, J., Xu, M., Lv, P., dan Zhao, X.: Robust lane detection using two-stage feature extraction with curve fitting. Pattern Recogn. **59**, 225–233 (2015)
15. Aly, M.: Real time detection of lane markers in urban streets. In: Intelligent Vehicles Symposium, pp. 7–12 (2008)

Rare Event Prediction Using Similarity Majority Under-Sampling Technique

Jinyan Li[1], Simon Fong[1(✉)], Shimin Hu[1], Victor W. Chu[2],
Raymond K. Wong[3], Sabah Mohammed[4], and Nilanjan Dey[5]

[1] Department of Computer Information Science,
University of Macau, Macau SAR, China
{yb47432, ccfong, yb72021}@umac.mo
[2] School of Computer Science and Engineering,
Nanyang Technological University, Singapore, Singapore
wchu@ntu.edu.sg
[3] School of Computer Science and Engineering,
University of New South Wales, Sydney, Australia
wong@cse.unsw.edu.au
[4] Department of Computer Science, Lakehead University, Thunder Bay, Canada
mohammed@lakeheadu.ca
[5] Department of Information Technology,
Techno India College of Technology, Kolkata, India
neelanjan.dey@gmail.com

Abstract. In data mining it is not uncommon to be confronted by imbalanced classification problem in which interesting samples are rare. Having too many ordinary but too few rare samples as training data, will mislead the classifier to become over-fitted by learning too much from majority class samples and become under-fitted lacking recognizing power for minority class samples. In this research work, a novel rebalancing technique that under-samples (reduce by sampling) the majority class size for subsiding the imbalanced class distributions without synthesizing extra training samples, is studied. This simple method is called Similarity Majority Under-Sampling Technique (SMUTE). By measuring the similarity between each majority class sample and its surrounding minority class samples, SMUTE effectively discriminates the majority and minority class samples with consideration of not changing too much of the underlying non-linear mapping between the input variables and the target classes. Two experiments are conducted and reported in this paper: one is an extensive performance comparison of SMUTE with the states-of-the-arts using generated imbalanced data; the other is the use of real data representing a case of natural disaster prevention where accident samples are rare. SMUTE is found to be working favourably well over other methods in both cases.

Keywords: Imbalanced classification · Under-Sampling · Similarity measure
SMUTE

© Springer Nature Singapore Pte Ltd. 2017
A. Mohamed et al. (Eds.): SCDS 2017, CCIS 788, pp. 23–39, 2017.
https://doi.org/10.1007/978-981-10-7242-0_3

1 Introduction

Classification is a popular data mining task. A trained classifier is a classification model which is inferred from training data that predicts the category of unknown samples. However, most of current classifiers were designed assuming that the distribution of dataset is balanced. Practically, many datasets found in real life are imbalanced, especially those that carry rare events. This gives rise to weakening the recognition power of the classifier with respect to minority class, and probably overfitting the model with too much training samples from majority class.

In essence, the imbalanced problem which degrades the classification accuracy is rooted at the imbalanced dataset, where majority class samples outnumber those of the minority class in quantity. E.g. the ratios of majority class samples and minority class samples at 20:1, 100:1, 1000:1, and even 10000:1 [1] are not uncommon. The reason for attracting the researcher's attention is that, almost always, the minority class is the prediction target which is of interest while the massive majority class samples are mediocre. The imbalanced classification problems often appear naturally in real-life applications, such as in bioinformatics dataset analysis [2], forecasting accidents and natural disasters [3], image processing [4] as well as assisting diagnosis and treatment through biomedical and diseases datasets [5, 6].

Conventional classifiers are designed to learn the relation between input variables and target classes, without regards to whichever class the samples come from. Training a classification model with imbalanced class dataset would result in biasing the classifier with a tendency of overfitting to the majority class samples and neglecting minority class samples. At the end, since the majority class samples dominate a large proportion in the training dataset, the classification model will still appear to be very accurate when being validated with the same dataset which contains mostly the majority samples for which the model was trained very well. However, when the classifier is being tested with the emergence of rare instances, the accuracy declines sharply. When such model is being used in critical situations, such as rare disease prediction, disaster forecast or nuclear facility diagnosis, the insensitivity of the trained model for accurately predicting the rare exceptions would lead to grave consequence.

The drawback of the classification model is due to the lack of proper training with the few rare samples available. When the model is tested with fresh samples from the minority class, it becomes inadequate. Knowing that "accuracy" is unreliable in situation like this, prior researchers adopted other evaluation metrics remedially to replace or supplement accuracy in order to justly assess the classification model and the corresponding rebalancing techniques. These metrics include AUC/ROC [7], G-mean [8], Kappa statistics [9], Matthews correlation coefficient (MCC) [10], and F-measure [11], etc. In general, researchers tried to solve the imbalanced problem of classification by re-distributing the data from the other major and minor classes through sampling techniques in the hope of making the classifiers robust. One common approach is to over-sample more instances from the minority class, even artificially.

Previously, an alternative and novel under-sampling method, namely Similarity Majority Under-Sampling Technique (SMUTE) for subsiding the imbalanced problem was suggested [12]. SMUTE works as a pre-processing mechanism for improving the imbalanced ratio in the training dataset. It adopts a filter strategy to select the majority class samples which are shown to work well in combination with the existing minority class samples. It works by referring to the similarity between the majority class samples and minority class samples. Then it screens off some majority class samples which are very similar to those minority class samples, according to the given under-sampling rate, in order to reduce the imbalanced ratio between two classes. It works by calculating the similarity of each majority class samples and its surrounding minority class samples. Then a distance value is obtained corresponding to each sample in the majority class, which is the sum of a given number of the most similar minority class samples' similarity to that majority class sample. Rank these majority class samples by their sum similarity in ascending order. Finally, the algorithm retains a given number of majority class samples (e.g. top k) using a filtering approach. Those majority class samples which are most dissimilar to the minority class samples are removed. This method could effectively segregate majority class samples and minority class samples in data space and maintain high distinguishing degree between each class, for the sake of keeping up with the discriminative power and high classification accuracy.

The remaining paper is organized as follows. Some previous approaches and papers for solving imbalanced problem are reviewed in Sect. 2. In Sect. 3, we elaborate our proposed method and the process. Then, the data benchmark, comparison algorithms, our experiments over the two case studies and results are demonstrated and discussed about in Sect. 4. Section 5 summarizes this paper.

2 Related Work

As introduced earlier, that imbalanced classification is a crucial problem for which effective solutions are in demand. Since the conventional classification algorithms were not originally designed to embrace training from imbalanced dataset, it triggered a series of problems, due to overfitting the majority class data and underfitting the minority class data. These problems include Data Scarcity [13], Data Noise [14], Inappropriate Decision Making and Inappropriate Assessment Criteria [15].

For overcoming this imbalanced data problem, current methods can be broadly divided into data level and algorithm level. The data level methods adopt resampling techniques to re-adjust the distribution of imbalanced dataset. At the algorithm level, the conventional classification algorithms are modified to favour the minority class samples through assigning weights on samples that come from different classes or ensemble techniques where the candidate models that are trained with minority class data are selected more often.

Prior arts suggested that rebalancing the dataset at the data level, by pre-processing is relatively simpler at data level, and as effective as biasing imbalanced classification [16]. Hence, sampling methods have been commonly used for addressing imbalanced classification by redistributing the imbalanced dataset space. Under-Sampling reduces the number of majority class samples and Over-sampling increases the amount of minority class samples. These two sampling approaches are able to reach an ideal class distribution ratio that maximizes the classifier performance. However, there is no rule of thumb on how much exactly to over-sample or under-sample so to achieve the best fit. A simple way is to randomly select majority class samples for downsizing and likewise for repeatedly upsizing minority class samples, randomly. Random under-sampling may miss out important samples by chances. Inflating rare samples without limit will easily cause over-fitting too. Synthetic Minority Over-sampling Technique (SMOTE) [17] is one of the most popular and efficient over-sampling methods in the literature. Each minority class sample replicates about several of its neighbour minority class samples to synthesise new minority class samples, for the purpose of rebalancing the imbalanced dataset. One weakness of this method is that the synthesized minority class samples may coincide with the surrounding sample of majority class sample [18]. For this particular weakness, researchers invented a number of modifications, extending SMOTE to better versions: such as, Borderline SMOTE [19], and MSMOTE [20] etc.

Basically, over-sampling will dilute the population of the original minority class samples by spawning extra synthetic samples. On the other hand, eliminating some majority class samples by under-sampling helps relieve the imbalanced classification problems too. But if this is done blindly or randomly, under-sampling may drop some meaningful and representative samples by chance, from the population of the majority class. Estabrooks and Japkowic concurrently used over-sampling and under-sampling with different sampling rates to obtain many sub-classifiers, like an ensemble method. The sub-classifiers are then integrated by the frame of mixture-of-experts in the following step [21]. The experimental results showed that this method is much better than the other ensemble methods. Balance Cascade [22] is a classical under-sampling method. Through iteration strategy, it is guided to remove the useless majority class samples gradually.

Ensemble learning and cost-sensitive learning are two typical techniques at algorithm level for solving imbalanced problem. They function by assigning different weights or votes or further iterations to counter-bias the imbalanced ratio, while conventional methods concern about increasing size of the minority class samples.

Ensemble learning gathers a number of base classifiers and then it adopts some ensemble techniques to incorporate them to enhance the performance of classification model. Boosting and Bagging [23], are the most frequently used approaches. AdaBoosting is a typical construct in boosting series methods. It adaptively assigns different and dynamic weights to each sample in iterations to change the tendentiousness of classifier [24]. Bagging implements several votations of sub-classifiers to promote the performance. These sub-classifiers classify repeatedly using the re-sampled dataset. A winning classifier which is most voted would be selected to produce the final results after several rounds of voting.

A lot of research works are focused on over-sampling the minority class samples either at the data level or tuning up the bias at the algorithm level. It was supposed that properly recognizing the minority class samples is more valuable than the majority class samples. The belief is founded on the consequence that misclassifying any minority class sample would often need to pay a high price in critical applications. Cost-sensitive learning followed this basic idea to assign different costs of misclassified class. Besides attempts to pushing down the misclassification rate, during the training of a cost-sensitive classifier, the classifier will be forced to boost a higher recognition rate for minority class samples, since keeping the total cost at the lowest is one of the optimization goals of this learning model.

3   Similarity Majority Under-Sampling Technique

In our new design, called SMUTE, the integrity of the minority class samples as well as their population size would be left untouched. This is believed that the minority class samples are better to be preserved as original as they are, in such a way that no more or no less amount should be intervened over them. Hence an appropriate classifier which is trained right from the original minority class samples, would offer the purist recognition power. Different from most of the popular class rebalancing techniques reviewed above, SMUTE manipulates only at the majority class samples, repopulating those majority instances which are found to be compatible (or similar) to the minority instances.

In the data space of an imbalanced dataset, majority class samples occupy most of it. The inequity causes the classifier to be insufficiently trained for identifying the minority class samples and the overwhelming majority class samples interfere the recognition power of the minority class samples. Consequently, the classifiers will bias majority class samples and it suffers from a pseudo high accuracy when it was tested with imbalanced dataset apart from those used in training. Over-sampling techniques reverse the bias of classifiers through synthesizing additional minority class samples. However, with the extra data generated, the data space will become more crowded, there might even be some overlaps between these samples that give rise to confusion to the training. Furthermore, over-sampling techniques increase the computational cost of model training because the overall training dataset size will increase with extra samples are synthesized and added, but the discrimination between the two classes of samples becomes vague with more overlaps.

Under-Sampling is a reasonable approach for reducing the disproportion between the two classes. Some under-sampling methods are introduced and used hereinbefore, such as instance selection and clustering method. The art of under-sampling is about how to reduce the majority class samples, in a way that the effective distinction between samples from different classes remains sharpened. It should ensure the data space does not get congested but the class samples are well distributed closely corresponding to the underlying non-linear relations inside the classifier. In SMUTE, majority class samples are selected free from deletion based on how "compatible" the majority samples to the minority samples are, while keeping the minority samples intact. Similarity measure is used here as a compatibility check which calculates the similarity distance between two data points in multi-dimensional space. Calculation methods vary for similarity measure. The most common measure is correlation which has been widely used in the similarity measure that adheres to four principles: (1) the similarity of their own is 0; (2) the similarity is a non-negative real number quantified as the distance apart; (3) Symmetry, if the similarity from A to B is equal to the similarity from B to A. (4) Triangular rule: the sum of both sides is greater than the third side of the similarity triangle.

Two steps of computation are involved in the proposed Similarity Majority Under-Sampling Technique.

1. Each majority class sample calculates the distances pairing between itself and each K minority class samples, and sum up these distances to a similarity score.
2. Given an under-sampling rate, [0, 1], select a subset of majority class samples which have the top percentage of high similarity scores; the disqualified samples (which have relatively low similarity scores) are discarded.

Figures 1, 2 and 3 show the several steps of the operation process of SMUTE. Given an imbalanced dataset $D = \{x_1, x_2, \ldots, x_n, y_1, y_2, \ldots, y_m\}$, here x_i is the ith majority class sample, y_j is the jth minority class sample. For x_i there are m example pairs (x_i, y_1), …, (x_i, y_m).

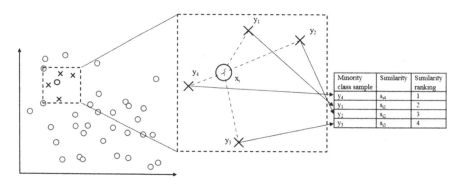

Fig. 1. First step of SMUTE: Calculating the similarity of a majority class samples with its surrounding minority class samples.

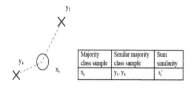

Fig. 2. Second step of SMUTE: Selecting a given number of minority class samples and calculating the sum similarity of each majority class sample

Majority class sample	Sum similarity	Similarity ranking
x_{89}	s_{89}'	1
x_3	s_3'	2
x_{30}	s_{41}'	3
…	…	…
x_1	s_1'	r_1
…	…	…
x_{56}	s_{56}'	80
x_{47}	s_{47}'	81
x_{39}	s_{39}'	82
…	…	…
x_{17}	s_{17}'	100

SMUTE result

Majority class sample	Sum similarity	Similarity ranking
x_{89}	s_{89}'	1
x_3	s_3'	2
x_{30}	s_{41}'	3
…	…	…
x_1	s_1'	r_1
…	…	…
x_{56}	s_{56}'	80

Fig. 3. Last step of SMUTE: Sorting and filtering a given number of majority class samples to obtain the new majority class dataset.

Then we obtain $S_i = \{s_{i1}, s_{i2}, \ldots, s_{ij}, \ldots, s_{im}\}$ by calculating similarity measure for each sample example pair, where s_{ij} is the similarity value between x_i and y_j. The minority class samples in S_i are ordered as their corresponding values of s_{ij} decreases.

From $S_i = \{s_{i1}, s_{i2}, \ldots, s_{ij}\}$, top w minority class samples with highest similarity are chosen to calculate a sum value S_1', where S_1' represents similarity level between majority class sample x_i and the minority class. Here w is a given value.

Then a ranking of $S' = \{S'_1, S'_2, \ldots S'_n\}$ in increasing order is computed. As a given under-sampling rate, N, a subset of majority class samples is filtered and selected.

In our previous experiment, eight common methods of calculating similarity are tried, together with 20 equidistant different under-sampling rates ranging from 100% to 0.5% to rebalance the imbalanced datasets. The eight similarity measures are Euclidean Distance [25], Manhattan Distance [26], Chebyshev Distance [27], Angle Cosine [28], Minkowski Distance [29], Correlation coefficient [30], Hamming Distance [31] and Jaccard similarity coefficient [32]. These eight similarity measures were respectively used under the hood of SMUTE. They are put under test for intellectual inquisition. SMUTE could evolve to a bagging type of algorithm, sorting out a version that offers the best performance. For a thorough performance assessment, 100 binary class imbalanced datasets from KEEL [24] were used for testing SMUTE with different versions of similarity. A standard C4.5 Decision Tree is used as the base classifier, which is subject to 10-cross validation method for recording the classification performances. For each dataset, each similarity measure method, and at each under-sampling rate will be repeatedly run 10 times before averaging them to a mean value. In addition to accuracy, Kappa statistic [25, 26] is chosen as the main evaluation metric because it indicates how reliable the accuracy the classifier in terms of generalizing its predictive power on other datasets.

The processed dataset is composed of original minority class samples and new majority class samples, it has lower imbalanced ratio. The following is the pseudo code of SMUTE.

SMUTE Pseudo code
1.   Input a imbalanced dataset and divide the dataset into two parts, minority class samples and majority class samples
2.   There are n majority class samples and m minority class samples
3.   **FOR** $i = 1, \ldots, n$ **DO**
2.   **FOR** $j = 1, \ldots, m$ **DO**
3.   Calculate the similarity value s_{ij} using similarity measure
4.   add s_{ij} to S_i
5.   **END FOR**
6.   Rank $S_i = \{ s_{i1}, s_{i2}, \ldots, s_{ij} \}$ in descending order
7.   Choose the top w minority class samples from the S_i ranking list, w is a give number
8.   Calculate s_i' the sum of the w minority class samples
9.   Add s_i' to S'
10. **END FOR**
11. Rank $S' = \{ s_1', s_2', \ldots, s_n' \}$ in increasing order
12. Choose n majority class samples according to the S' ranking list, n is a given number.
13. Combine the new majority class samples and the original minority class samples into a new dataset.

As shown in Figs. 1, 2, 3 and 4 of the simulation runs [12], it seems that the similarity measure of Angle Cosine is more suitable for SMUTE than the correlation coefficient versions. In summary, it is possible to implement the most effective version of SMUTE using Angle Cosine after testing out various options. It is apparently useful for easing the imbalanced classification problem. Since it is shown to be superior, Angle Cosine, again will be used in the following experiments for comparing the classification performances of SMUTE and the other existing rebalancing methods.

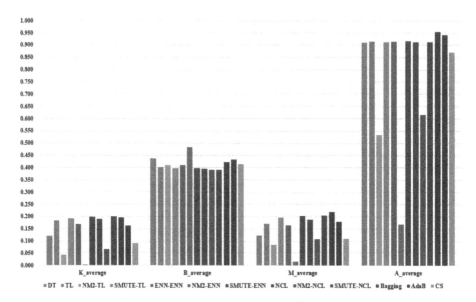

Fig. 4. The total average values of each evaluation metrics of each algorithm for all the synthetic datasets.

The Cosine similarity is independent of the amplitude of the vector, only in relation to the direction of the vector. The Angle Cosine is in the range $[-1,1]$. The larger the Angle Cosine value means the narrower the angle and the greater similarity between the two vectors, vice versa. When the direction of the two vectors coincides, the Angle Cosine takes the maximum value 1, and when the direction of the two vectors is exactly at opposite, the Angle Cosine takes the minimum value -1.

$$\cos(\theta) = \frac{\sum_{i=1}^{n} x_i y_i}{\sum_{i=1}^{n} x_i^2 \sum_{i=1}^{n} y_i^2} \tag{1}$$

4   Experiments

Minority class samples in real life are rare. Therefore, in this paper, the experiment adopted strict tenfold cross validation to implement the testing part. Each imbalanced dataset is randomly divided into ten parts. Each part is separate and rotated as testing dataset, and the remaining nine parts in the composition of the corresponding training dataset, respectively. The sampling approaches and algorithm level techniques are acting on the training dataset to build a recovered classification model for the testing dataset.

4.1 Case Study of Synthetic Imbalanced Datasets

There are seven competitive methods for tackling imbalanced classification. As a baseline reference, the Decision Tree algorithm which is directly built from the imbalanced datasets without any pre-processing of rebalancing is used. Four classical and typical under-sampling methods, three conventional methods in algorithms level, cost-sensitive, Adaboosting and Bagging are used in the benchmarking test.

- Decision Tree (DT): Decision tree directly adopts tenfold cross-validation to classify the original imbalanced datasets.
- Tomek-links (TL): It aims at removing the noise and boundary points in majority class samples [33].
- Edited Nearest Neighbour (ENN): The basic idea of ENN is to remove the samples whose class is different from two class samples of the three nearest three samples [34].
- Neighbourhood cleaning rule (NCL): Based on ENN algorithms to remove more samples [35].
- NearMiss-2 (NM2): The previous experimental results demonstrate that NearMiss-2 could give competitive results in Near Miss series under-sampling algorithms [36]. Based on K-NN algorithm, it selects the majority class samples whose average distance from the three farthest minority class samples is the smallest [37].
- Bagging: Bagging+DT. It adopts 20 base classifiers in an ensemble.
- AdaBoost.M1 (AdaBM1): AdaBM1+DT.AdaBoost.M1 uses AdaBoosting for addressing binary imbalanced classification problem, its maximum iteration number is 100.
- Cost-sensitive (CTS): CTS+DT. Cost matrix assigned 100 costs to the misclassified minority class samples, and the cost of misclassified majority class samples is 1. The costs of correct classified classes are 0.

The mean of the ten times repeated test results is deemed as the final result. In this experiment, we used the generator of Random Tree Generator in MOA to randomly generate 20 artificial imbalanced datasets, all the features are numeric type [38]. Table 1 contains the information of the 20 imbalanced datasets. The imbalanced ratio between majority class and minority class is ranging from 10.45:1 to 58.89:1.

For the experimentation, the simulation software is programmed in MATLAB version 2014b. The simulation computing platform is CPU: E5-1650 V2 @ 3.50 GHz, RAM: 62 GB.

Tomek-links, ENN and NCL don't have a specific given under-sampling rate as Near Miss-2 and SMUTE. Therefore, in our experiment, Near Miss-2 and SMUTE respectively adopted the same and corresponding under-sampling rates of Tomek-links, ENN and NCL. They are named NM2-TL, NM2-ENN, NM2-NCL, SMUTE-TL, SMUTE-ENN, SMUTE-NCL.

In the experimentation, Kappa statistics is used to supplement the accuracy measure for evaluating the goodness of the classification model. We also used Balanced Error Rate (BER) [6] and Matthews's correlation coefficient (MCC) [39] to assist the former two indicators to evaluate the solved imbalanced classification model. The experimental results in Fig. 4 show total average values of each evaluation metrics of each

Table 1. Information of testing datasets generated by MOA

Majority	Minority	#Features	Imb.ratio (maj/min)
742	27	8	27.48
559	31	15	18.03
1856	146	13	12.71
1590	27	16	58.89
1343	63	12	21.32
719	51	9	14.10
753	47	18	16.02
1021	40	18	25.53
1442	138	11	10.45
1544	145	13	10.65
1040	90	13	11.56
596	42	8	14.19
1889	52	13	36.33
1028	50	10	20.56
1099	36	10	30.53
1189	90	11	13.21
811	71	12	11.42
917	38	12	24.13
1034	55	13	18.80
1389	72	15	19.29

algorithm over all the testing datasets. These metrics of original datasets, which was directly classified by Decision tree reflect that the imbalanced classification model obtain a very high value of accuracy. But its Kappa statistics, BER and MCC are poor. NCL defeated ENN and TL in every indicator. The performance of TL is the middle of the three and ENN is the worst one. When NM2 and SMUTE used the under-sampling rates of these three methods, this diagram demonstrates that NM2 is the worst method in the five under-sampling techniques. However, SMUTE overcomes all the other under-sampling methods, regardless of the under-sampling rates used from the first methods. The performance of SUMTE-ENN and SMUTE-NCL are similar. In the three algorithm level methods, which are shown on the left side, bagging obtains the best performances. Since the cost matrix of cost-sensitive learning is not specific, its performance is even worse than the original results.

With respect to Kappa statistics and BER, SMUTE is a little better than Bagging. Bagging methods exceed SMUTE in the indicator of MCC and Accuracy. However, Fig. 5 displays the total standard deviation of each evaluation metrics of each algorithm for all datasets, in order to assess the stability of each method. SMUTE could obtain a significantly lower standard deviation value than Bagging.

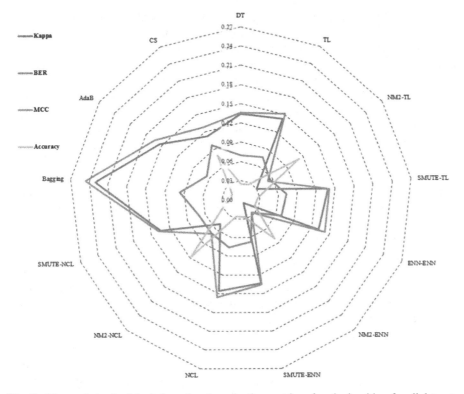

Fig. 5. The total standard deviation of each evaluation metrics of each algorithm for all datasets.

4.2 Case Study of Real Life Imbalanced Dataset

The same computing platform is used, and three popular rebalancing filtering tools in Weka are used to compare with the efficacy of SMUTE. The two filtering tools are SMOTE and RESAMPLE, which inflates the minority samples and reduces the majority samples respectively. A combination of SMOTE and RESAMPLE is used together to compete with SMUTE as well. These two rebalancing techniques represent some of the most popular tools implemented and available on an open source Java-based machine learning benchmarking program called Weka. In particular, we want to evaluate the efficacy of SMUTE in a case real life dataset such as mining hazards. The dataset of the case study being considered here is called seismic bumps[1] which is taken from UCI data repository. Seismic bump is a training dataset that records mining shifts information and seismic activities that are picked up by the installed sensors for a coal mine. It is known that coal mining is a dangerous industry and the operations are always associated with dangers of cave and underground mine collapses etc. Such imminent occupational threat is required to be detected and pre-dicted in the hope of saving lives, when some early warning could be foretold. On par

---

[1] http://archive.ics.uci.edu/ml/datasets/seismic-bumps.

with the difficulty of predicting earthquake, this kind of natural hazards demand for the most prediction software model and the most reliable monitoring hardware. Advanced acoustic monitoring systems are able to continuously check over the rock mass, earth movements, in the count of seismic bumps within a specific interval. The prediction model is far from perfect, due to unknown seismic occurrence complexity. Another contributing factor is due to the very high disproportion between the available records of high-energy (10^4 J) phenomena which is known as the minority class, and a large number of low-energy seismic activities known as the majority class. The imbalance ratio of the two classes of samples makes statistical techniques incompetent for reliable seismic hazard prediction.

The dataset which is imbalanced in nature, consists of rows as summary statement within a cycle of 8 h shift recording about the measured seismic activity in the rock mass. Each record associates with a binary state of 0 or 1 which is interpreted as "hazardous" or "hazardless". A high energy seismic bump is deemed to occur in the next shift if the class label has the value 1, and vice-versa. The records dataset infer some causality relation between some recorded high energy tremors that have occurred in history and the seismo-acoustic activity measured in the previous shifts. The dataset attributes could be found from the UCI website (which is referenced at the footnote). A both accurate and reliable prediction model is there much needed for this real life case, which is able to rectify the imbalanced classification data mining problem. With an accurate and reliable model ready, mine disaster can be reduced by conducting preventive rock burst via distressing shooting, a way of releasing the high energy or by evacuating mine workers from the threatened mine. Therefore, being able to rectify the imbalanced problem is a matter of great practical urgency in real life application. The data set is highly imbalanced as there are only 170 rare event samples of class 1 out of 2584 samples.

The dataset is first put under test using three rebalancing techniques, SMOTE, RESAMPLE and the combination of the two. Three typical classification algorithms are used to couple with these pre-processing filters, they are decision tree (J48), Naïve Bayes (NB) and Neural Network (NN). The performance evaluation metrics used are the same as those used in Sect. 4.1. Particularly, Kappa statistic is the main focus of our performance evaluation as it is know that having high accuracy is meaningless because the training and testing are on the same dataset; it is relatively more important that the induced model would be able to generalize to test well on unseen data – that is where exactly the practical prediction model is to be applied, to predict and test on possibly some totally unseen seismic activities in order to decide whether or not the next shift is going to be hazardous. The training is from the original training dataset and testing is by the dataset which contains all the 170 rare samples and 170 samples randomly selected from the majority class. The results are presented in Table 2, and visually presented in Fig. 6.

As it can be seen from the Table and Figures, the performances vary when different rebalancing methods and classifiers are used. In general, all the classifiers can be benefited from the rebalancing methods in improving all the performance indicators. In particular, decision tree seems to be most susceptible to imbalanced problem but also it gains hugely when the data rebalanced. The Kappa for decision tree has increased from 0 to 0.9235. But the maximum accuracy achievable is limited at 96.1765%.

Table 2. Performance comparison of seismic bump model by various classifiers and rebalancing methods

Original	Algorithm	Accuracy%	Kappa	BF	Gain
	J48	50	0	0.333	0
	NB	65.5882	0.3118	0.812	16.60573
	NN	56.7647	0.1353	0.54	4.147343
SMOTE	Algorithm	Accuracy%	Kappa	BF	Gain
	J48	70.8824	0.4176	0.815	24.1244
	NB	68.2353	0.3647	0.863	21.47611
	NN	66.4706	0.3294	0.752	16.46535
Resample	Algorithm	Accuracy%	Kappa	BF	Gain
	J48	94.7059	0.8941	0.994	84.16849
	NB	70	0.4	0.947	26.516
	NN	77.9412	0.5588	0.983	42.81313
Resample+SMOTE	Algorithm	Accuracy%	Kappa	BF	Gain
	J48	96.1765	**0.9235**	0.997	88.55254
	NB	69.1176	0.3824	0.962	25.42621
	NN	85	0.7	0.996	59.262

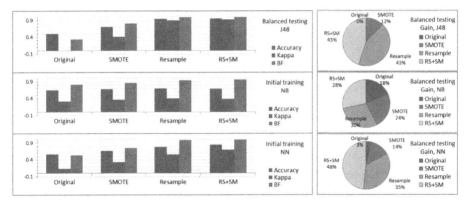

Fig. 6. Performance comparison of seismic bump model by various classifiers and rebalancing methods in bar charts and pie charts.

For testing out SMUTE, in the subsequent part of the experiment, 50 different retention ratios of majority class are varied, from 100% to 2%. The idea is to test how the retention ratio as a model variable influence the resultant Kappa of decision tree. The imbalanced ratio of training dataset is 2414:170, and the imbalance ratio of testing data is 50:1. The other parameter, the K neighborhoods, is randomly selected. The result is shown in Figs. 7 and 8 for views of all the performance indicators and Kappa respectively. As it can be clearly seen from Fig. 8, when SMUTE is tuned properly, at

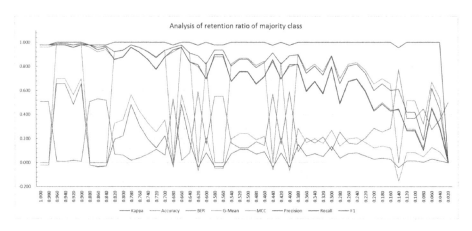

Fig. 7. Performance of SMUTE over different retention ratios showing all performance indicators

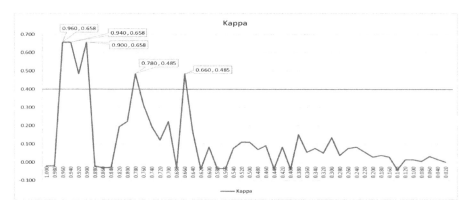

Fig. 8. Performance of SMUTE over different retention ratios showing only Kappa statistic

the right ratios, Kappa can be maximized at 0.658 which is in a moderate-to-strong region. This indicates SMUTE can be a good candidate outperforming SMOTE, in achieving a relatively reliable and very accurate classifier at precision performance between 98% and 100% most of the times.

5   Conclusion

In this paper, we further studied a novel class rebalancing method for easing the imbalanced classification problems, called Similarity Majority Under-sampling Technique (SMUTE) with case studied. SMUTE reduces the majority class samples by computing, checking and ranking the similarity between each majority class sample and minority class samples. The majority class samples that are most similar to those of minority class samples are retained. The motivation by doing so is that we prefer to keep the amount of minority class samples intact without reducing any of them or generating any extra synthesis samples. For reducing the majority samples in order to try even out their ratio, preference is given to those majority samples and the minority samples are alike. Through this simple pre-processing technique, three advantages with respect to rebalancing the imbalanced dataset are attained: (1) The minority samples which are the rare events are well preserved in number; (2) No artificial samples are generated from either majority or minority class, so the overall training dataset is capped with no increase of artificial samples. The training time hence will not be prolonged; (3) Because we filter off those majority samples other than those which are similar (compatible) to the minority samples, the classification performance generally will be kept up. The experimental results show that SMUTE could exceed the performance of some state-of-the-art down-sampling method at the same under-sampling rates. SMUTE is believed to be useful in data mining, particularly as a pre-processing approach for fixing training dataset that contains only few but important minority class samples. It works by preserving all the available minority class samples (which are supposed to be precious), and reducing the overly large population of majority class samples, keeping only those majority class samples that are similar to the rare samples by data distances. In the case study of mining hazard prediction, SMUTE can effectively increase the performance of a decision tree to 0.658 Kappa which lies within a moderate-to-strong reliability region and high precision at approximately 98% and 100%. It demonstrated its efficacy in improving a serious problem in seismic hazard prediction.

Acknowledgement. The authors are thankful to the financial support from the research grants, (1) MYRG2016-00069, titled 'Nature-Inspired Computing and Metaheuristics Algorithms for Optimizing Data Mining Performance' offered by RDAO/FST, University of Macau and Macau SAR government. (2) FDCT/126/2014/A3, titled 'A Scalable Data Stream Mining Methodology: Stream-based Holistic Analytics and Reasoning in Parallel' offered by FDCT of Macau SAR government. Special thanks go to a Master student, Jin Zhen, for her kind assistance in programming and experimentation.

References

1. Weiss, G.M., Provost, F.: Learning when training data are costly: the effect of class distribution on tree induction. J. Artif. Intell. Res. **19**, 315–354 (2003)
2. Li, J., et al.: Improving the classification performance of biological imbalanced datasets by swarm optimization algorithms. J. Supercomputing **72**(10), 3708–3728 (2016)
3. Cao, H., et al.: Integrated oversampling for imbalanced time series classification. IEEE Trans. Knowl. Data Eng. **25**(12), 2809–2822 (2013)
4. Kubat, M., Holte, R.C., Matwin, S.: Machine learning for the detection of oil spills in satellite radar images. Mach. Learn. **30**(2–3), 195–215 (1998)
5. Li, J., et al.: Solving the under-fitting problem for decision tree algorithms by incremental swarm optimization in rare-event healthcare classification. J. Med. Imaging Health Inform. **6**(4), 1102–1110 (2016)

6. Li, J., et al.: Adaptive swarm cluster-based dynamic multi-objective synthetic minority oversampling technique algorithm for tackling binary imbalanced datasets in biomedical data classification. BioData Mining 9(1), 37 (2016)
7. Chawla, N.V.: C4. 5 and imbalanced data sets: investigating the effect of sampling method, probabilistic estimate, and decision tree structure. In: Proceedings of the ICML (2003)
8. Tang, Y., et al.: SVMs modeling for highly imbalanced classification. IEEE Trans. Syst. Man Cybern. Part B (Cybern.) 39(1), 281–288 (2009)
9. Li, J., Fong, S., Yuan, M., Wong, R.K.: Adaptive multi-objective swarm crossover optimization for imbalanced data classification. In: Li, J., Li, X., Wang, S., Li, J., Sheng, Q. Z. (eds.) ADMA 2016. LNCS, vol. 10086, pp. 374–390. Springer, Cham (2016). https://doi.org/10.1007/978-3-319-49586-6_25
10. Stone, E.A.: Predictor performance with stratified data and imbalanced classes. Nat. Methods 11(8), 782 (2014)
11. Guo, H., Viktor, H.L.: Learning from imbalanced data sets with boosting and data generation: the databoost-im approach. ACM Sigkdd Explor. Newslett. 6(1), 30–39 (2004)
12. Li, J., et al.: Similarity majority under-sampling technique for easing imbalanced classification problem. In: 15th Australasian Data Mining Conference (AusDM 2017), Melbourne, Australia, 19–25 August 2017, Proceedings. Australian Computer Society (2017)
13. Weiss, G.M.: Learning with rare cases and small disjuncts. In: ICML (1995)
14. Weiss, G.M.: Mining with rarity: a unifying framework. ACM Sigkdd Explor. Newslett. 6 (1), 7–19 (2004)
15. Arunasalam, B., Chawla, S.: CCCS: a top-down associative classifier for imbalanced class distribution. In: Proceedings of the 12th ACM SIGKDD International Conference on Knowledge Discovery and Data Mining. ACM (2006)
16. Drummond, C., Holte, R.C.: C4. 5, class imbalance, and cost sensitivity: why under-sampling beats over-sampling. In: Workshop on Learning from Imbalanced Datasets II. Citeseer (2003)
17. Chawla, N.V., et al.: SMOTE: synthetic minority over-sampling technique. J. Artif. Intell. Res. 16, 321–357 (2002)
18. Li, J., et al.: Adaptive multi-objective swarm fusion for imbalanced data classification. Inf. Fusion 39, 1–24 (2018)
19. Han, H., Wang, W.-Y., Mao, B.-H.: Borderline-SMOTE: a new over-sampling method in imbalanced data sets learning. In: Huang, D.-S., Zhang, X.-P., Huang, G.-B. (eds.) ICIC 2005. LNCS, vol. 3644, pp. 878–887. Springer, Heidelberg (2005). https://doi.org/10.1007/11538059_91
20. Hu, S., et al.: MSMOTE: improving classification performance when training data is imbalanced. In: Second International Workshop on Computer Science and Engineering, WCSE 2009. IEEE (2009)
21. Estabrooks, A., Japkowicz, N.: A mixture-of-experts framework for learning from imbalanced data sets. In: Hoffmann, F., Hand, David J., Adams, N., Fisher, D., Guimaraes, G. (eds.) IDA 2001. LNCS, vol. 2189, pp. 34–43. Springer, Heidelberg (2001). https://doi.org/10.1007/3-540-44816-0_4
22. Quinlan, J.R.: Bagging, boosting, and C4. 5. In: AAAI/IAAI, vol. 1 (1996)
23. Sun, Y., Kamel, M.S., Wang, Y.: Boosting for learning multiple classes with imbalanced class distribution. In: Sixth International Conference on Data Mining, ICDM 2006. IEEE (2006)
24. Alcalá, J., et al.: Keel data-mining software tool: data set repository, integration of algorithms and experimental analysis framework. J. Multiple-Valued Logic Soft Comput. 17 (2–3), 255–287 (2010)

25. Li, J., Fong, S., Zhuang, Y.: Optimizing SMOTE by metaheuristics with neural network and decision tree. In: 2015 3rd International Symposium on Computational and Business Intelligence (ISCBI). IEEE (2015)
26. Viera, A.J., Garrett, J.M.: Understanding interobserver agreement: the kappa statistic. Fam. Med. **37**(5), 360–363 (2005)
27. Cha, S.-H.: Comprehensive survey on distance/similarity measures between probability density functions. City **1**(2), 1 (2007)
28. Nguyen, H.V., Bai, L.: Cosine similarity metric learning for face verification. In: Kimmel, R., Klette, R., Sugimoto, A. (eds.) ACCV 2010. LNCS, vol. 6493, pp. 709–720. Springer, Heidelberg (2011). https://doi.org/10.1007/978-3-642-19309-5_55
29. Santini, S., Jain, R.: Similarity measures. IEEE Trans. Pattern Anal. Mach. Intell. **21**(9), 871–883 (1999)
30. Ahlgren, P., Jarneving, B., Rousseau, R.: Requirements for a cocitation similarity measure, with special reference to Pearson's correlation coefficient. J. Am. Soc. Inf. Sci. Technol. **54** (6), 550–560 (2003)
31. Xu, Z., Xia, M.: Distance and similarity measures for hesitant fuzzy sets. Inf. Sci. **181**(11), 2128–2138 (2011)
32. Choi, S.-S., Cha, S.-H., Tappert, C.C.: A survey of binary similarity and distance measures. J. Syst. Cybern. Inform. **8**(1), 43–48 (2010)
33. Kotsiantis, S., Kanellopoulos, D., Pintelas, P.: Handling imbalanced datasets: a review. GESTS Int. Trans. Comput. Sci. Eng. **30**(1), 25–36 (2006)
34. Tomek, I.: An experiment with the edited nearest-neighbor rule. IEEE Trans. Syst. Man Cybern. **6**, 448–452 (1976)
35. Bekkar, M., Alitouche, T.A.: Imbalanced data learning approaches review. Int. J. Data Mining Knowl. Manag. Process **3**(4), 15 (2013)
36. Mani, I., Zhang, I.: kNN approach to unbalanced data distributions: a case study involving information extraction. In: Proceedings of Workshop on Learning from Imbalanced Datasets (2003)
37. He, H., Garcia, E.A.: Learning from imbalanced data. IEEE Trans. Knowl. Data Eng. **21**(9), 1263–1284 (2009)
38. Bifet, A., et al.: Moa: massive online analysis. J. Mach. Learn. Res. **11**(May), 1601–1604 (2010)
39. Ding, Z.: Diversified ensemble classifiers for highly imbalanced data learning and their application in bioinformatics (2011)

Image Feature Classification
and Extraction

Pattern Recognition of Balinese Carving Motif Using Learning Vector Quantization (LVQ)

I Made Avendias Mahawan[(✉)] and Agus Harjoko

Department of Computer Science and Electronics, University of Gadjah Mada,
Yogyakarta, Indonesia
made.avendias.m@mail.ugm.ac.id, aharjoko@ugm.ac.id

Abstract. Bali is a world tourism destination with its cultural uniqueness, one of the Balinese cultural products that need to be maintained is the art of Balinese carvings in traditional buildings and sacred buildings, to inherit the culture it needs a management, documentation and dissemination of information by utilizing technology. Digital image processing and pattern recognition can be utilized to preserve arts and culture, the technology can be utilized to classify images into specific classes. Balinese carving is one of the carvings that have many variations, if these carvings are analyzed then required an appropriate method for feature extraction process to produce special features in the image. So they can be recognized and classified well and provide information that helps preserve Bali. The aim of this research is to get the right feature extraction method to recognize and classify Bali carving pattern image based on the accuracy of HOG feature extraction method with PCA trained using LVQ. The results of the test data obtained the best accuracy of HOG is 90% with cell size 32×32 and block size 2×2, PCA obtained 23.67% with threshold 0.01 and 0.001, from training input with learning rate = 0.001 and epoch = 1000.

Keywords: Balinese carving motif · Learning Vector Quantization (LVQ)
Histogram of Oriented Gradient (HOG) · Canny edge detection
Principal Component Analysis (PCA)

1 Introduction

The era of globalization has made Indonesia one of the fastest growing countries in technology, Indonesia as a developing country faces a serious threat in the cultural field because does not have competitive power equivalent to developed countries and begins to erode the values of Indonesian cultural identity [1]. Bali is the worlds best tourist destination with its unique culture, one of Balinese cultures that needs to be maintained is the Balinese sculpture from traditional and sacred buildings. Digital technology is one way to maintain, inherit and introduce the art and culture of Bali to the world, in an effort to maintain and inherit the culture hence required a management, documentation and dissemination of information by utilizing technology that can be poured in the form of games, educational software, digital music or animated films and others [2], other digital technology areas that can be utilized for preservation of art and culture is the processing of digital images and pattern recognition, one of the problems in the field of

© Springer Nature Singapore Pte Ltd. 2017
A. Mohamed et al. (Eds.): SCDS 2017, CCIS 788, pp. 43–55, 2017.
https://doi.org/10.1007/978-981-10-7242-0_4

pattern recognition is the classification of images into a particular class. Balinese carving are very complex and rich with variations on each species, it is a combination of one motif and another, if the pattern of Balinese carvings poured into digital images and analyzed, it would be difficult to make recognition and classification, then it required a method of feature extraction to help obtain the special characteristics of the input image so that later can facilitate the process of recognition and classification to provide information related to Balinese carving motif and contribute to preserve the art and culture of Bali.

This study utilizes a form of Balinese carvings that are dominant containing arches so that the approach used for feature extraction is the gradient of the line. The Histogram of Oriented Gradient (HOG) method is a method of feature extraction that utilizes the distribution of image gradients at a particular orientation by utilizing cells and blocks, but the problem with HOG is to obtain accurate recognition and classification of the selection of cell sizes and blocks must be done manually and randomly selecting cell and block size variations [3]. Another method that can do the gradient calculation is Canny edge detection method, the advantage of this method is to detect the weak edge properly because this method will remove the noise with gaussian filter before detection the edge, but the constraint that occurs is the size of the resulting image before and after the edge detection is the same so that in the introduction process will take a long time if the input image used has a large size, this constraint requires a support method that can accelerate the process of introduction and classification of Balinese carvings, Principal Component Analysis (PCA) is an appropriate method for reducing image size and minimizing memory usage in the data training process so that it takes less time for recognition and classification [4]. Features or characteristics derived from both the feature extraction methods of HOG and PCA are further trained using artificial neural network Learning Vector Quantization (LVQ) that performs competitive and supervised learning and intelligently classifies data into specific classes according to training data, the results of recognition with best accuracy will be chosen as the right method for the pattern recognition of Balinese carving.

2   Methodology

2.1   Digital Image Processing

Digital image is a function of two dimensional light intensity f (x, y), where x and y denote spatial coordinates. The f value at each point (x, y) shows the color level (value) of the image at that point [5].

2.2   Image Acquisition

The result of image acquisition consists of six types of carving motif, karang gajah, karang goak, karang tapel, mas-masan motif, kakul motif, and patra cina. The process of image acquisition used a digital camera with posisition that the main object can meet at least 80% of the size of an image of the acquisition.

2.3 System Overview

The comparison of Balinese carvings recognition used HOG and PCA methods. Overview of the pattern recognition with HOG and PCA methods shown in Figs. 1 and 2.

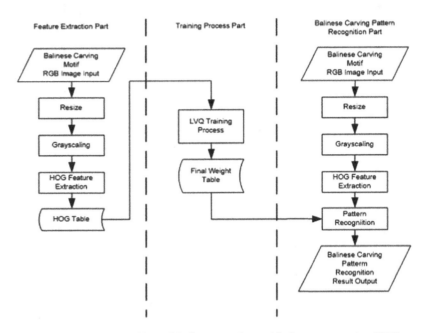

Fig. 1. Pattern recognition of Balinese carvings with feature extraction HOG

The system is designed into three main parts, the establishment of feature data, the data training section, and the pattern recognition of Balinese carvings. In the character forming data section is divided into three main processes, the first process is preprocessing shown in Fig. 1 consists of the process of image size changes or resize, grayscaling and in Fig. 2 added Canny edge detection process. The second process of feature extraction, in Fig. 1 used feature extraction by the HOG method, in Fig. 2 used feature extraction by the PCA method. The third process is the process of storing feature data into tables.

The training section consists of a training process with LVQ and the process of storing the final training weight on the final weight table. The recognition part is the same part as the formation of characteristic data but added recognition process using data from the final weight table to obtain the output data.

2.4 Preprocessing

Preprocessing serves to prepare the image or signal in order to produce better characteristics in the next stage. The process include resize the image, grayscaling or

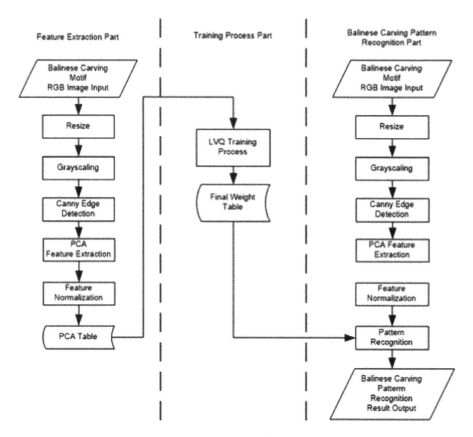

Fig. 2. Pattern recognition of Balinese carvings with feature extraction PCA

changes the image to grayscale, and edge detection process, the process for resizing the image is performed using an interpolation method that obtains the average value of a given region, this process can be illustrated in Fig. 3.

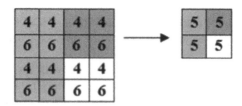

Fig. 3. Rezising image using interpolation

Grayscaling is the process of converting the image composition to one channel value with a range of one value being between 0–255, the process of converting RGB image to grayscale can be calculated by Eq. (1) [6].

$$0.2989 * R + 0.5870 * G + 0.1140 * B \tag{1}$$

Further preprocessing is edge detection using the Canny method, the steps for the Canny edge detection process as follows.

1. Reduce input image noise with gaussian filter.
2. Calculate the intensity of the gradient Gx and Gy with the sobel operator and determine the direction of the edge. The equation used to calculate the gradient is shown in Eq. (2) and to calculate the edge direction is represented by Eq. (3) [5].

$$G(x, y) = \sqrt{(G_x)^2 + (G_y)^2} \tag{2}$$

$$\theta = tan^{-1}\left(\frac{G_y}{G_x}\right) \tag{3}$$

3. Directions based on the obtained angle are:

$$Degree(\theta) = \begin{cases} 0, & 0 \leq angle < 22,5 || angle \geq 157,5 \\ 45, & 22,5 \leq angle \leq 67,5 \\ 90, & 67,5 \leq angle < 112,5 \\ 135, & 112,5 \leq angle < 157,5 \end{cases}$$

4. Streamline the edge with non-maximum suppression.
5. Binaries with two thresholding T1 and T2.

2.5   Feature Extraction

The feature extraction process used two ways with HOG method and PCA method as a comparison. The HOG method describes the intensity distribution or direction of the edge gradient of an object in the image. The steps for the HOG feature extraction process are as follows [7].

Calculate the gradient value of each pixel input image, it is grayscale image using filter [−1 0 1] and [−1 0 1]^T that shown in Eqs. (2) and (3).

1. Divide the image into several cells with size n × n pixels.
2. Determining the number of orientation bins used in histogram creation or also called spatial orientation binning.
3. Then the cells grouped into larger sizes and named block.
4. The subsequent process of overlapping normalization of blocks can be calculated by Eq. (4).

$$v = \frac{v_{i,j}}{\sqrt{||v_{i,j}||^2 + \epsilon^2}} \tag{4}$$

The result of the HOG is a feature used for the recognition process, the blocks produce a feature vector that will be used as a descriptor, the vector normalization process aims to obtain a data that is normal distribution, v is a normalized vector, $v_{i,j}$ is the Vector obtained from the binning process, ϵ is a small value constant.

The feature extraction with the PCA method is a transformation process that reduces a dimension value of data, PCA represents and selects the optimal base of an image vector given by the eigenvalues and transforms by reducing the image vector of a high-dimensional space into a lower dimension space. The process of feature extraction with PCA is as follows.

1. Change the image matrix of edge detection results into column vectors in a dataset.
2. Looking for the zero mean of the dataset.
3. Calculating the covariance matrix.
4. Looking for the eigenvalues of the covariance matrix.
5. Calculate the principal component.

The PCA feature extraction results will be normalized to make the data at a certain interval, the feature normalization process can be calculated by Eq. (5) [8].

$$X_{normal} = \frac{X - X_{min}}{X_{max} - X_{min}} * (S - R) + R \tag{5}$$

X_{normal} is normalized features, X is a not yet normalized feature, X_{max} is maximum feature, X_{min} is the minimum feature, S is the upper limit of normalization ($S = 1$), R is the lower limit of normalization ($R = 0$).

2.6   Training Process

Training process use to obtain the final weight that will be used as a reference in the pattern recognition process. The steps for training the data on LVQ are as follows [9].

1. The first step is to determine each output class, determine the initial weight, and set the learning rate α.
2. Compare each input with each defined weight by measuring the distance between each weight w_0 and input x_p. The equation shown in Eq. (6).

$$\|x_p - w_0\| \tag{6}$$

3. The minimum value of the comparison result will determine the class of input vectors and the new weight $\left(w_0'\right)$ can be calculated using Eqs. (7) and (8).
   For input and weight that have the same class calculated by Eq. (7).

   - For input and weight that have the same class calculated by Eq. (7).

$$w_0' = w_0 + \alpha(x - w_0) \tag{7}$$

   - For input and weight that have different class calculated by Eq. (8).

$$w_0' = w_0 - \alpha(x - w_0) \qquad (8)$$

4. The new weight replaces the initial weight (step 2), calculations will continue until the weight value does not change with new input, this requires a very large memory to perform calculations, for that performing LVQ calculations determined maximum epoch.

2.7 Pattern Recognition

The process of recognition is done by calculating the minimum value of the calculation between the final weight and the value of the characteristics obtained from the extraction process of the test image feature, the method used is eucledian distance. The equations used for calculating eucledian distance are shown in Eq. (9).

$$d = \sqrt{\sum_{k=1}^{n} (x_k - w_k)^2} \qquad (9)$$

d the distance between the feature and the final weight, the distance is used to measure the similarity degree between the feature and the final weight, x_k is a feature derived from feature extraction, w_k is the final weight.

3 Results

Results on this research related to training process analysis, testing process analysis, and comparison results of Balinese carving pattern recognition method.

3.1 Training Process Results with LVQ

Training process needs to make the artificial neural network system can learn based on case/pattern until the artificial neural network system can recognize the pattern, if the output produced by the network does not meet the target then the artificial neural network will do the renewal of the weight. The training process will stop if epoch = maximum epoch or previous weight (w_0) = new weights (w_0). Tables 1 and 2 are the training process results of Balinese carving pattern using PCA and HOG feature extraction that was taught by LVQ.

In PCA data obtained the smallest Mean Square Error (MSE) result on upper threshold 0.01 and lower threshold 0.001 with learning rate (alpha) value 0.0010 and epoch 1000.

In HOG data obtained the smallest MSE on the variation of cell 8 × 8 and block 2 × 2 with alpha/learning rate value 0.0010 and epoch 1000, compared with PCA then HOG obtained greater MSE.

The training data will be retested on the basis of the final weights obtained in the training process to get the accuracy. Tables 3, 4 and 5 are test results of training process using PCA and HOG.

Table 1. Training process results with PCA

No	Feature extraction	Up threshold	Low threshold	Alpha	Max epoch	MSE
1	PCA	0.01	0.001	0.0010	1000	2.9463e−13
2	PCA	0.01	0.001	0.0025	1000	2.4815e−11
3	PCA	0.01	0.001	0.0050	1000	1.6307e−09
4	PCA	0.01	0.001	0.0075	1000	2.1968e−07
5	PCA	0.01	0.001	0.0100	1000	1.5407e−06
6	PCA	0.1	0.01	0.0010	1000	9.9373e−09
7	PCA	0.1	0.01	0.0025	1000	6.1007e−07
8	PCA	0.1	0.01	0.0050	1000	2.5490e−06
9	PCA	0.1	0.01	0.0075	1000	5.2965e−06
10	PCA	0.1	0.01	0.0100	1000	8.1809e−06
11	PCA	0.2	0.1	0.0010	1000	5.7268e−09
12	PCA	0.2	0.1	0.0025	1000	5.5475e−07
13	PCA	0.2	0.1	0.0050	1000	3.7159e−06
14	PCA	0.2	0.1	0.0075	1000	9.2270e−06
15	PCA	0.2	0.1	0.0100	1000	1.6278e−05

Table 2. Training process results with HOG

No	Feature extraction	Cell (n)	Block (n)	Alpha	Max epoch	MSE
1	HOG	8	2	0.0010	1000	8.2492e−12
2	HOG	8	2	0.0025	1000	2.2864e−11
3	HOG	8	2	0.0050	1000	2.1923e−11
4	HOG	8	2	0.0075	1000	4.4222e−11
5	HOG	8	2	0.0100	1000	1.3116e−10
6	HOG	16	2	0.0010	1000	5.4179e−11
7	HOG	16	2	0.0025	1000	1.5671e−10
8	HOG	16	2	0.0050	1000	1.5772e−10
9	HOG	16	2	0.0075	1000	2.6711e−10
10	HOG	16	2	0.0100	1000	6.2313e−10
11	HOG	32	2	0.0010	1000	1.0150e−10
12	HOG	32	2	0.0025	1000	1.1398e−10
13	HOG	32	2	0.0050	1000	3.6382e−10
14	HOG	32	2	0.0075	1000	4.8919e−10
15	HOG	32	2	0.0100	1000	8.8085e−10

The highest accuracy obtained by the PCA method was 22% with the upper and lower threshold values of Canny edge detection respectively 0.01 and 0.001 with recognizable images totaling 132 and not recognizing amounting to 468. Increases were obtained when the feature size was increased to 8192 (50% size of the input image). Table 4 is a table of training process accuracy using PCA with feature size 8192.

Table 3. Accuracy of training process with PCA features size 4096

No	Feature extraction	Up threshold	Low threshold	Alpha	Known imagery	Unknown imagery	Accuracy %
1	PCA	0.01	0.001	0.0010	132	468	22.00
2	PCA	0.01	0.001	0.0025	130	470	21.67
3	PCA	0.01	0.001	0.0050	100	500	16.67
4	PCA	0.01	0.001	0.0075	100	500	16.67
5	PCA	0.01	0.001	0.0100	100	500	16.67
6	PCA	0.1	0.01	0.0010	130	470	21.67
7	PCA	0.1	0.01	0.0025	111	489	18.50
8	PCA	0.1	0.01	0.0050	110	490	18.33
9	PCA	0.1	0.01	0.0075	106	494	17.67
10	PCA	0.1	0.01	0.010	105	495	17.50
11	PCA	0.2	0.1	0.0010	124	476	20.67
12	PCA	0.2	0.1	0.0025	123	477	20.50
13	PCA	0.2	0.1	0.0050	114	486	19.00
14	PCA	0.2	0.1	0.0075	113	487	18.83
15	PCA	0.2	0.1	0.010	113	487	18.83

Table 4. Accuracy of training process with PCA features size 8192

No	Feature extraction	Up threshold	Low threshold	Alpha	Known imagery	Unknown imagery	Accuracy %
1	PCA	0.01	0.001	0.0010	146	454	24.33
2	PCA	0.1	0.01	0.0010	134	467	22.17
3	PCA	0.2	0.1	0.0010	135	465	22.50

Table 5. Accuracy of training process with HOG

No	Feature extraxtion	Cell (n)	Block (n)	Alpha	Known imagery	Unknown imagery	Accuracy %
1	HOG	8	2	0.0010	581	19	96.83
2	HOG	8	2	0.0025	564	36	94.00
3	HOG	8	2	0.0050	548	52	91.33
4	HOG	8	2	0.0075	536	64	89.33
5	HOG	8	2	0.0100	527	73	87.83
6	HOG	16	2	0.0010	579	21	96.50
7	HOG	16	2	0.0025	558	42	93.00
8	HOG	16	2	0.0050	545	55	90.83
9	HOG	16	2	0.0075	537	63	89.5
10	HOG	16	2	0.0100	537	63	89.5
11	HOG	32	2	0.0010	600	0	100.00
12	HOG	32	2	0.0025	586	14	97.67
13	HOG	32	2	0.0050	576	24	96.00
14	HOG	32	2	0.0075	575	25	95.83
15	HOG	32	2	0.0100	569	31	94.83

The feature size that changed to 8192 has an accuracy increase of 24.33% with upper threshold 0.01, lower threshold 0.001, and alpha/learning rate 0.001. Table 5 is a table of training process accuracy using HOG.

The highest accuracy obtained from HOG method is 100% with variation of cell size 32 × 32, block 2 × 2, and alpha/learning rate 0.0100 with all images are recognized.

3.2 Testing Process Results

The best five accuracies training data will be used for reference to the testing process, as for the best five accuracy of the PCA training data that will be tested again shown in Table 6.

Table 6. Accuracy of testing process with PCA

No	Feature extraction	Up threshold	Low threshold	Alpha	Known imagery	Unknown imagery	Accuracy %
1	PCA	0.01	0.001	0.0010	71	229	23.67
2	PCA	0.01	0.001	0.0025	58	242	19.33
3	PCA	0.1	0.01	0.0010	40	260	13.33
4	PCA	0.2	0.1	0.0010	51	249	17.00
5	PCA	0.2	0.1	0.0025	50	250	16.67

The highest accuracy obtained in the test from the upper threshold 0.01, the lower threshold 0.001 and alpha value 0.0010 with the recognition accuracy is 23.67%. Improvement efforts were made by changing the feature size from 4096 to 8192, the accuracy from the feature size 8192 shown in Table 7.

Table 7. Accuracy of testing process with PCA features size 8192

No	Feature extraction	Up threshold	Low threshold	Alpha	Known imagery	Unknown imagery	Accuracy %
1	PCA	0.01	0.001	0.0010	55	245	18.33
2	PCA	0.1	0.01	0.0010	40	260	13.33
3	PCA	0.2	0.1	0.0010	49	251	16.33

In the features size 8192 obtained the highest accuracy 18.33% with the recognizable image amounted to 55 and the unrecognized amounted to 245. The highest accuracy obtained from the upper threshold value 0.01 and the lower threshold value 0.001 with the same alpha value 0.001, in this case accuracy not increase when using feature size 8192 with testing process.

The next test is using HOG, the best five accuracy of the HOG training data that tested again with the testing process shown in Table 8.

Table 8. Accuracy of testing process with HOG

No	Feature extraction	Cell (n)	Block (n)	Alpha	Known imagery	Unknown imagery	Accuracy %
1	HOG	8	2	0.0010	191	109	63.67
2	HOG	16	2	0.0010	213	87	71.00
3	HOG	32	2	0.0010	270	30	90.00
4	HOG	32	2	0.0025	229	71	76.33
5	HOG	32	2	0.0050	217	83	72.33

The highest accuracy of the HOG testing process was obtained from the cell size 32 × 32, image blocks 2 × 2, and alpha/learning rate 0.001 with 270 recognizable images and 30 unrecognizable images.

3.3 Comparison Analysis for Recognition Patterns of Balinese Carving with HOG and PCA

The best accuracy of testing process with HOG method reached 90%, but the the best accuracy of testing process with PCA only reached 23.67%. This is because PCA used to reduce dataset can not guarantee that the accuracy obtained from matrix reduction results will increase, but PCA reduction can support systems in accelerating computation. Research by [4] mentions that PCA is a method to reduce image dimension that can reduce memory usage and shorten time during data training process.

Another factor that causes the low accuracy obtained from the recognition of Balinese carving patterns with the Canny edge detection and PCA approach is the lack of accurate edge detection of the engraving object, this is because the background is too dirty (the background has many other object or the same light intensity that causes the background detected as an object). [5] states that the edge of an object formed from point one to the next point in the image can be detected by the difference in the intensity of light, if the degree of difference in light intensity is high then the edge can be detected clearly, and if the low light intensity the resulting edge will be unclear. [10] in his research describes PCA as a statistical technique used to convert the dimensions of space from high dimension to low dimension and finds a pattern in the data using standard deviation concepts, mean values, variance, covariance and algebraic matrices from mathematical field statistics, in this case the image is a data that has a high dimension that required a technique or method that can be used to reduce dimensions. As well as the research that has been done [4] the effect of edge detection is very high on recognition accuracy. The separation of objects and backgrounds plays an important role in obtaining good edge detection results.

The HOG method as a descriptor can produce maximum accuracy because the HOG calculates gradient values in a particular orientation, [3] in his research explaining that the selected cell size and block may affect detection quality and recognition time, the best results obtained are the results of an experimental approach based on correction gamma, gradient filter type, and cell size and block size. The best cell size and block determination on HOG does not apply to the whole case, it depends on the dataset. In the case of Balinese carvings pattern recognition the best accuracy of HOG is obtained from the largest cell size of 32×32 and 2×2 blocks with minimum overlaping blocks, feature length generated from 32×32 cell size and 2×2 block with 128×128 image size is 1×324, this size is the smallest of three variation of cell size and block. The 8×8 and 2×2 blocks feature 1×8100, 16×16 and 2×2 block features 1×1764 size, the smallest feature size can help speed up computing and provide shorter time in the data training process.

4   Conclusion

Based on the analysis of training and testing results the recognition of Balinese carvings pattern, it can be taken some conclusions as follows.

1. The test results of the training data with total data is 600 images show that HOG can produce better accuracy than PCA, HOG can recognize the whole data on cell size 32×32 and block 2×2, while PCA only able to recognize the most amount of 146 image at features size 8192 with upper threshold 0.01 and lower threshold 0.001 for Canny edge detection.
2. The test results of the testing data with total data is 300 images with each motif amounting to 50 images shows that HOG produces a better accuracy than PCA that reached 90% accuracy with 32×32 cell size and 2×2 block, while the accuracy with PCA only reached 23.67% with features size 4096 (25% of image size), upper threshold and lower threshold value Canny detection is 0.01 and 0.001.
3. The testing process using the training data or testing data of the HOG method produces better accuracy than the PCA method because the feature extraction of HOG method performs gradient calculations on a particular orientation and does not depend on the edge of the image; the HOG feature extraction utilizes gradients on the orientation of specific cells and blocks.
4. In the PCA method the acquisition of accuracy is influenced by the edge detection process, in the case of Balinese carving pattern recognition found a problem that the separation of objects with a background unperfect give a very big influence on edge detection results. The PCA feature extraction also does not provide assurance to improve accuracy, but PCA as a method to reduce the image can accelerate computing and reduce memory usage.

Based on the results obtained, it can be suggested some things for further research such as adding a method that can separate the object and background well so that the edge detection process produces maximum results, can compare other methods to perform feature extraction e.g. comparison of HOG and SIFT.

References

1. Safril, M.A.: Revitalisasi Identitas Kultural Indonesia di Tengah Upaya Homogenisasi Global. Jurnal Global & Strategis (online), FISIP Universitas Airlangga. Surabaya, Indonesia (2011)

2. Sitokdana, M.N.N.: Digitalisasi Kebudayaan Indonesia. Seminar Nasional Teknologi Informasi dan Komunikasi (SENTIKA), Yogyakarta, Indonesia (2015)
3. Kachouane, M., Sahki, S., Lakrouf, M., Ouadah, N.: HOG based fast human detection. In: 24th ICM International Conference on Microelectronics, Algeria, pp. 1–4 (2012)
4. Chanklan, R., Chaiyakhan, K., Hirunyawanakul, A., Kerdprasop, K., Kerdprasop, N.: Fingerprint recognition with edge detection and dimensionality reduction techniques. In: Proceedings of the 3rd International Conference on Industrial Application Engineering, Japan (2015)
5. Gonzalez, R.C., Woods, R.E.: Digital Image Processing. 2/E Publishing. Company, Inc., USA (2002)
6. Güneş, A., Kalkan, H., Durmuş, E.: Optimizing the color-to-grayscale conversion for image classification. Sig. Image Video Process. **10**(5), 853–860 (2015). Springer, London
7. Dalal, N., Triggs, B.: Histograms of oriented gradients for human detection. In: IEEE Computer Society Conference on Computer Vision and Pattern Recognition (CVPR 2005), San Diego, CA, USA, vol. 1, pp. 886–893 (2005)
8. Gajera, V., Gupta, R., Jana, P.K.: An effective multi-objective task scheduling algorithm using Min-Max normalization in cloud computing. In: 2016 2nd International Conference on Applied and Theoretical Computing and Communication Technology (iCATccT), Bangalore, pp. 812–816 (2016)
9. Ruslianto, I., Harjoko, A.: Pengenalan Karakter Plat Nomor Mobil Secara Real Time. IJCCS (Indonesian J. Comput. Cybern. Syst.) **7**(1), 35 (2013). Yogyakarta, Indonesia
10. Subbarao, N., Riyazoddin, S.M., Reddy, M.J.: Implementation of FPR for save and secure internet banking. Global J. Comput. Sci. Technol. **13**(9 Version 1.0) (2013). USA

Feature Extraction for Image Content Retrieval in Thai Traditional Painting with SIFT Algorithms

Sathit Prasomphan[(✉)]

Department of Computer and Information Science, Faculty of Applied Science,
King Mongkut's University of Technology North Bangkok,
1518 Pracharat 1 Road, Wongsawang, Bangsue, Bangkok 10800, Thailand
ssp.kmutnb@gmail.com

Abstract. This research presents a novel algorithm for feature extraction in Thai traditional painting by using keypoint generated from SIFT algorithms. The proposed algorithms aim to retrieve knowledge inside an image. Content Based Image Retrieval (CBIR) technique was applied. The generated keypoint and descriptors were used as input in neural network for training. Neural network technique was used for image classification to get details of image. The algorithms were tested with the corridor around the temple of the Emerald Buddha. The experimental results show that the proposed framework can efficiently give the correct descriptions to the image compared to using the traditional method. The proposed algorithms can be used to classify the type of Thai traditional painting which can provide the accuracy about 70–80% in average.

Keywords: Content Based Image Retrieval · SIFT algorithms
Feature extraction

1 Introduction

Thai traditional painting [1] is one kind of arts that can be shown the cultural of country. It showed the feeling of life and mind of Thais. The contents inside an image are created from the former of nation history also it showed the national characteristics and the special style of nation [1]. Thai traditional painting are shown on the wall of building, for example, the temple and palace or special things such as silk, furniture [2]. Most of image's contents are shown the history of the people life, the history of the Buddhist or the famous literature. Example of the well known Thai traditional painting are the painting surround the Emerald Buddha or also known as in the other name, Wat Phra Kaew. A story of the Ramayana literary was painted on the balcony around the temple. There are 178 rooms built since the reign of King Rama I by writing and repairing during the reign of King Rama I and King Mongkut. Then, the coming of the reign of King Chulalongkorn, the painting were renovated on the occasion of 100 years Rattanakosin era. In this painting, several characteristics from different character were occurred. Those who are not interested in historical studies may not understand the characters or the manner in which the characters appear on those wallpapers [3]

© Springer Nature Singapore Pte Ltd. 2017
A. Mohamed et al. (Eds.): SCDS 2017, CCIS 788, pp. 56–67, 2017.
https://doi.org/10.1007/978-981-10-7242-0_5

especially the foreign tourists who interest in the painting. The nature of the canvas painting is difficult to extract within the image due to the large number of stripes and the appearance of the image.

There are several algorithms for an image content retrieval for generating description inside an image [4–6]. For example, Socher et al. [7] introduced a method for finding relationship between image and sentence by using recursive neural network [8]. A model for generating language and description for an image was proposed by Karpathy and Li [9] and Farhadi et al. [10]. Both of these two methods were shown the rules for generating relationship between language and image. [9] used the recurrent neural network which was tested with several databases such as Flickr8K, Flickr30K and MSCOCO. In [10], they used the scoring method for calculating the relationship between sentence and image. They suggested the model for simulating the relationship between image space, meaning space, and sentence space.

A large-scale image database and World-Wide Web Image needs an image search engine for a person who need to collect images from this large database. One of mechanism for searching is the technique calls content-based image retrieval. Color, texture or shapes were used inside the algorithm for searching a required image. A low-level image features and high-level image semantics has a big gap for this technique. Researchers have been interested in the experiments for studying image retrieval. Several researches were introduced. Kulikarnratchai and Chitsoput [11] presented the research "Image retrieval by weight distribution with Gaussian distribution for a color histogram in the HSV color model", the paper presented an image retrieval model with HSV (Hue, Saturation, Value), which using color histograms to the image index. The method presented in the article was compared by using the color histogram. The color to be compared was weighted differently according to the weight distribution of the Gaussian for making the higher retrieval performance. Sangswang [12] offered research name "Image search by histograms comparison using vector models". The research, proposed an image retrieval procedure with color histogram using vector modeling. The system can convert the image into a histogram and store it in the database. The image/histogram search was performed by comparing the histogram between images and histogram of the image in the database using the similarity in the vector model. They measured the similarity of the image with approximated values. The similarity value closest to 1 indicates that the image is similar to the one most wanted. Kerdsantier and Sodanil [13] presented the research title "Image retrieval with color histogram by using fuzzy set theory", the process is to convert RGB color features into color features based on the HSV model, which is similar to the perception of the human eye. The color values are determined by the H (Hue) value, which is the main color indicator and uses S (Saturation) value, and V (Brightness value) value. Color considered being able to more clearly attribute for using in the fuzzy set theory. Then bring the image features of interest to compare with the features of the image in the database. The results will be visualized in a database with color features similar to the underlying image. Facktong [14] introduced research "Retrieval of digital amulets by feature extraction and the nearest neighbor techniques". The paper presented digital amulets retrieval using the nearest neighbor method. There are methods of taking pictures of amulets, templates, and test images from digital cameras. The research focused on the retrieval of powdered amulets. Features used to extract image data. Gray surface matrix analysis was performed using a gray matrix. Collecting statistic values come from 10 surface texture analyses and 44 systematic powders testing using a total of 1,400 images. The image is divided into 880 images prototype, and image has not been recognized 440 images. The accuracy was 98.87%. Based on the research and the problem, our objective is interested in developing an algorithm for retrieving contents in the images, focusing on the Thai traditional painting images in the temple of the Emerald Buddha to find descriptions and stories which is appeared in the images for easy to search. Most of research in image retrieval is based on its content. The following feature is importance to describe the image: color, texture and shape. The current issue which is difficult to perform is how to add the caption of image as same as the aspect of human ability. Content based retrieval systems use query images to retrieve required images. It brings the important characteristics that have been compared to the key characteristics of each image stored in the database. If any image is as important or similar to a source image. It will be displayed as a search result for the user. Most research uses color, which is one of the key characteristics of the image that is often used in image retrieval. Based on content using color histograms is a popular way of describing the color distribution of images, but in the case of large databases, there may be some images that are different but there is a similar color histogram. To fix such problems, in [2, 5, 6] proposed to index the image with the essential characteristics of color. By using human color perception to colorize the uniform, in combination with the Gaussian distribution technique, it can be computed in order to obtain the color correlation in the histogram of each image. By applying colorimetric data in the histogram, it can be applied to enhance the retrieval of images from the image database.

From that reason, if we were able to take a photo and take that photo to search for a similar image and access the details that were visible through the photos, we could better understand the history. Also, if a poem associated with the picture that appeared in the corridor around the temple is known, it will benefit to the viewer or interested person. For this reason, the researcher is interested in developing a technique of image retrieval based on content based image retrieval, applied to murals, the corridor around the temple of the Emerald Buddha.

The remaining of this paper is organized as follows. At first, we show the characteristics of Thai traditional painting. Next, the process for generating image description based on SIFT algorithms and neural network is discussed in Sect. 2. Results and discussions are discussed in Sect. 3. Finally, the conclusions of the research are presented in Sect. 4.

2 Proposed Algorithms

In this section we first briefly review the fundamental of characteristics of Thai traditional painting. Next, the image content retrieval is described. After that, image content description generation is introduced based on recognizing keypoint. Finally, neural network is shown for learning the image description and generating that description to the Thai traditional painting image.

A. Characteristics of Thai Traditional Painting

Thai traditional painting is very popular on the walls of Buddhist buildings and high-rise buildings. Tiranasar [15] explains the concept of Thai traditional art that are mostly relevance with the behaviors of Thai and their life which have a relation with the religions and the natural environment. Five categories of Thai traditional paintings were explained. (1) The relationship between emotions of human and the painting, for example, sad, love, and anger. (2) The relationship between behaviors among human, for example the kings and their people. In these categories, different levels of relationship between them were displayed. (3) The relationship between people and the environment. (4) The relationship between human and their belief. (5) The relationship between human and the moral value such as good and bad in the scene of previous life of the Lord Buddha.

Other categories of Thai paintings can be divided into four levels: idealistic, combination of idealistic and realistic, realistic, and surrealistic. At the idealistic level is considered the stories of the Lord Buddha, kings and heavenly beings. This level is the highest level. At the second level is the combination of idealistic and realistic. They focus on the change in tradition from the past to present which express the contents and philosophy of religion. In this level it includes high status people. For the realistic level, the society and environment in that era was expressed. The emotion of people, the way of life of them, the history and the atmosphere in that particular time is included. For the lowest level which is surrealistic, the artists used their imagination to create the work to draw the emotion of horror and melancholy to warn the viewers to be aware of the results of the wrong doings [15].

The examples of famous Thai traditional painting are the painting inside the temple of the Emerald Buddha as shown in the Figs. 1, 2 and 3.

Fig. 1. Example of Thai traditional painting (Giant).

B. Scale Invariant Feature Transforms (SIFT) Algorithms

Scale-invariant feature transform (SIFT) [16, 17] is a powerful framework to retrieve image in a large database which was published by David Lowe. Keypoint in an image was extracted by using the framework. The meaning of keypoint is the interesting points in each image. Keypoint refers to a pixel in the image which is changed the orientation from two-dimensional of brightness levels surrounded a keypoint. Local features in images were detected and described. The main advantage of this algorithm

Fig. 2. Example of Thai traditional painting (Monkey).

Fig. 3. Example of Thai traditional painting (Human).

is the features are invariant to image scale and rotation. To perform reliable recognition, it is important that the features extracted from the training image be detectable even under changes in image scale, noise and illumination. The scale or the orientation angle position of image was not included directly for the comparison between image in SIFT algorithm. Accordingly, the calculated features can be used to compare the features in more easily and accurately, more precisely. To detect a keypoint from an input image, a series of x, y coordinates were obtained firstly. Next, the keypoint *description* or *descriptor vector*s were calculated. These vectors were calculated from the brightness of the pixels in the area surrounded keypoint.

To calculate the most similar image between the query image and reference images in database, *descriptor vector*s were used. Following are the major steps for generating the set of image features:

1. **Scale-space extrema detection:** This process is for detecting points of interest, which is termed keypoint in the SIFT. The most importance feature of image which getting from the framework does not depend on the size or orientation of an image. Gaussian function in each octave is performed for burring image. Several burring scale are created by burring normal scale and increasing scale parameter which effect to the burring image. Octave size in each step will be half reduced.

2. **Keypoint localization:** Scale-space extrema detection generated too many keypoint candidates. Locating maxima/minima in Difference of Gaussian image is used for getting keypoint and is used for finding sub pixel maxima/minima. After that, a data fit to the nearby data for accurate location, scale, and ratio of principal curvatures is performed [17]. This information allows points to be rejected that have low contrast or are poorly localized along an edge.

3. **Orientation assignment:** Collected magnitude and orientation of the gradient of the area around the keypoint to determine a keypoint direction. Next, take the magnitude and direction of the gradient of a pixel around the keypoint to create histograms, which the x-axis is degrees, and the y-axis is the magnitude of the gradient of the pixel. There is a condition that if any keypoint has a peak of the histogram more than 80% of the maximum peak, it will break into a new keypoint.

4. **Keypoint descriptor:** 16×16 windows are created. Next, the window is divided to 4×4 windows with 16 sets. Magnitude and gradient was calculated. After that, create histogram with 8 bins from these windows. The 128 feature vectors are obtained. The 128 feature vectors are used to the matching process.

The process in each step can be shown in Fig. 4.

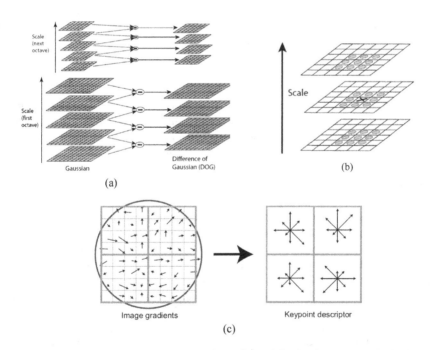

Fig. 4. (a) Gaussian function in each octave is performed for burring image. (b) Scale-space extrema detection generated too many *keypoint* candidates. Locating maxima/minima in Difference of Gaussian image is used for getting *keypoint* and is used for finding sub pixel maxima/minima. (c) 16×16 windows are created. Next, the window is divided to 4×4 windows with 16 sets. Magnitude and gradient was calculated. After that, create histogram with 8 bins from these windows. The 128 feature vectors are obtained [11].

C. Image Based Retrieval

In the past, techniques used in image retrieval rely on image caption retrieval. It is not a true characteristic of the image and is not suitable for large databases at present because image definitions depend on each user. As the database grows larger, defining individual images based on just a few words is quite difficult. In addition, some image databases may consist of several types of image data in the same group, such as geographic imagery databases, which may consist of multiple ground images, multiple tree types, and so on. It also requires human beings to define the image. Image specificity is a feature that can be obtained with image processing algorithms. The basic characteristics of the image consist of three parts: color, shape and texture. *Color* is a unique feature that plays an important role in the image retrieval system. Because color is remarkable, it can be clearly seen from the picture. It can also help in identifying things. *Shape* is the specificity of the image that describes the shape and appearance, as well as the size of the object within the image. This makes it possible to distinguish between differently shaped objects. *Texture* is a unique feature used to describe the roughness, resolution, or complexity of an object within an image. Each image may be composed of objects with different surface characteristics. Surface analysis helps to better distinguish objects. Image indexing is the implementation of a visual identity, such as a histogram of colors that creates a specific vector of each image in the database and stores it. The individual characteristics of the individual images are in the form of numeric values, n values or n-dimensional vectors, which are the characteristic vectors of the image. These characteristic vectors are used as image indices, which represent the points in the n dimensional space [18–21].

Feature extraction is the basis of content-based image retrieval [19]. Features may include both text-based features and visual features (color, texture, shape, faces). Within the visual feature scope, the features can be further classified as general features and domain. The former includes color, texture, and shape features while the latter is application-dependent and may include, for example, human faces and finger prints [22].

D. Image content Generator System

In this section, we describe the process of image content generator system as show in Fig. 5. The following steps will perform for retrieving the description of image content:

Pre-processing: The training set of Thai traditional painting image is stored in database. The query image which wants to know its descriptions will be taken from camera. After that gray scale image was obtained by transforming the RGB image to gray scale image. The quality of image is enhanced by using the image enhancement algorithms. After that, edge detection is performed. The edge that pass through or near to the interested point is checked by measure the different of intensity of nearest points or finding the line surrounding the object inside image. The clearly edge of image can be easily used for the classification process. The edge is formed by the difference of intensity from one point to other points. The high different of these points affect to the accurate edge, if not, the edge is blurred. Too many algorithms were used to find edge. In this research, Laplacian was used to detect edge of image.

Feature extraction: This process used for retrieving the identity of each image. SIFT algorithms is used for getting keypoint inside image. The process for feature extraction by using SIFT algorithms was explained in the previous section.

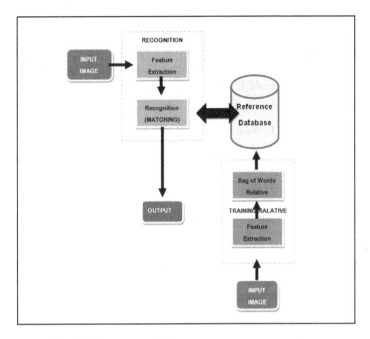

Fig. 5. The process of image content generator system.

Matching: For matching the most similar image between query image and the reference images, neural network was used. In this paper, neural networks have been created in order to distinguish for keypoint of an image. The architecture of neural network is described as following: one input layer, one hidden layer, and one output layer was created. 128 input attributes from the keypoint and descriptors generated from SIFT algorithms was used. Feed-forward multilayer neural network with back propagation learning algorithms was used. The training set was trained by setting the stopping criteria and the network parameters.

Set of inputs and desired output belongs to training patterns were fed into the neural network to learn the relationship of data. Inside hidden layer is the process for adjusting weight which connected to each node of input. To calculate the error between desired output and its calculated output, the root mean square error (RMSE) is calculated. The predefined RMSE value was set. If the RMSE value does not satisfy the predefined values, it will propagate error back to the former layer. This will be done from the direction of the upper layer towards the input layer. This algorithm will adjust weight from initial weight until it gives the satisfied mean square error.

Generating description: In the process of generating description after matching process, the algorithm was used descriptions of that image inside the database to show and set description to the input image. Accordingly, inside the image database, instead of contains only the reference image, it will contain the descriptions of the reference image as well.

3 Experimental Setup

A. Experimental setup
In this research, images used in the experiments were mural images around the balcony, the temple of the Emerald Buddha. The requirements were as follows (Table 1).

Table 1. Number of samples in the experiments

Character type	Number of image
Human	67
Giant	44
Monkey	46
All	**301**

1. Picture used in the experiment were murals around the temple of the Emerald Buddha.
2. Images used in the experiment had at least 80% of the characters in the image.
3. Pictures used in the experiment must be taken vertically.
4. The image used in the experiment must be a .jpeg or .png extension.

B. Performance indexed
Tables 2, 3, and 4 is the confusion matrix. The confusion matrix is graphical tools that reflect the performance of an algorithm. Each row of the matrix represents the instances of a predicted class, while each column represents the instances of an original class. Thus, it is easy to visualize the classifier's errors while trying to accurately predict each original class' instances. Percentage of the test data being classified to the original image was shown inside the table.

Table 2. Confusion matrix of the proposed algorithms (SIFT with neural network)

Character type	Image recognition		
	Human	Giant	Monkey
Human	**80.67**	4.43	7.35
Giant	4.07	**79.35**	3.29
Monkey	7.43	3.79	**82.47**

C. Comparing algorithms
In this paper we used the following algorithms for comparing: SIFT with Euclidean distance, SIFT with neural network, SIFT with nearest neighbors, and using only support vector machine (SVM) for classifying images. We used SIFT algorithms to compare the efficiency of generating descriptions which used only the descriptors with several classification method, for example, Euclidean distance, k-nearest neighbors and neural network. The accuracy of the proposed method was measured by the accuracy of classification result.

Table 3. Confusion matrix of SIFT with K-NN

Character type	Image recognition		
	Human	Giant	Monkey
Human	**60.82**	18.43	21.98
Giant	18.93	**62.78**	23.29
Monkey	21.98	23.29	**63.72**

Table 4 Confusion matrix of SIFT with Euclidean distance

Character type	Image recognition		
	Human	Giant	Monkey
Human	**70.28**	15.68	14.38
Giant	15.68	**75.22**	12.11
Monkey	14.38	12.11	**73.74**

We compared each of this method with our proposed algorithms which used the combination of SIFT algorithms and neural network algorithms.

D. Experimental results

The experimental results for Image Content Retrieval in Thai traditional painting by using keypoints generated from SIFT will be shown. The classification results by using the proposed algorithms to classify the type of characters were used. Accordingly, the accuracy of our image description generator depends on the classification accuracy. The classification accuracy were shown in Tables 2, 3, 4 and 5. Table 2 shows the confusion matrix of the proposed algorithms which is the combination between SIFT algorithms and neural network. It gives the accuracy 80.67%, 79.35% and 82.47% in Human, Giant, and Monkey. Table 3 shows the confusion matrix of using the combination between SIFT algorithms and k-nn algorithm. This algorithm gives the accuracy 60.82%, 62.78% and 63.27% in Human, Giant, and Monkey. Table 4 shows the confusion matrix of using the combination between SIFT algorithms and Euclidean distance. This algorithm gives the accuracy 70.28%, 75.22%, and 73.64% in Human, Giant, and Monkey. Table 5 shows the confusion matrix of using the SVM. This algorithm gives the accuracy 72.98%, 69.63%, and 69.78% in Human, Giant, and Monkey. The experimental results confirms that using keypoint of image which is generated from the SIFT algorithms and using the neural network for training the keypoint to get the type of character of image can be successfully used for generating

Table 5. Confusion matrix of the SVM

Character type	Image recognition		
	Human	Giant	Monkey
Human	**72.98**	13.72	12.61
Giant	13.61	**69.63**	20.29
Monkey	12.61	20.95	**69.78**

the description of image as shown in table. The proposed algorithms can be used to classify the type of Thai Traditional Painting which can provide the accuracy about 70-80% in average.

4 Conclusion

This research presents novel algorithms for generating image description via keypoint descriptor from SIFT algorithms. The keypoint descriptors of an image are used to distinguish identity of an image. We extract the important feature from Thai Traditional Painting image and provide the new architecture for neural network to reduce the difficulty of the classification process and reduce number of feature to be sent to neural network compared with the other method. We have presented a new approach to feature extraction based on analysis of keypoint descriptor of Thai traditional painting. The algorithm was tested with mural paintings around the balcony of the temple of the Emerald Buddha. The experimental results show that the proposed framework can efficiently find the correct descriptions compared to using the traditional method.

Acknowledgment. This research was funded by King Mongkut's University of Technology North Bangkok. Contract no. KMUTNB-60-GEN-016.

References

1. Special Committee on Thai Book Development: Our Thailand. 2nd edn, p. 44. National Identity Office, Secretariat of the Prime Minister (1992). ISBN 974-7771-27-6
2. Chareonla, C.: Buddhist Arts of Thailand. Buddha Dharma Education Association Inc (1981)
3. Bantukwattanatham, S.: Thai Traditional Painting, vol. 610. Siamese Heritage, March (2001)
4. Young, P., Lai, A., Hodosh, M., Hockenmaier, J.: From image descriptions to visual denotations: new similarity metrics for semantic inference over event descriptions. TACL **2**, 67–78 (2014)
5. Hodosh, M., Young, P., Hockenmaier, J.: Framing image description as a ranking task: data, models and evaluation metrics. J. Artif. Intell. Res. **47**, 853–899 (2013)
6. Su, H., Wang, F., Li, Y., Guibas, L.J.: 3D-assisted image feature synthesis for novel views of an object. CoRR (2014). http://arxiv.org/abs/1412.0003
7. Socher, R., Karpathy, A., Le, Q.V., Manning, C.D., Ng, A.Y.: Grounded compositional semantics for finding and describing images with sentences. TACL **2**, 207–218 (2014)
8. Zaremba, W., Sutskever, I., Vinyals, O.: Recurrent neural network regularization. arXiv preprint arXiv:1409.2329 (2014)
9. Karpathy, A., Li, F.: Deep visual-semantic alignments for generating image descriptions. In: The 2015 IEEE Conference on Computer Vision and Pattern Recognition (CVPR), pp. 1–14 (2015)
10. Farhadi, A., Hejrati, M., Sadeghi, M.A., Young, P., Rashtchian, C., Hockenmaier, J., Forsyth, D.: Every picture tells a story: generating sentences from images. In: Daniilidis, K., Maragos, P., Paragios, N. (eds.) ECCV 2010. LNCS, vol. 6314, pp. 15–29. Springer, Heidelberg (2010). https://doi.org/10.1007/978-3-642-15561-1_2

11. Kulikarnratchai, P., Chitsoput, O.: Image retrieval using color histogram in HSV color sampler. In: 29th Electrical Engineering Symposium (EECON-29), pp. 1029–1032 (2006)

12. Sangswang, A.: Image search by histogram comparison using vector models. In: The 2nd National Conferences Benjamit Academic, 29 May 2012

13. Kerdsantier, P., Sodanil, M.: Color retrieval system with color histogram using fuzzy set theory. In: The 6th National Conference on Computing and Information Technology (NCCIT2010), pp. 698–703, June (2010)

14. Facktong, P.: Retrieval of digital amulets by extraction techniques and nearest neighbors. J. Inform. Technol. **9**(2), 34–40 (2013)

15. Tiranasar, A.: Thai traditional art and art education. In: The 2nd Asia-Pacific Art Education Conference, Hong Kong, 28–30 December 2004

16. Lowe, D.G.: Object recognition from local scale-invariant features. In: Proceedings of 7th International Conference on Computer Vision (ICCV 1999), Corfu, Greece, pp. 1150–1157 (1999)

17. Lowe, D.G.: Distinctive image features from scale-invariant keypoints. Int. J. Comput. Vis. **60**(2), 91–110 (2004)

18. Baeza-Yates, R., Berthier, R.N.: Modern Information Retrieval. Addison Wesley, Boston (1999)

19. Manjunath, B.S., Ohm, J.R., Vasudevan, V.V., Akio, Y.: Color and texture descriptors. IEEE Trans. Circ. Syst. Video Technol. **11**(6), 703–715 (2001)

20. World Wide Program: HSV to RGB color model source code algorithm C Programming CS1355-graphics and multimedia lab (2010). http://worldwyde-programs.blogspot.com/2010/05/hsv-to-rgb-color-model-source-code.html

21. Eakins, J., Graham, M.: Content-based image retrieval. University of Northumbria at Newcastle. Accessed 10 Mar 2014

22. Kaur, S., Banga, V.K., Kaur, A.: Content based image retrieval. In: International Conference on Advances in Electrical and Electronics Engineering (ICAEE 2011) (2011)

Modeling of the Gaussian-Based Component Analysis on the Kernel Space to Extract Face Image

Arif Muntasa[(⊠)] and Indah Agustien Siradjuddin

Informatics Engineering Department, Engineering Faculty, University of
Trunojoyo, Ry Telang Po. Box 2 Kamal, Madura, Bangkalan, Indonesia
arifmuntasa@trunojoyo.ac.id

Abstract. One of the methods to extract the image characteristics on the feature
space is Kernel Principal Component Analysis. Gaussian is a model to transform
the image to the features spaces by using kernel trick. In this paper, the new
model is proposed to add the image features to be more dominant, so that the
main image features can be raised. Two databases were used to verify the
proposed method, which are the YALE and the CAI-UTM. Three scenarios
have been applied with different training samples. The results demonstrated that
the proposed method can recognize the face image 87.41% for two training sets,
90.83% for three training sets, and 92.38% for four training sets on the YALE
database. On the CAI-UTM database, the proposed method could classify
correctly by 83.75%, 85.57%, and 87.33% for two, three, and four training sets
respectively. The comparison results show that the results of the proposed
approach outperformed to other methods.

Keywords: Gaussian kernel · Component analysis · Feature selection

1 Introduction

Image analysis has been developed to solve and aid live problem, one of them is a
biometric field. Many algorithms were created and developed by researchers to obtain
better results i.e. Eigenface and its extension [1–3], Fisherface [4–7], Laplacianfaces
[8–10], Factor analysis, Independent Component Analysis [11], metric multidimensional scaling [12], Isomap [13], Hessian, Laplacian Eigenmaps [14], and other
methods.

Eigenface is the simplest method to extract the features of the image by reducing
the dimensionality of the image. The Eigenface can reduce the dimensionality of the
image up to some data used for the training sets. The Eigenface is obtained from the
projection of the eigenvector and the training sets, whereas the eigenvector is derived
from the covariance matrix of the training sets. Eigenface can reduce the dimension of
the image up to the size of the image dimension minus the amount of data, when the
covariance matrix is based-orthogonal. However, the Eigenface can only map the
feature linearly, since it can only solve linear model in high dimensionality, while
non-linear distribution cannot be solved by the Eigenface. In fact, the used data

© Springer Nature Singapore Pte Ltd. 2017
A. Mohamed et al. (Eds.): SCDS 2017, CCIS 788, pp. 68–80, 2017.
https://doi.org/10.1007/978-981-10-7242-0_6

distribution cannot be predicted whether linear or non-linear [8, 15]. Besides, the produced features of the Eigenface depend on the amount of training sets utilized. If the number of training sets is more than the image dimensionality, then reduction dimensionality cannot be applied to obtain the features.

Fisherface has improved the weakness of the Eigenface methods, where the features can be reduced so that the number of features to become as the amount of classes. Unlike the Eigenface, Fisherface can distinguish the information, though it is not based-orthogonal, therefore the Eigenface is not better than the Fisherface. However, the Fisherface has also the weakness, i.e. inability to separate non-linear distributed features [4]. The Fisherface will also fail to obtain the features, when the within-scatter class is a singular matrix.

One of the methods that can solve non-linear data distribution is Isomap. It is one graph based-method that used to reduce the dimensionality. However, an Isomap produces a graph with topological instability, though the methods have been improved by reducing several data points on the graph. Another weakness of an Isomap is non-convex manifold because it will succeed the process is not complete.

Another non-linear method is Local Linear Embedding (LLE). It is a graphical method that similar to Isomap method, where Isomap tries to preserve the characteristics of the local data. The weakness of non-convex of the Isomap can be solved by preserving of the local data characteristics. However, several researches have reported that the LLE cannot success to visualize the data points. LLE has delivered the constraints of the data points so that they spread to undesired areas [16].

Two graphical based-methods has also weakness, therefore in this research, we proposed another approach to reduce the image dimensionality. A transformation from the image to the feature spaces is proposed to project the eigenvector of the Gaussian kernel matrix. Gaussian kernel found the distance between the points to the others of the training sets. Gaussian considers standard deviation as inner scale to record the data distribution deviation, while Gaussian kernel is a way to map image to feature spaces by Gaussian equation. The principal component is just applied the Gaussian kernel matrix to obtain the Eigenvalue and Eigenvector. In this case, the average, zero mean, and covariance matrixes are not necessary to be calculated, because the image samples have been mapped into feature spaces by Gaussian Kernel. The results of the Eigenvector are delivered to calculate the projection matrix as the image features. The image features will be further process to classify the face image.

2 Proposed Approach

Kernel trick is method to convert from image to feature spaces. On the feature spaces, there are four models to conduct kernel trick, which are Linear, Gaussian, Polynomial, and Polyplus [15, 17, 18]. Gaussian model is one of the kernel trick model considered the distance, mean, and variant of an object. Therefore, Gaussian kernel trick model can extract the more dominant features than the others, i.e. Linear, Polynomial, and Polyplus. Suppose the training and the testing sets are represented by X and Y, where X has m samples and n image dimensionalities, while Y is row matrix with n columns as described in the Eq. (1).

$$\mathcal{X} = \begin{pmatrix} x_{1,1} & x_{1,2} & x_{1,3} & \cdots & x_{1,n} \\ x_{2,1} & x_{2,2} & x_{2,3} & \cdots & x_{2,n} \\ x_{3,1} & x_{3,2} & x_{3,3} & \cdots & x_{3,n} \\ \vdots & \vdots & \vdots & \ddots & \vdots \\ x_{m,1} & x_{m,2} & x_{m,3} & \cdots & x_{m,n} \end{pmatrix} \tag{1}$$

For each testing set is written as follows,

$$\mathcal{Y} = \begin{pmatrix} y_{1,1} & y_{1,2} & y_{1,3} & \cdots & y_{1,n} \end{pmatrix} \tag{2}$$

2.1 Gaussian Kernel Matrix

Gaussian is one of models used to map from image to feature spaces [8]. Gaussian-based component analysis on the kernel can be obtained by calculation $K_x(X, X^T)$ and $K_y(Y, Y^T)$, i.e. Gaussian kernel matrix for the training and testing sets. To obtain Gaussian kernel matrix $K_x(X, X^T)$, the distance between point and others is calculated as follows,

$$\mathcal{A}_{i,1} = \sum\nolimits_{j=1}^{n} \left(\mathcal{X}_{i,j} \cdot \mathcal{X}_{i,j} \right) \tag{3}$$

The value of i has range as follows: $i \in 1, 2, 3, \cdots, m$. If the Eq. (3) is calculated from $i = 1$ until $i = m$, then the Eq. (5) can be also written as follows,

$$\mathcal{A}_{m,1} = \begin{pmatrix} \sum_{j=1}^{n} \left(\mathcal{X}_{1,j} \cdot \mathcal{X}_{1,j} \right) \\ \sum_{j=1}^{n} \left(\mathcal{X}_{2,j} \cdot \mathcal{X}_{2,j} \right) \\ \vdots \\ \sum_{j=1}^{n} \left(\mathcal{X}_{m,j} \cdot \mathcal{X}_{m,j} \right) \end{pmatrix} \tag{4}$$

Furthermore, Eq. (4) is duplicated to be m columns, so that the matrix dimensionality will be m rows and m columns. The effect of column duplication is the same values for each column as shown in Eq. (5)

$$\mathcal{A}_{m,m} = \underbrace{\begin{pmatrix} \sum_{j=1}^{n} \left(\mathcal{X}_{1,j} \cdot \mathcal{X}_{1,j} \right) & \sum_{j=1}^{n} \left(\mathcal{X}_{1,j} \cdot \mathcal{X}_{1,j} \right) & \cdots & \sum_{j=1}^{n} \left(\mathcal{X}_{1,j} \cdot \mathcal{X}_{1,j} \right) \\ \sum_{j=1}^{n} \left(\mathcal{X}_{2,j} \cdot \mathcal{X}_{2,j} \right) & \sum_{j=1}^{n} \left(\mathcal{X}_{2,j} \cdot \mathcal{X}_{2,j} \right) & \cdots & \sum_{j=1}^{n} \left(\mathcal{X}_{2,j} \cdot \mathcal{X}_{2,j} \right) \\ \vdots & \vdots & \ddots & \vdots \\ \sum_{j=1}^{n} \left(\mathcal{X}_{m,j} \cdot \mathcal{X}_{m,j} \right) & \sum_{j=1}^{n} \left(\mathcal{X}_{m,j} \cdot \mathcal{X}_{m,j} \right) & \cdots & \sum_{j=1}^{n} \left(\mathcal{X}_{m,j} \cdot \mathcal{X}_{m,j} \right) \end{pmatrix}}_{\text{Equation (4) is duplicated to be } m \text{ columns}} \tag{5}$$

$$\mathcal{B}_{m,m} = \left(\underbrace{\begin{pmatrix} \sum_{j=1}^{n}\left(\mathcal{X}_{1,j}\cdot\mathcal{X}_{1,j}\right) & \sum_{j=1}^{n}\left(\mathcal{X}_{1,j}\cdot\mathcal{X}_{1,j}\right) & \cdots & \sum_{j=1}^{n}\left(\mathcal{X}_{1,j}\cdot\mathcal{X}_{1,j}\right) \\ \sum_{j=1}^{n}\left(\mathcal{X}_{2,j}\cdot\mathcal{X}_{2,j}\right) & \sum_{j=1}^{n}\left(\mathcal{X}_{2,j}\cdot\mathcal{X}_{2,j}\right) & \cdots & \sum_{j=1}^{n}\left(\mathcal{X}_{2,j}\cdot\mathcal{X}_{2,j}\right) \\ \vdots & \vdots & \ddots & \vdots \\ \sum_{j=1}^{n}\left(\mathcal{X}_{m,j}\cdot\mathcal{X}_{m,j}\right) & \sum_{j=1}^{n}\left(\mathcal{X}_{m,j}\cdot\mathcal{X}_{m,j}\right) & \cdots & \sum_{j=1}^{n}\left(\mathcal{X}_{m,j}\cdot\mathcal{X}_{m,j}\right) \end{pmatrix}}_{\text{The transpose result of Equation (5)}} \right)^{T} \Bigg\} m$$

$$\tag{6}$$

Moreover, the Eq. (5) is transposed, and the transpose result is represented by using B as represented in Eq. (7).

The distance of the training sets can be computed by using the simple operation as shown in Eq. (8), while the value of C can be calculated by using the Eq. (9)

$$\mathcal{D}_{m,m} = \left| \mathcal{A}_{m,m} + \mathcal{B}_{m,m} - \mathcal{C}_{m,m} \right| \tag{7}$$

$$\mathcal{C} = 2 \times \mathcal{X} \times \mathcal{X}^{T} \tag{8}$$

Based on the Eq. (8), the Gaussian kernel matrix can be obtained by using the Eq. (10), while the variable of σ represents the standard deviation. The value of standard deviation can be defined as positive integer, but usually the values of standard deviation utilized are 1, 2, or 3.

$$\mathcal{K}(\mathcal{X}, \mathcal{X}^{T}) = exp\left(-\frac{\mathcal{D}_{m,m}}{2 \times \sigma^{2}} \right) \tag{9}$$

2.2 Sharpen of Gaussian-Based Component Analysis

In order to calculate the kernel on the feature space (G), the matrix $\mathbb{I}$ is required, where all of elements are 1. Moreover, the feature space can be processed by the simple operation as written in the following equation

$$\mathcal{G} = \mathcal{K} - (\mathbb{I} \times \mathcal{K}) - (\mathcal{K} \times \mathbb{I}) + (\mathbb{I} \times \mathcal{K} \times \mathbb{I}) + (\mathcal{K} \times \mathbb{I} \times \mathcal{K}) \tag{10}$$

The operation of $(\mathcal{K} \times \mathbb{I} \times \mathcal{K})$ is utilized to sharpen the features of the object, so that the object is easier to be recognized. The addition of the operation in the Eq. (12) has indicated that the dominant features can be maximally extracted. The result of Eq. (12) is furthermore applied to gain the eigenvalues and eigenvectors as written in Eq. (12)

$$Det(\mathcal{G} - I \times \lambda) = 0 \tag{11}$$

The variable of I represents the identity matrix which is the matrix with zero elements except the main diagonal with 1 value. The calculation result of the Eq. (13) produces the eigenvalues and eigenvectors as shown in Eqs. (13) and (14)

$$\lambda = \begin{pmatrix} \lambda_{1,1} & 0 & 0 & \cdots & 0 \\ 0 & \lambda_{2,2} & 0 & \cdots & 0 \\ 0 & 0 & \lambda_{3,3} & \cdots & 0 \\ \vdots & \vdots & \vdots & \ddots & \vdots \\ 0 & 0 & 0 & \cdots & \lambda_{m,m} \end{pmatrix} \tag{12}$$

$$\Lambda = \begin{pmatrix} \Lambda_{1,1} & \Lambda_{1,2} & \Lambda_{1,3} & \cdots & \Lambda_{1,m} \\ \Lambda_{1,2} & \Lambda_{2,2} & \Lambda_{2,2} & \cdots & \Lambda_{2,m} \\ \Lambda_{1,3} & \Lambda_{3,2} & \Lambda_{3,3} & \cdots & \Lambda_{3,m} \\ \vdots & \vdots & \vdots & \ddots & \vdots \\ \Lambda_{1,m} & \Lambda_{m,2} & \Lambda_{m,3} & \cdots & \Lambda_{m,m} \end{pmatrix} \tag{13}$$

The eigenvalues as represented in Eq. (13) can be also represented by using row matrix as shown in Eq. (15).

$$\lambda = \begin{pmatrix} \lambda_{1,1} & \lambda_{2,2} & \lambda_{3,3} & \cdots & \lambda_{m,m} \end{pmatrix} \tag{14}$$

These values of λ are not decreasingly ordered yet, therefore these values must be decreasingly ordered. The sorting result is represented by using $\left(\hat{\lambda} \right)$ as shown in the following equation,

$$\hat{\lambda}_{1,1} \geq \hat{\lambda}_{2,2} \geq \hat{\lambda}_{3,3} \geq \hat{\lambda}_{4,4} \geq \cdots \hat{\lambda}_{m,m} \tag{15}$$

The change of column position is also effect of column position of the eigenvectors. The sorting results of the eigenvectors $\left(\hat{\Lambda} \right)$ are composed based on the index found of the sorting results of the eigenvalues as written in Eq. (17).

$$\hat{\Lambda} = \begin{pmatrix} \hat{\Lambda}_{1,1} & \hat{\Lambda}_{1,2} & \hat{\Lambda}_{1,3} & \cdots & \hat{\Lambda}_{1,m} \\ \hat{\Lambda}_{1,2} & \hat{\Lambda}_{2,2} & \hat{\Lambda}_{2,2} & \cdots & \hat{\Lambda}_{2,m} \\ \hat{\Lambda}_{1,3} & \hat{\Lambda}_{3,2} & \hat{\Lambda}_{3,3} & \cdots & \hat{\Lambda}_{3,m} \\ \vdots & \vdots & \vdots & \ddots & \vdots \\ \hat{\Lambda}_{1,m} & \hat{\Lambda}_{m,2} & \hat{\Lambda}_{m,3} & \cdots & \hat{\Lambda}_{m,m} \end{pmatrix} \tag{16}$$

2.3 Projection of the Gaussian-Based Component Analysis

To obtain the features using Gaussian kernel, for both the training $\mathcal{P}_x$ and testing Gaussian kernel $\mathcal{P}_y$ can be simply represented.

$$\mathcal{P}_x = \mathcal{K}_x \left(\mathcal{X}, \mathcal{X}^T \right) \times \hat{\Lambda} \tag{17}$$

In this case, $\mathcal{K}_x(\mathcal{X}, \mathcal{X}^T)$ is the Gaussian kernel matrix for the training sets, it was obtained by using Eq. (10), while the matrix values $\hat{\Lambda}$ was also calculated by using Eq. (17).

$$\mathcal{P}_y = \mathcal{K}_y(\mathcal{Y}, \mathcal{Y}^T) \times \hat{\Lambda} \tag{18}$$

$\mathcal{K}_y(\mathcal{Y}, \mathcal{Y}^T)$ is the Gaussian kernel matrix for the testing sets. The difference between $\mathcal{K}_x(\mathcal{X}, \mathcal{X}^T)$ and $\mathcal{K}_y(\mathcal{Y}, \mathcal{Y}^T)$ is used as the input. $\mathcal{K}_x(\mathcal{X}, \mathcal{X}^T)$ applies training sets as the input (see Eq. (1)), whereas $\mathcal{K}_y(\mathcal{Y}, \mathcal{Y}^T)$ applies the testing sets as input (see Eq. (2)).

The calculation result of Eqs. (18) and (19) can be written in the following matrix as seen in Eq. (20) for $\mathcal{P}_x$ and in Eq. (21) for $\mathcal{P}_y$.

$$\mathcal{P}_x = \left. \begin{pmatrix} P_{1,1} & P_{1,2} & P_{1,3} & \cdots & P_{1,m} \\ P_{1,2} & P_{2,2} & P_{2,2} & \cdots & P_{2,m} \\ P_{1,3} & P_{3,2} & P_{3,3} & \cdots & P_{3,m} \\ \vdots & \vdots & \vdots & \ddots & \vdots \\ P_{1,m} & P_{m,2} & P_{m,3} & \cdots & P_{m,m} \end{pmatrix} \right\} m \tag{19}$$

$$\underbrace{\hphantom{P_{1,1} \quad P_{1,2} \quad P_{1,3} \quad \cdots \quad P_{1,m}}}_{m}$$

$$\mathcal{P}_y = \begin{pmatrix} \hat{P}_{1,1} & \hat{P}_{1,2} & \hat{P}_{1,3} & \cdots & \hat{P}_{1,m} \end{pmatrix} \tag{20}$$

2.4   Features Selection and Similarity Measurements

As mentioned in Eqs. (20) and (21), the column dimensionality of them is m. It is indicated that the number of features produced is m, for both the training and the testing sets. However, features generated will not be used all for measurement. Therefore the features produced must be selected to be applied on similarity measurements. The selection results can be shown in Eqs. (22) and (23).

$$\mathcal{P}_x = \left(\begin{array}{cccc|ccc} P_{1,1} & P_{1,2} & \cdots & P_{1,t} & P_{1,t+1} & \cdots & P_{1,m} \\ P_{2,1} & P_{2,2} & \cdots & P_{2,t} & P_{2,t+1} & \cdots & P_{2,m} \\ P_{3,1} & P_{3,2} & \cdots & P_{3,t} & P_{3,t+1} & \cdots & P_{3,m} \\ \vdots & \vdots & \ddots & \vdots & \vdots & \ddots & \vdots \\ P_{m,1} & P_{m,2} & \cdots & P_{m,t} & P_{m,t+1} & \cdots & P_{m,m} \end{array} \right) \tag{21}$$

$$\underbrace{\hphantom{P_{1,1} \quad P_{1,2} \quad \cdots \quad P_{1,t}}}_{t \text{ training features are used}} \qquad \underbrace{\hphantom{P_{1,t+1} \quad \cdots \quad P_{1,m}}}_{(m-t) \text{ features are not used}}$$

$$\mathcal{P}_y = \left(\underbrace{\hat{P}_{1,1} \quad \hat{P}_{1,2} \quad \cdots \quad \hat{P}_{1,t}}_{t \text{ testing features are used}} \middle| \underbrace{\hat{P}_{1,t+1} \quad \hat{P}_{1,t+2} \quad \cdots \quad \hat{P}_{1,m}}_{(m-t) \text{features are not used}} \right) \tag{22}$$

The first column represented the more dominant feature than the second, the third, until the mth columns. It means that the bigger the column index, the feature less dominant. The feature selection is also intended to reduce the computation time when the similarity measurements are applied to classify the face images. In order to classify the face image, the simple method is applied, which is the city block method as shown in Eq. (23)

$$\mathcal{D}_{m,1} = |\mathcal{P}_x - \mathcal{P}_y| \tag{23}$$

3 Experimental and Discussion

In order to evaluate the proposed approach, two face databases have been prepared, which are the YALE and the CAI-UTM database. The YALE face database is a small database, which has only fifteen people, where for each people has eleven different poses [19], sample of face image can be seen in see Fig. 1. Though, it is small face database, but the images have many illumination, accessories, and expressions variation. The second face database is the CAI-UTM, where it has a hundred people, and for each people has ten different poses as shown in Fig. 2 as a sample of image. Therefore, the second face database has a thousand images.

Fig. 1. The yale sample

Fig. 2. CAI-UTM sample

3.1 The Results on the Yale Database

In this paper, three scenarios are implemented to evaluate the proposed approach, which are using two, three, and four face images as the training sets. Each scenario will be conducted five times experiments with different images indexes (five-fold cross validation). For each scenario, the features will be selected based on the training sets applied. For the first scenario, 20 until 29 features are selected, 30 until 39 features for the second scenario, and 40–49 features for the last scenario.

The results of the first scenario can be seen in Fig. 3. The accuracy of each experiment and the average accuracy were described in the figure. The investigation results show that the accuracy depends on the images used as the training sets. The use of face image training with illumination and expression can produce higher accuracy than the others. In this case, the second experiment has delivered the highest accuracy. The use of features also influences the accuracy produced. Based on Fig. 3, the more features applied, the higher accuracy delivered. It is also shown on the average accuracy obtained, where the line of accuracy tend increase proportional to number of features applied. The results of the first scenario, maximum accuracy produced is 87.41%, while the average accuracy is 80.54%.

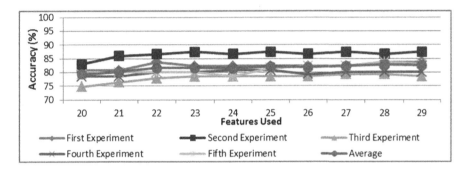

Fig. 3. The first scenario on the YALE

The similar results are also shown in the second scenario, where the use of the illumination and expression face images as the training sets has delivered the better results than the others. On the second scenario has shown the best performance, it is influenced by the sample images used, which is illumination and expression images has been applied as the training sets. The worst result has occurred on the fourth experiment, the investigation results show that the use of normal images as training sets will delivered the lower accuracy than the illumination and expression face images as the training sets, even the more features applied, the less the accuracy obtained.

Based on Fig. 4, the maximum accuracy is 90.83%, the results show that the third scenario is better than the second scenario. This can also be seen from the obtained average accuracy, the third scenario has delivered higher average accuracy than in the second scenario, which is 87.22%.

Four images as the training sets are applied on the last scenario, the worst result (more than 84%) is still better than the average accuracy of the first scenario (less than 81%). The average accuracy of the last scenario (more than 88%) is still better than the maximum accuracy of the first scenario (less than 88%). Similar to the third scenario, the fourth experiment has delivered the worst result on the last scenario, it is caused by the sample used has not been representative of the training sets. The result of the last

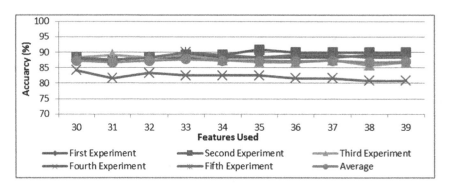

Fig. 4. The second scenario on the YALE

scenario can be shown in Fig. (5). The average accuracy also tends to increase in proportion to number features used. Maximum and average accuracies achieved are 92.38% and 88.91%.

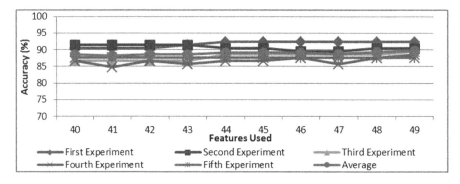

Fig. 5. The third scenario on the YALE

The results were also compared to other approaches for all of scenarios, which are Eigenface, Fisherface, and Laplacianfaces. The performance results of the first and the second scenarios show that the proposed approach outperformed to other methods, while the last scenario shows that Fisherface and Laplacianfaces are better than the proposed approach for accuracy average, but the proposed approach outperformed to the others for the maximum accuracy as seen in Fig. 6.

3.2 The Results on the CAI-UTM Database

Different database was also applied to evaluate the performance of the proposed approach, which is the CAI-UTM. A thousand images are prepared from a hundred people [20]. The Difference with the previous database, in this face database, the number of used features is more than previous database. Since the number of captured sample people is more. For the first scenario, 185 until 199 features are selected, 285

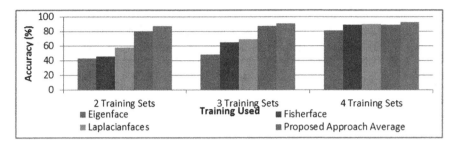

Fig. 6. Performance of the proposed approach was compared to other methods

until 299 features for the second scenario, and 385–399 features for the last scenario. Each scenario is tested using 5-fold Cross Validation.

The first scenario results can be displayed in Fig. 7. Six different line models were described in Fig. 7. However, four line models are almost overlapping, because they produce similar similarity for each feature, include the average accuracy line. As shown in Fig. 6, the proposed approach delivered the stable accuracy result for each features.

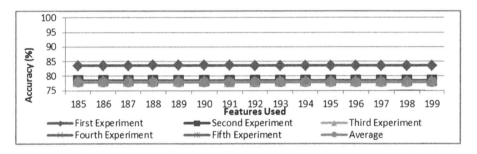

Fig. 7. The first scenario on the CAI-UTM

The best performance has occurred on the first experiment, while the worst results are in the last experiment. By using representative face image samples, the proposed approach system can recognize face image model. Representative face image samples are face image with different models, i.e. open smile, close smile or other expressions. But if the samples used are normal face without expressions, then the proposed approach sometimes cannot recognize the face with other expressions, such as surprise or the others. Performance maximum and average of the proposed approach are 83.75% and 77.90%.

The second scenario is evaluated by using three image samples. The obtained results are shown in Fig. 8. The result shows that the fewer features have represented the characteristics of an image. The second and fifth experiments have delivered the highest performance, because they have applied the image with different poses that represent the other poses, while the other experiments used the images with similar poses, therefore the produced features do not represent overall image poses on the same class. The maximum (85.57%) and average (84.01%) performance of the second

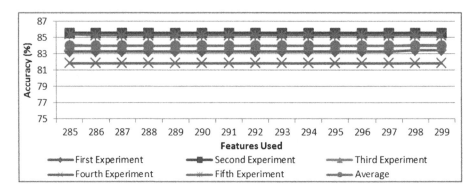

Fig. 8. The second scenario on the CAI-UTM

experimental results is better than the first experiment results, it indicated that the accuracy is affected by number of training samples and poses applied.

The results of the last scenario show the similar trend to the first and the second scenario, i.e. the number of features clearly influenced the performance of the proposed approach. The final scenario results are also evident that the use of training sets with diverse samples is able to produce better accuracy than similar poses as described in Fig. 9. The results of the use of diverse training sets can be seen in the third and fifth experiments, whereas the similar use of the images as the training sets can be seen in the first, second and fourth experiments. The experimental results also proved that the use of the amount of training data affects the resulting accuracy, which are 87.33% for the maximum and 86.43% for the average.

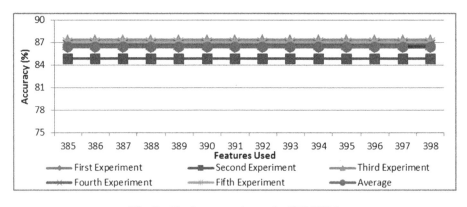

Fig. 9. The last scenario on the CAI-UTM

4   Conclusion

In this paper, the proposed approach has proved that modeling of the Gaussian-based Component Analysis on the kernel space can extract the image features by reducing the dimensionality of the training sets. The proposed approach can recognize the face image with small sample training sets and even it is better than other methods, i.e. Eigenface, Fisherface, and Laplacianfaces. The proposed approach was also evaluated by using local face image database, i.e. CAI-UTM database, where the evaluation results show that the proposed approach is able to recognize facial image more than 87%.

Acknowledgement. This research has been covered by Ministry of Research Technology and Higher Education and facilitated by Laboratory of Computational Artificial Intelligence, University of Trunojoyo Madura, 2017.

References

1. Zou, H., Hastie, T., Tibshirani, R.: Sparse principal component analysis. J. Comput. Graph. Stat. **15**(2), 265–286 (2006)
2. Lu, H., Plataniotis, A.N.V.K.N.: MPCA: multilinear principal component analysis of tensor objects. IEEE Trans. Neural Netw. **19**(1), 18–39 (2008)
3. Huang, S.M., Yang, J.F.: Improved principal component regression for face recognition under illumination variations. IEEE Trans. Sig. Process. Lett **19**(4), 179–182 (2012)
4. Muntasa, A.: New modelling of modified two dimensional fisherface based feature extraction. Telkomnika (Telecommun. Comput. Electron. Control) **12**(1), 115–122 (2014)
5. Muntasa, A., Siradjuddin, I.A., Wahyuningrum, R.T.: Multi-criteria in discriminant analysis to find the dominant features. Telkomnika **14**(3), 1113–1122 (2016)
6. Ye, J., Li, Q.: A two-stage linear discriminant analysis via QR-decomposition. IEEE Trans. Pattern Anal. Mach. Intell. **27**(6), 929–941 (2005)
7. Zheng, W.-S., Lai, J.H., Li, S.Z.: 1D-LDA vs. 2D-LDA: when is vector-based linear discriminant analysis better than matrix-based? Pattern Recogn. **41**(7), 2156–2172 (2008)
8. Muntasa, A.: The Gaussian orthogonal laplacianfaces modelling in feature space for facial image recognition. Makara J. Technol. **18**(2), 79–85 (2014)
9. Niu, B., Yang, Q., Shiu, S.C.K., Pal, S.K.: Two-dimensional Laplacianfaces method for face recognition. Pattern Recogn. **41**(10), 3237–3243 (2008)
10. He, X., Yan, S., Hu, Y., Niyogi, P., Zhang, H.-J.: Face recognition using Laplacianfaces. IEEE Trans. Pattern Anal. Mach. Intell. **27**(3), 328–340 (2005)
11. Bartlett, M.S., Movellan, J.R., Sejnowski, T.J.: Face recognition by independent component analysis. IEEE Trans. Neural Netw. **13**(6), 1450–1464 (2002)
12. Biswas, S., Bowyer, K.W., Flynn, P.J.: Multidimensional scaling for matching low-resolution face images. IEEE Trans. Pattern Anal. Mach. Intell. **34**(10), 2019–2030 (2012)
13. Yang, M.-H.: Face recognition using extended isomap. In: 2002 International Conference on Image Processing on Proceedings, Rochester, NY, USA. IEEE (2002)
14. Belkin, M., Niyogi, P.: Laplacian eigenmaps for dimensionality reduction and data representation. Neural Comput. **15**(6), 1373–1396 (2003)
15. Muntasa, A.: The human facial expression classification using the center Kernel subspace based the ridge regression. J. Comput. Sci. **11**(11), 1054–1059 (2015)
16. Memisevic, R., Hinton, G.: Improving dimensionality reduction with spectral gradient descent. Neural Netw. **18**(1), 702–710 (2005)
17. Pumomo, M.H., Sarjono, T.A., Muntasa, A.: Smile stages classification based on Kernel Laplacian-lips using selection of non linear function maximum value. In: VECIMS 2010 - 2010 IEEE International Conference on Virtual Environments, Human-Computer Interfaces and Measurement Systems on Proceedings, pp. 151–156. IEEE, Taranto-Italy (2010)

18. Nadil, M., Souami, F., Labed, A., Sahbi, H.: KCCA-based technique for profile face identification. EURASIP J. Image Video Process. **2017**(2), 1–13 (2017)
19. Yale Face Image Database. http://www.face-rec.org/databases/. Accessed 2000
20. CAI-UTM, CAI-UTM Face Image Database (2016)

Classification of Batik Kain Besurek Image Using Speed Up Robust Features (SURF) and Gray Level Co-occurrence Matrix (GLCM)

Fathin Ulfah Karimah and Agus Harjoko[✉]

Department Computer Science and Electronics,
Universitas Gadjah Mada, Yogyakarta, Indonesia
fathin.ulfah.k@mail.ugm.ac.id, aharjoko@ugm.ac.id

Abstract. Indonesian Batik has been endorsed as world cultural heritages by UNESCO. The batik consists of various motifs each of whom represents characteristics of each Indonesian province. One of the motifs is called as Batik Kain Besurek or shortly Batik Besurek, originally from Bengkulu Province. This motif constitutes a motif family consisting of five main motifs: *Kaligrafi, Rafflesia, Relung Paku, Rembulan, and Burung Kuau.* Currently most Batik Besureks reflect a creation developed from combination of main motifs so that it is not easy to identify its main motif. This research aims to classify Indonesian batik according to its image into either batik besurek or not batik besurek as well as reidentifying its more detailed motif for the identified batik besurek. The classification is approached through six classes: five classes in accordance with classification of Batik Besurek and a class of not Batik Besurek. The preprocessing system converts images to grayscale and followed by resizing. The feature extraction uses GLCM method yielding six features and SURF method yielding 64 descriptors. The extraction results are combined by assigning weight on both methods in which the weighting scheme is tested. Moreover, the image classification uses a method of k-Nearest Neighbor. The system is tested through some scenarios for the feature extraction and some values k in k-NN to classify the main motif of Batik Besurek. So far the result can improve system performance with an accuracy of 95.47% according to weighting 0.1 and 0.9 for GLCM and SURF respectively, and k = 3.

Keywords: Classification · Batik Kain Besurek · SURF · GLCM · k-NN

1 Introduction

Batik is Indonesian cultural heritage which philosophycally posseses visceral meaning and high value. It has also been acknowledged by United Nations Educational, Scientific, and Culture Organization (UNESCO) as one of world cultural heritage. Batik is so various kinds, almost every region in Indonesia has a different batik [1]. One of those is Batik Kain Besurek which is originally from Bengkulu City [2]. The word "Besurek" is Bengkulu language meaning "mail". Therefore Batik Kain Besurek means a batik made from fabric with writing like mail. Batik kain besurek has 5 main motifs,

© Springer Nature Singapore Pte Ltd. 2017
A. Mohamed et al. (Eds.): SCDS 2017, CCIS 788, pp. 81–91, 2017.
https://doi.org/10.1007/978-981-10-7242-0_7

those are *Kaligrafi, Rafflesia, Relung Paku, Rembulan, and Burung Kuau* (see Fig. 1). However batik kain besurek is also an art so that a motif of batik kain besurek can sometime be created by craftsmen through combining some motifs. Hence classifying batik kain besurek based on their motifs needs to observe them thoroughly and therefore the way of classifying is a kind of problem to solve. The classification is also beneficial in term of inventorying them in manageable way. On the other hand the so many kind of batik in Indonesia have made many Indonesian people so hard to recognise between batik kain besurek and not batik kain besurek. Identifying and classifying motifs of batik constitute problems related to digital image processing.

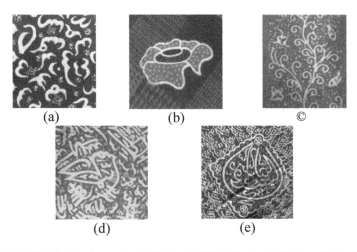

(a)   (b)   ©

(d)   (e)

Fig. 1. Motif of batik kain besurek. (a) Kaligrafi, (b) Raflesia, (c) Relung Paku, (d) Burung Kuau, (e) Rembulan

Digital image processing [3] is a process imposed to images for obtaining some certain results as required. By using the digital images processing, the image of batik such as Batik Kain Besurek can also be processed to identify whether the image is right as the image of Batik Kain Besurek and to classify its motif with respect to a certain class of motif. This manner could be one of methods to solve the above-mentioned problem. Research on batik image processing or Content Based Batik Image Retrieval (CBBIR) is a research focused on image processing on the basis of motif characteristic and so many researchers have worked on it. Research on CBBIR that used four extracts of texture features, those are mean, energy, entropy and standard deviation, has resulted in performance 90–92% [4]. The use of treeval and treefit in term of decision tree has been utilized in research [5] to optimize CBBIR, and the result has shown the percentage 80–85%. CBBIR has also been used in research [6] where feature extraction is Gray Level Co-occurrence Matrix (GLCM) and the obtained sensitivity is 82%. This paper proposes development of extraction through other feature for a better result. Combining feature extraction of GLCM and other method has been done such as with wavelet feature [7]. GLCM and wavelet have been combined and the result has shown percentage 94.87%, more accurate performance in compared with its previous methods.

Another feature extraction that has ever been used for classification is Speed Up Robust Features (SURF). This feature extraction has made description based on yielded interest points by using wavelet as having been done in [7].

Based on the term of references above, this paper proposes another way not only to identify a batik as batik kain besurek or not but also to classify the image of the batik according to its motif. The identification of batik in this research uses threshold to differentiate a batik as a member of batik besurek class or not. The feature extraction uses the Gray Level Co-occurrence Matrix (GLCM) method as having been used by previous researchers and is combined with the Speed Up Robust Features (SURF) method. The combination of both methods gives better results [8]. GLCM constitutes one of methods to extract texture features in order to know how often the combination of pixel brightness value for different position happens for an image [9]. SURF constitutes an algorithm which is able to detect keypoint and produce description for every point. Keypoint itself is a point whose value is persistent for the change of scale, rotation, blurring, 3-dimensional transformation, lighting, and shape. Thus, the algorithm works well for images with noise, detection error, as well as geometric and photometric change [10]. In term of classification, we have used k-Nearest Neighbor Method [11] where Euclidean Distance has been used to determine distance of feature extraction result obtained through GLCM and Hausdorff Distance has been used to determine distance of feature extraction result obtained from SURF. Result of distance calculation for the feature extraction method is then combined by using weighting scheme. Afterward we could find the best weight for each of both distance according to proposed testing scenario.

2 Related Work

2.1 Research Materials

Data used in this research are both primary data and secondary data. The primary data are batik kain besurek images consisting of 117 images each of them has been collected three times (actual size, scaled down size, and rotated size) so that all images of batik kain besurek are 351 images. All the images are categorized into two groups, those are the one encompassing 70% of images stored as database for reference of classification (training) and the other one covering 30% of images for testing. The images of batik kain besurek have been taken by using a camera of size 1200 pixels × 1200 pixels. Meanwhile the secondary data comprise 20 images of not batik besurek which have been collected from internet. These latter images have been used for testing.

2.2 System Design

The system design is made to show the work flow of research. This research consists of two main parts, those are training and testing (see Fig. 2). Flow diagram is divided into two, part 1 and part 2. In both parts, the process starts with image acquisition and preprocessing. The next process is the feature extraction process for obtaining the characteristics of each image. In this system, two methods for feature extraction are

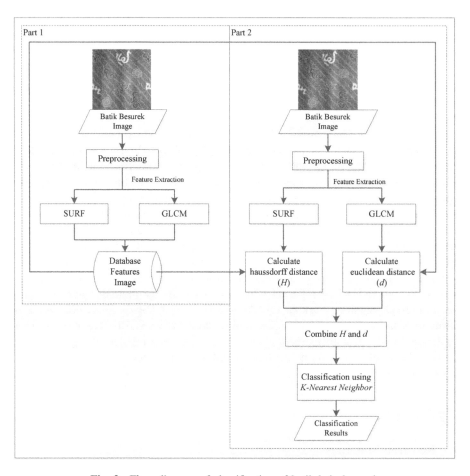

Fig. 2. Flow diagram of classification of batik kain besurek

GLCM [12] and SURF [13, 14]. In Part 1, the feature extraction results of both methods are stored into the system database. While part 2 will be a process of measuring the distance between the feature extraction of the image test and image database to obtain the results of classification.

Preprocessing
The preprocessing of this research comprises the conversion of images to be smaller for its scale and to be greyish for its color. Beside resizing its scale to be smaller, resizing is also to reduce the amount of image pixel so that the duration of system computation can be lessened. Every image input needs to make smaller to the size of about 200×200. Subsequently the image inputs need to convert through RGB to grayscale.

Speed Up Robust Features (SURF)
Feature extraction through SURF consists of 5 main steps as shown by Fig. 3. A SURF process is started by shaping integral image [15, 16] through reducing computational

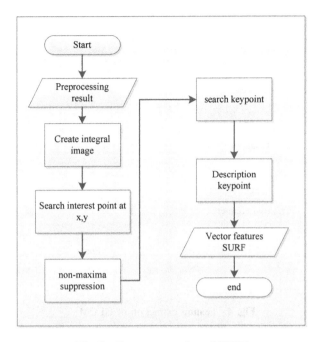

Fig. 3. Feature extraction of SURF

time [14]. Afterward a keypoint as a candidate of image feature needs to search through Fast Hessian [14]. In order to get tenacity of the feature in term of its scale, then the search of feature must be applied to a scale space in order to get image pyramid. The scale space is obtained by iterating convolution to image inputs through filter of discrete Gaussian kernel [17, 18]. In addition, a process of sub-sampling is carried out repeatedly to reduce the image size [14]. By the process, an image pyramid is shaped, meaning that the image size in the scale space decreases continuously.

After finding a feature candidate of the image in the scale space, a keypoint candidate is searched by applying non-maxima suppression method [14]. The keypoint candidate is searched by utilising extreme respons of hessian matrix. Subsequently the obtained keypoint candidate is examined again by determining its deviation. A point can be classified as a keypoint if the extreme value is smaller than the given threshold [19]. The final process is to describe the obtained keypoint. The process is done by searching pixel distribution of neighbor around the keypoint. The description aims to make the keypoint more endure with respect to an angle rotation. The obtained result of processing feature extraction through SURF is a feature vector describing image input.

Gray Level Co-occurrence Matrix (GLCM)
GLCM method of this research is the one with 4 directions ($0°$, $45°$, $90°$ and $135°$) and the neighborhood distance of d = 1 pixel. The direction and distance d use the result of research with best accuration [20]. Figure 4 constitutes steps of GLCM used in this research. The process starts from converting images to become greyish of 8 levels. This conversion is performed to reduce computational time. Then co-occurrence matrices for

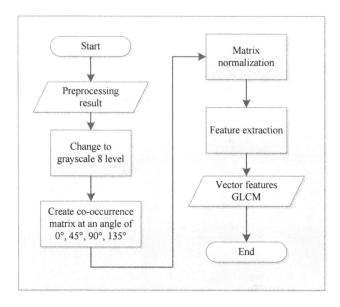

Fig. 4. Feature extraction of GLCM

4 directions are formed. The formations of 4 co-occurrence matrices are done by finding out the frequency of pixel emergence with their neighbors. All the four co-occurrence matrices then are summed up to get one co-occurrence matrix. The result is then d with respect to all matrix elemens. The feature of normalized matrix is then extracted with harlick features [21], those are contrast, correlation, inverse difference moment, angular second moment, entrophy and variance. The six parameters are the most relevant parameters of feature extraction of GLCM [12].

Classification

Features obtained from extraction through GLCM and SURF could be used as references for measuring a closest distance between object and accustomed data. The measurement is applied for different closest distances of feature extraction obtained from GLCM and SURF. In term of applying GLCM, the measurement uses an euclidean distance. Method of Euclidean Distance (ED) is the most often method used to determine distance of 2 vectors. ED can be calculated through the formula as follows

$$d_{ij} = \sqrt{\sum_{k=1}^{n} \left(x_{ik} - x_{jk} \right)^2} \tag{1}$$

Moreover, in term of applying SURF, the measurement uses hausdorff distance. Method of Hausdorff Distance (HD) is introduced by Filix Huasdorff [22]. The method works by determining a maximum distance of closest points of two-point sets. HD can be calculated through the formula as follows

$$H(\mathcal{A}, \mathcal{B}) = \max(h(\mathcal{A}, \mathcal{B}), h(\mathcal{B}, \mathcal{A})) \tag{2}$$

The classification process starts from combining the results of the distance calculation of both feature extraction methods. The combination of both methods is done by two steps, those are (1) multiplying each of the calculated feature extraction distance by the determined weight and (2) then summing up their results. If the result of the two step calculation is greater than the threshold, then the test image has been identified as a non-besurek class and the classification process has been completed. Whereas if the result is smaller than threshold, then the test image is identified to be one class of 5 batik besurek class and the process of classification of k-NN will be processed. The classification using the k-Nearest Neighbors method will involve the nearest neighbor image based on the smallest distance calculation value of k that has been determined. The class with the smallest distance is the result of image classification.

3   Experimental Results

This section explains on testing of research. The test is a weighting method combining feature extraction method, and value of k on k-NN. After having tested in accordance with the given scenario, the results can be compared to find out the best results for each test. The results of this research were compared based on the resulting performance of the confusion matrix on each test that has been done.

Before comparison of system test results for the best feature extraction method (SURF, GLCM or combination of both methods), we search the best weights for combined GLCM and SURF feature extraction. The test results on this weight search are shown in Fig. 5. The weighting results show that the weight of 0.9 for the SURF extraction method and 0.1 for GLCM provide the best system performance of the other

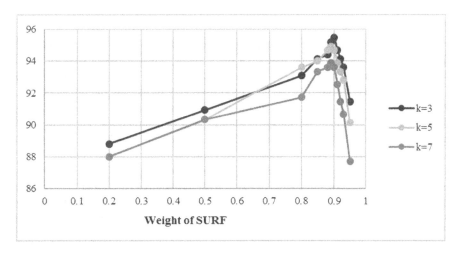

Fig. 5. The performance system to find the best weight for combined of SURF and GLCM

tested weighted ratios. Thus, the weights of 0.9 and 0.1 are weighted for combining the feature extraction methods.

After testing of both weights of extraction method, the next test are for feature extraction method and k value on k-NN. In Fig. 6, comparison graphs of the accuracy of the feature extraction methods and classification methods which are used in this research is presented. In testing of feature extraction method, it shows that the combination of GLCM feature extraction method and SURF is the method with highest accuracy result on every variant of k-NN value. On testing the variation of k, the value k = 3 is the highest accuracy value among all other k value variation. The highest accuracy on the value of k = 3 applies to all the tested featured extraction methods. Thus, the combination of both methods can improve the accuracy of the system. Moreover, the system can identify the image of batik besurek or batik non besurek and batik besurek classification according to the motifs well.

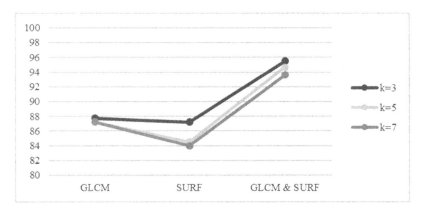

Fig. 6. Comparison of the system using feature extraction GLCM, SURF and combination of both

System performance is shown according to its sensitivity, specificity, precision and accuracy based on the confusion matrix of the system. The four best testing of SURF and GLCM combined for feature extraction and k = 3 for k-NN are shown in Table 1.

The performance according to sensitivity is useful for measuring the proportion of a predefined class according to actual conditions. This system has the highest percentage of sensitivity that is 100% for class of relung paku and rembulan. Thus, all tested images in the class of relung paku and rembulan are well predicted. This is evidenced by 0 in False Negative (FN) column for the class of relung paku and rembulan. The FN column shows the number of incorrect predictably tested images. While the smallest sensitivity for the class of rafflesia is 70.37% with the number of correct (TP column) predictably tested images are 19 images and the number of wrong (FN column) predicted images are 8 images. The average sensitivity of the results of this study is 88.19%.

Table 1. The testing on the combination of SURF and GLCM with 3-NN

No	Class name	TP	FN	FP	TN	Sensitivity (%)	Precision (%)	Specificity (%)	Accuracy (%)
1.	Raflesia	19	8	0	98	70.37	100.00	100.00	93.60
2.	Kaligrafi	27	4	2	92	87.10	93.10	97.87	95.20
3.	Burung Kuau	11	1	2	111	91.67	84.61	98.23	97.60
4.	Relung Paku	20	0	5	100	100.00	80.00	95.24	96,00
5.	Rembulan	15	0	6	104	100.00	71.43	94.54	95.20
6.	Non besurek	16	4	2	103	80.00	88.89	98.09	95.20
Average						88.19	86.34	97.33	95.47

The performance based on specificity is the proportion of incorrectly predefined class with respect to the actual conditions. This performance is calculated to measure how well the training image of the class which is independent of other classes. The average specificity performance in the six classes is 97.33% and the specificity performance for each class is greater than 94%. The highest performance of specificity is found in the class of raflesia, 100%. This shows that the class of raflesia does not affect the results of other class classifications.

Predicting data as member of a class does not mean that the data is assured correctly for the class. It is possible that the data are member of other classes. The possibility that we can do is by determining the performance of precision. Precision is a measure of the performance determined by calculating the proportion of a predictable class according to the true condition of all correctly predictable data. The highest precision performance is shown in the class of raflesia with a percentage of 100%. It shows that the test data of the other 5 classes is unpredictable as the raflesia class. While the lowest precision is related to the class of rembulan with a percentage of 71.43%. The average of system precision is 86.34%. Finally the last measure of performance is accuracy. This accuracy is the percentage of correct predictably data of a class. The average percentage of accuracy is 95.47%. The accuracy for each class is above 93%.

4 Conclusion

In this paper, we presented comparison GLCM and SURF as well as the combination of both methods for feature extraction. Based on the results of research, it can be concluded that the combination of GLCM and SURF methods can improve the accuracy of classification of the batik besurek's main motif. It is verified that the combination of both methods has resulted in the highest accuracy performance related to the testing scenario. The highest average accuracy for the combined methods of feature extraction is 95.47%. The accuracy is obtained by combining the GLCM and SURF methods with the weight of 0.1 for GLCM method and the weight of 0.9 for the SURF method, whereas regarding the classification using k-NN, the value k in the test yielding the highest accuracy is k = 3. Testing with the highest accuracy performance is a highest performance test besides other measure of performance such as sensitivity, specificity and precision. It means that the highest performance is the convincing result of testing. Class of relung paku and rembulan have made sure of 100% as the highest sensitivity for the performance testing of the combined feature extraction methods. It suggests that the combination of both GLCM and SURF methods can help to identify the motif of relung paku and rembulan very well.

References

1. Intangible Cultural Heritage UNESCO. https://ich.unesco.org/en/RL/indonesian-batik-00170 . Accessed 12 Sept 2017
2. Pemerintah Provinsi Bengkulu. http://bengkuluprov.go.id/batik-basurek-motif-asli-khas-bengkulu/. Accessed 12 Sept 2017
3. Gozales, R.C., Woods, R.E.: Digital Image Processing, 2nd edn. Prentice Hall, Upper Saddle River (2002)
4. Rangkuti, A.H., Harjoko, A., Putro, A.E.: Content based batik image retrieval. J. Comput. Sci. **10**(6), 925–934 (2014)
5. Rangkuti, A.H., Rasjid, Z.E., Santoso, D.: Batik image classification using treeval and treefit as decision tree function in optimizing content based image retrieval. In: International Conference on Computer Science and Computational Intelligence, Jakarta, pp. 577–583. Elsevier (2015)
6. Ernawati, E., Andreswari, D., Karimah, F.U.: Rancang bangun aplikasi pencarian citra batik besurek dengan gray level co-occurrence matrix dan Euclidean distance. Jurnal Teknologi Informasi **11**(1), 64–77 (2015)
7. Parekh, R.: Improve texture recognition using combined GLCM dan wavelet features. Int. J. Comput. Appl. **29**(10), 41–46 (2011)
8. Benavides, C., Villegas, J., Román, G., Avilés, C.: Face classification by local texture analysis through CBIR and SURF points. IEEE Lat. Am. Trans. **14**(5), 2418–2424 (2016)
9. Wibawanto, H.: Identifikasi citra massa kritik berdasarkan fitur gray level co-occurence matrix: In: Seminar Nasional Aplikasi Teknologi Informasi (2008)
10. Alfanindya, A., Hashim, N., Eswaran, C.: Content based image retrieval and classification using speeded-up robust features (SURF) and grouped bag-of-visual-words (GBoVW). In: 3rd International Conference on Technology, Informatics, Management, Engineering & Environment, Bandung. IEEE (2013)
11. Singh, S., Haddon, J., Markou, M.: Nearest neighbor strategies for image understanding. In: Proceedings of Workshop on Advanced Concepts for Intelligent Vision, Baden-Baden (1999)
12. Baraldi, A., Parmiggiani, F.: An investigation of the textural characteristics associated with gray level coocurence matrix statistical parameters. IEEE Trans. Geosci. Remote Sens. **33**(2), 293–304 (1995)
13. Bay, H., Fasel, B., Gool, L.V.: Interactive museum guide: fast and robust recognition of museum object. In: Proceedings of the First International Workshop on Mobile Vision (2006)
14. Bay, H., Ess, A., Tuytelaars, T., Gool, L.: Speeded-up robust features (SURF). Comput. Vis. Image Underst. **110**, 346–359 (2008). Elsevier

15. Kovesi, P.: Fast almost-Gaussian filtering. In: Proceedings of the Australian Pattern Recognition Society Conference on Digital Image Computing: Techniques and Applications, Sydney (2010)
16. Derpanis, K.G.: Integral image-based representations. Departement of Computer Science and Engineering, York University (2007)
17. Computer Aided Medical Procedures and Augmented Reality Technische Universität München. http://campar.in.tum.de/Chair/HaukeHeibelGaussianDerivatives. Accessed 23 May 2017
18. Golub, G.H., VanLoan, C.F.: Matrix Computations, 3rd edn. The Johns Hopkins University Press, Maryland (1996)
19. Yunmar, R.A., Harjoko, A.: Sistem identifikasi relief pada situs bersejarah menggunakan perangkat mobile android (studi kasus candi borobudur). Indones. J. Comput. Cybern. Syst. **8**(2), 133–144 (2014)
20. Listia, R., Harjoko, A.: Klasifikasi massa pada citra mammogram berdasarkan gray level coocurence matrix (GLCM). Indones. J. Comput. Cybern. Syst. **8**, 59–68 (2014)
21. Haralick, M.R., Shanmugam, K., Dinstein, I.: Textural features for image classification. IEEE Trans. Syst. Man Cybern. 610–621 (1973)
22. Jesorsky, O., Kirchberg, K.J., Frischholz, R.W.: Robust face detection using the Hausdorff distance. In: Bigun, J., Smeraldi, F. (eds.) AVBPA 2001. LNCS, vol. 2091, pp. 90–95. Springer, Heidelberg (2001). https://doi.org/10.1007/3-540-45344-X_14

Applications of Machine Learning

Model Selection in Feedforward Neural Networks for Forecasting Inflow and Outflow in Indonesia

Suhartono[✉], Prilyandari Dina Saputri, Farah Fajrina Amalia,
Dedy Dwi Prastyo, and Brodjol Sutijo Suprih Ulama

Department of Statistics, Institut Teknologi Sepuluh Nopember,
Kampus ITS Keputih Sukolilo, Surabaya 60111, Indonesia
{suhartono,dedy-dp,brodjol_su}@statistika.its.ac.id,
prildisa@gmail.com, farahfajrinaamalia@gmail.com

Abstract. The interest in study using neural networks models has increased as they are able to capture nonlinear pattern and have a great accuracy. This paper focuses on how to determine the best model in feedforward neural networks for forecasting inflow and outflow in Indonesia. In univariate forecasting, inputs that used in the neural networks model were the lagged observations and it can be selected based on the significant lags in PACF. Thus, there are many combinations in order to get the best inputs for neural networks model. The forecasting result of inflow shows that it is possible to testing data has more accurate results than training data. This finding shows that neural networks were able to forecast testing data as well as training data by using the appropriate inputs and neuron, especially for short term forecasting. Moreover, the forecasting result of outflow shows that testing data were lower accurate than training data.

Keywords: Forecasting · Inflow · Outflow · Nonlinear · Neural network

1 Introduction

In recent years, neural networks are one of the most popular methods in forecasting. Neural networks applied in vary fields such as energy, financial and economics, environment, etc. [1–4]. The interest in study using neural networks models has increased as they are able to capture non linear pattern and have a great accuracy.

In financial and economics, neural networks are used to predict the movement of stock price index in Istanbul [5]. This method was compared to Support Vector Machine (SVM). They found that neural networks model were significantly better than SVM model. This result shows that neural networks have a great performance in financial data.

Forecasting inflow and outflow plays an important role to achieve the stability of economics in Indonesia. The currency that exceeded the demand will lead to inflation. On the other hand, the currency that less than the demand will lead to the declining of economic growth. In Indonesia, the suitability of the currency was maintained by Bank Indonesia. In order to guarantee the availability of currency, Bank Indonesia needs to plan the demand and supply of currency. The forecasting results of inflow and outflow will be used as the indicator in determining the demand of currency in the next period.

© Springer Nature Singapore Pte Ltd. 2017
A. Mohamed et al. (Eds.): SCDS 2017, CCIS 788, pp. 95–105, 2017.
https://doi.org/10.1007/978-981-10-7242-0_8

Thus, an accurate forecast of inflow and outflow in Indonesia will lead to the suitability of the demand and supply currency.

The study about forecasting inflow and outflow has been done by using many methods, such as ARIMAX, SARIMA, time series regression, etc. [6–9]. By using classical methods, there are many assumptions that must be fulfilled. One of them is the homoscedasticity of the variance residuals. Unfortunately, the previous study found that the residuals from ARIMAX model didn't fulfill the assumption [6, 8]. Thus, using neural networks for forecasting inflow and outflow in Indonesia seems to have promising opportunity, since this method was free from assumption.

The selection of the input were one of the most important decisions for improve the forecast accuracy [10]. In univariate forecasting, inputs that used in the model were the lagged observations. In this study, selections of inputs are based on the significant lags in PACF [11]. Thus, there are many combinations in order to get the best inputs for neural networks model. The model selection will be done by using cross validation. Cross validation is the most generally applicable methods for model selection in neural networks [12]. The forecasting results from FFNN model will be compared with the widely used classical methods, i.e. ARIMA and ARIMAX.

2   Methods

2.1   Data

The data used in this study is secondary data obtained from Bank Indonesia. The data used are inflow and outflow data from January 2003 to December 2016. The data will be divided into training data and testing data. The training data are inflow and outflow data from January 2003 to December 2015, while testing data are inflow and outflow data from January 2016 to December 2016.

2.2   Autoregressive Integrated Moving Average (ARIMA)

The ARMA model is a combined model of the Autoregressive and Moving Average process. The AR process is a process that describes Y_t that influenced by the previous condition $(Y_{t-1}, Y_{t-2}, \ldots Y_{t-p})$ and has a white noise a_t. The MA process is a process which shows that the estimated value of Y_t influenced by error at the time t and the previous error $(a_{t-1}, a_{t-2}, \ldots a_{t-q})$. Non-stationary time series data can be differenced on a certain order to produce a stationary data. The general equation of the ARIMA model (p, d, q) can be written as follows [13].

$$\phi_p(B)(1 - B)^d Y_t = \theta_0 + \theta_q(B)a_t, \tag{1}$$

where:

$\phi_p(B) = (1 - \phi_1 B - \ldots - \phi_p B^p)$
$\theta_q(B) = (1 - \theta_1 B - \ldots - \theta_q B^q)$
$(1 - B)^d =$ differencing order,
$a_t =$ error at the time t.

2.3 Autoregressive Integrated Moving Average with Exogenous Variable (ARIMAX)

ARIMAX model is a development of the ARIMA model. In the ARIMAX model, there is used an additional variable known as exogenous variable. The exogenous variables used can be dummy variables (non-metric) or other time series variables (metrics). In this study, exogenous variables used are dummy variables i.e. trend, monthly seasonal, and calendar variations effects. The general equation of the ARIMAX model can be written as follows [14].

$$Y_t = \beta_0 + \beta_1 V_{1,t} + \beta_2 V_{2,t} + \ldots + \beta_h V_{h,t} + N_t, \tag{2}$$

$$N_t = \frac{\theta_q(B)}{\phi_p(B)} a_t, \tag{3}$$

where:

$V_{h,t}$ = dummy variable,
$\phi_p(B) = (1 - \phi_1 B - \ldots - \phi_p B^p)$,
$\theta_q(B) = (1 - \theta_1 B - \ldots - \theta_q B^q)$,
N_t = residual of the time series regression process,
a_t = residual of the ARIMAX process.

2.4 Feedforward Neural Network

Neural networks are one of nonlinear regression methods and widely applied in pattern recognition. The idea in building neural networks models is motivated by their similarity to working biological system, which consist of large number of neurons that work in parallel and have the capability to learn. Neural networks are able to process vast amounts of data and make accurate predictions [15].

In time series forecasting, the most popular neural networks model are Feedforward Neural Network. In FFNN, the process starts from inputs that are received by the nodes, where these nodes are grouped in input layers. Information received from the input layer proceeds to the layers in the FFNN up to the output layer. Layers between input and output are called hidden layers. The input that used in neural network for forecasting are the previous lagged observations and the output describing the forecasting results. The selection of input variables are based on the significant lags in PACF [11].

The accuracy of neural networks model is determined by three components, i.e. the network architecture, training methods or algorithms, and activation functions. FFNN with p input and one hidden layer that consist of m neuron can be described as the figure below.

The model of FFNN in Fig. 1 can be written as follows:

$$f(\mathbf{x}_t, \mathbf{v}, \mathbf{w}) = g_2\left\{ \sum_{j=1}^{m} v_j g_1\left[\sum_{i=1}^{p} w_{ji} x_{it} \right] \right\}, \tag{4}$$

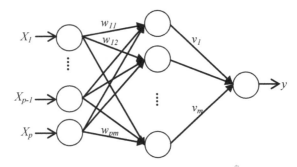

Fig. 1. Neural networks architecture

where **w** is the weights that connect the input layer to the hidden layer, **v** is the weights that connect the hidden layer to the output layer, $g_1(\cdot)$ and $g_2(\cdot)$ is the activation function, while w_{ji} and v_j are the weights to be estimated. The widely used activation function is tangent hyperbolic with the function below:

$$g(x) = \tanh(x) \tag{5}$$

2.5   Model Evaluation

The purpose of model evaluation is to know the performance of model in forecasting future period. Model evaluation will be based on the accuracy of forecasting result using *Root Mean Square Error* (RMSE). Model with the smallest RMSE will be selected as the best model in neural network. RMSE can be calculated by using the following formula [13]:

$$RMSE = \sqrt{\frac{1}{L}\sum_{l=1}^{L} e_l^2} \tag{6}$$

where $e_l = Y_{n+l} - \widehat{Y_n}(l)$.

3   Results

3.1   Forecasting Inflow and Outflow in Indonesia

The growth of inflow and outflow in Indonesia can be shown at the time series plot in Fig. 2. It shows that both inflow and outflow in Indonesia tends to increase every year. In other words, inflow and outflow in Indonesia have a trend pattern. In general, inflow and outflow data in Indonesia has a seasonal pattern [16]. However, the time series plot can't captured this pattern clearly. Thus, to determine the monthly pattern of the inflow and outflow, it will be used PACF plot in the next step.

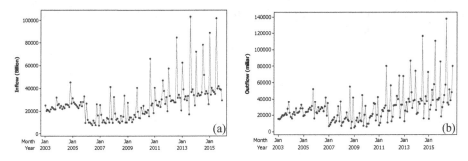

Fig. 2. Time series plot data of inflow (a) and outflow (b)

Selection of inputs that used in the neural network can be performed using lag plot. If nonlinear pattern is formed at a certain lag, then this lag will be used as input to the neural network models. Lag plot of inflow and outflow in Indonesia can be seen at the following figure.

Based on the lag plot in Fig. 3, the correlations with lagged observation until lag 15 tend to have the same pattern. Therefore, input determination based on the lag plot would be difficult to do. Thus, input determination will be done by using PACF from stationary data. If at certain lag the partial autocorrelation was significant, then this lag will be used as input to the neural network models. First, we will check the stationary of the data by using Augmented Dickey Fuller test, as shown in Table 1.

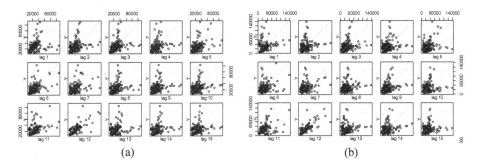

Fig. 3. Lag plot of inflow (a) and outflow (b)

Table 1 shows that the inflow data is not stationary in mean. Thus, the determination of the input for the inflow data will be based on the PACF of the differenced inflow data, while for outflow data, Augmented Dickey Fuller test shows that outflow data has been stationary in mean. Thus, the determination of inputs for outflow data

Table 1. Stationary test for inflow and outflow data

Data	Dickey-Fuller	p-value	Conclusion
Inflow	−2.238	0.476	Not stationary
Outflow	−3.824	0.019	Stationary

will be done based on PACF from outflow data. The PACF of differenced inflow data and PACF from outflow data can be seen at the Fig. 4.

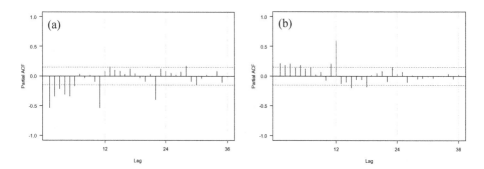

Fig. 4. PACF of stationary inflow data (a) and outflow data (b)

Figure 4 shows that data inflow and outflow in Indonesia have a seasonal pattern. In the PACF plot of differenced inflow data, lag 11 has the highest partial correlation. Thus, the high correlation of the inflow data is at lag 11 and 12, while in the outflow data, lag 12 has the highest partial correlation.

Significant lags in differenced inflow data are at lags 1, 2, 4, 5, and 11. Thus, the input combinations to be used are lag 1, 2, 3, 4, 5, 6, 11, and 12. In outflow data, the input combinations to be used are lag 1, 2, 3, and 12. Having known some input combinations, these combinations will be checked in linearity using the terasvirta test. Forecasting with neural network will be done if there is a nonlinear relationship between the input and the output. The results of the linearity testing for all possible inputs are shown at Table 2.

Table 2. Linearity test for all possible inputs for inflow and outflow models

Inflow			Outflow		
Input's lag	χ^2	p-value	Input's lag	χ^2	p-value
1	6.43	4.01×10^{-2}	12	9.91	7.03×10^{-3}
12	16.68	2.39×10^{-4}	1 and 12	56.71	6.82×10^{-10}
1 and 2	14.12	4.91×10^{-2}	2 and 12	39.72	1.43×10^{-6}
11 and 12	67.74	4.21×10^{-12}	1, 2, and 12	91.49	1.33×10^{-12}
1, 2, 11 and 12	159.77	0.00	1, 3, and 12	73.08	2.86×10^{-9}
1, 2, 3, 4, 5, 6, 11, and 12	Inf	0.00	1, 2, 3, and 12	102.04	8.84×10^{-10}

Based on Terasvirta test, all possible combinations input have a nonlinear relationship with the output. This is indicated by p-value for all combinations are less than α ($\alpha = 0.05$). Under null hypotheses that the input has linear relationship with the output, this null hypothesis was rejected. Thus, all combinations of inputs will be analyzed using the neural network in order to obtain the best input with the optimal number of neurons.

In forecasting using neural networks models, preprocessing of the data also plays an important role. One of the most widely used types of preprocessing is transformation. In general, transformations can improve the accuracy of forecasting especially for short-term forecasting. The better forecasting results are generally obtained by using logarithm natural transformation [17]. Thus, this transformation will be used in order to improve the accuracy of forecasting results. The comparison of RMSE obtained for training data and testing data shown at Table 3. In order to make it easier to compare, the RMSE value in Table 3 then be illustrated as Fig. 5.

The selections of optimal input are based on the input that has the smallest RMSE on the testing data. Figure 5 shows that for inflow models, the input with smallest RMSE is combination of lag 11 and 12 with 4 number of neuron. Meanwhile for outflow data, the input with smallest RMSE is combination of lag 1, 2, and 12 with 5 number of neuron. However, Fig. 5 also shows that the increasing number of neurons does not always produce smaller RMSE, especially on testing data. Therefore, in the selection of the number of neurons, trial and error need to be done to determine the optimal number of neurons that produce the smallest RMSE. Thus, the forecasting with neural network will be done by using the transformed data using natural logarithm, and different input and neuron for inflow and outflow. The comparison of actual and forecast data for the inflow and outflow in Indonesia shown as Fig. 6.

Figure 6 shows that the forecast for training data can follow the actual data for inflow and outflow data. Moreover, for inflow data, the accuracy for testing data has smaller RMSE than training data. RMSE for training data is 10409.87 billion, while RMSE for testing data is 6560.62 billion. Whereas, forecasting results for testing data in outflow tend to be less able to capture the actual data patterns. RMSE for the training data is 9042.16 billion, while RMSE of testing data is 21566.85 billion. The value of RMSE in testing data is larger than the training data. It is in line with the previous study that found neural network models which can capture patterns well in training data can also produce forecasting less accurate in testing data [18].

3.2 Comparison of FFNN Methods with Other Classical Methods

The accuracy of FFNN in forecast the data of Inflow and Outflow in Indonesia will be compared with classical method which is widely used in forecasting, i.e. ARIMA and ARIMAX. ARIMA model that fulfilled white noise assumption for inflow data is ARIMA $(2, 1, [23, 35]) (1, 0, 0)^{12}$, while ARIMA model for outflow data is ARIMA $([2, 3, 23], 0, [35]) (1, 0, 0)^{12}$. In the ARIMAX model, dummy variables used are trend, seasonal, and calendar variations effects. The comparison of RMSE for FFNN, ARIMA, and ARIMAX models can be shown as follows.

Table 4 shows that the best method for forecasting inflow data is FFNN, whereas for outflow data, the smallest RMSE is produced by ARIMAX method. This can be caused by the use of different inputs on the FFNN model and ARIMAX model. Based on Fig. 6(b), it can be seen that the FFNN model can't capture the effect of calendar variation so that the value of RMSE becomes large.

In ARIMAX model there is dummy variable that is calendar variations effects so that ARIMAX model will be able to capture the pattern of calendar variation in outflow data. However, compared to the ARIMA method, with the parsimonious inputs, the

Table 3. Comparison of RMSE for all possible inputs

Inflow				Outflow			
Input lag	Neuron	RMSE training	RMSE testing	Input lag	Neuron	RMSE training	RMSE testing
1	1	14143.91	28898.81	12	1	11703.26	28241.51
	2	22215.66	27086.60		2	11766.28	28979.40
	3	14266.64	29339.61		3	12806.41	29151.20
	4	14133.84	29068.68		4	12287.39	29129.95
	5	14563.87	26951.62		5	11567.62	28283.98
	10	15038.23	25980.72		10	11292.59	28334.56
	15	14154.05	28997.70		15	12751.28	29225.10
12	1	10871.00	14148.57	1 and 12	1	14653.29	30294.51
	2	11015.59	10919.43		2	10660.71	27789.08
	3	10516.49	11755.03		3	9598.81	25075.43
	4	11249.39	11739.50		4	10615.72	26766.65
	5	11275.52	9915.43		5	10100.57	27372.06
	10	11855.57	10518.17		10	11838.60	23831.50
	15	10732.50	11644.34		15	12612.10	26941.21
1 and 2	1	13562.32	27175.61	2 and 12	1	11526.31	26101.39
	2	13451.37	26923.92		2	11753.04	27320.93
	3	21500.65	22702.19		3	11600.49	24242.59
	4	15872.27	23861.09		4	11714.06	28441.15
	5	13404.75	26736.95		5	11495.97	27781.79
	10	13624.41	26655.08		10	11167.16	26931.99
	15	13469.46	26412.88		15	11884.34	28023.63
11 and 12	1	11160.52	9351.40	1, 2 and 12	1	12006.61	27205.56
	2	10577.98	18359.11		2	11266.77	26221.86
	3	9560.15	15585.81		3	11636.90	26548.47
	4	10409.87	**6560.62**		4	13738.92	29349.27
	5	9668.54	13062.22		5	9042.16	**21566.85**
	10	9637.45	10465.48		10	9549.50	25354.33
	15	8232.85	10779.40		15	9473.79	25289.16
1, 2, 11, and 12	1	10743.57	19329.36	1, 3 and 12	1	12810.72	27453.98
	2	11344.87	11356.93		2	13024.16	29289.44
	3	11997.43	8759.97		3	12016.07	28383.20
	4	10262.76	15117.49		4	10615.99	28469.27
	5	9173.02	10138.90		5	10295.00	26808.46
	10	9571.04	10089.72		10	11249.07	26378.39
	15	8754.30	11735.84		15	10363.29	25116.26
1, 2, 3, 4, 5, 6, 11, and 12	1	12432.70	22895.15	1, 2, 3 and 12	1	12203.34	27810.09
	2	9929.39	11359.77		2	9660.77	25247.26
	3	9891.63	8493.18		3	9240.43	22234.38
	4	8539.83	13215.84		4	10447.53	24038.35
	5	11106.36	9379.35		5	9463.79	23241.48
	10	10175.12	7727.62		10	9407.18	25092.63
	15	8462.65	8659.24		15	8264.03	22848.71

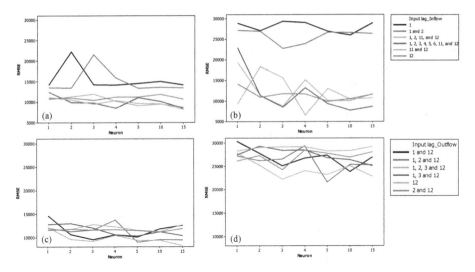

Fig. 5. RMSE comparison for selection input and neuron in training inflow (a), Testing inflow (b), Training outflow (c), and Testing outflow (d) Models

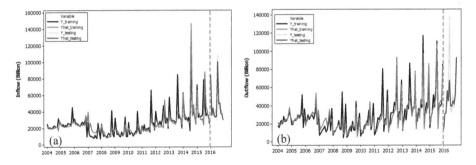

Fig. 6. Comparison of actual and forecast data for inflow (a) and outflow (b)

Table 4. Comparison of RMSE for FFNN, ARIMA, and ARIMAX models

Model	Inflow		Outflow	
	RMSE training	RMSE testing	RMSE training	RMSE testing
FFNN	10409.87	**6560.62**	9042.16	21566.85
ARIMA	6531.24	11697.04	7975.92	27212.73
ARIMAX	6551.84	15577.45	7969.11	**19381.40**

FFNN model has smaller RMSE values for testing data, both for inflow and outflow data. While for training data, RMSE of ARIMA model is lower than FFNN.

4 Discussion

Model selection in forecasting inflow and outflow in Indonesia are based on the significant lags in PACF. Thus, there are many combinations in order to get the best inputs for neural networks model. The model selection has been done by using cross validation. The forecasting results of inflow shows that it is possible that testing data have more accurate results than training data. This finding shows that neural networks were able to predict testing data as well as training data by using the suitable inputs and neuron, especially for short term forecasting. Meanwhile, by using FFNN the forecasting result of outflow shows that testing data were lower accurate than training data. Some previous studies showed that a neural network models can't capture the trend and seasonal patterns well [19, 20]. The best way to overcome these problems are to do detrend and deseasonal. So that, preprocessing by using detrend and deseasonal might prompting to improve the accuracy of forecasting.

All possible inputs that used in this study only based on the significant lag in PACF. There are many other methods to get the combination of inputs, such as by using stepwise linear regression [21], using lag from ARIMA models [18], and based on the increment of R^2 of FFNN model when added an input variable [22]. Additional research might examine the comparison between selection inputs by using those methods.

Acknowledgments. This research was supported by DRPM-DIKTI under scheme of "Penelitian Berbasis Kompetensi", project No. 532/PKS/ITS/2017. The authors thank to the General Director of DIKTI for funding and to anonymous referees for their useful suggestions.

References

1. Maier, H.R., Dandy, G.C.: Neural networks for the prediction and forecasting of water resources variables: a review of modelling issues and applications. Environ. Model Softw. **15**, 101–124 (2000)
2. Kandil, N., Wamkeue, R., Saad, M., Georges, S.: An efficient approach for short term load forecasting using artificial neural networks. Electr. Power Energy Syst. **28**, 525–530 (2006)
3. Azad, H.B., Mekhilef, S., Ganapathy, V.G.: Long-term wind speed forecasting and general pattern recognition using neural networks. IEEE Trans. Sustain. Energy **5**, 546–553 (2014)
4. Claveria, O., Torra, S.: Forecasting tourism demand to Catalonia: neural networks vs. time series models. Econ. Model. **36**, 220–228 (2014)
5. Kara, Y., Boyacioglu, M.A., Baykan, O.K.: Predicting direction of stock price index movement using artificial neural networks and support vector machines: the sample of the Istanbul stock exchange. Expert Syst. Appl. **38**, 5311–5319 (2014)
6. Guler, H., Talasli, A.: Modelling the daily currency in circulation in Turkey. Central Bank Repub. Turkey **10**, 29–46 (2010)
7. Nasiru, S., Luguterah, A., Anzagra, L.: The efficacy of ARIMAX and SARIMA models in predicting monthly currency in circulation in Ghana. Math. Theory Model. **3**, 73–81 (2013)
8. Rachmawati, N.I., Setiawan, S., Suhartono, S.: Peramalan Inflow dan dan Outflow Uang Kartal Bank Indonesia di Wilayah Jawa Tengah dengan Menggunakan Metode ARIMA, Time Series Regression, dan ARIMAX. Jurnal Sains dan Seni ITS, 2337–3520 (2015)

9. Kozinski, W., Swist, T.: Short-term currency in circulation forecasting for monetary policy purposes – the case of Poland. Financ. Internet Q. **11**, 65–75 (2015)
10. Hill, T., Marquezb, L., O'Connor, M., Remusa, W.: Artificial neural network models for forecasting and decision making. Int. J. Forecast. **10**, 5–15 (1994)
11. Crone, S.F., Kourentzes, N.: Input-variable specification for neural networks - an analysis of forecasting low and high time series frequency. In: International Joint Conference on Neural Networks, pp. 14–19 (2009)
12. Anders, U., Korn, O.: Model selection in neural networks. Neural Netw. **12**, 309–323 (1999)
13. Wei, W.W.S.: Time Series Analysis Univariate and Multivariate Methods. Pearson, New York (2006)
14. Lee, M.H., Suhartono, S., Hamzah, N.A.: Calendar variation model based on ARIMAX for forecasting sales data with Ramadhan effect. In: Regional Conference on Statistical Sciences, pp. 349–361 (2010)
15. Sarle, W.S.: Neural networks and statistical models. In: Proceedings of the Nineteenth Annual SAS Users Group International Conference (USA 1994), SAS Institute, pp. 1538–1550 (1994)
16. Apriliadara, M., Suhartono, Prastyo, D.D.: VARI-X model for currency inflow and outflow forecasting with Eid Fitr effect in Indonesia. In: AIP Conference Proceedings, vol. 1746, p. 020041 (2016)
17. Proietti, T., Lutkepohl, H.: Does the Box-Cox transformation help in forecasting macroeconomic time series? Int. J. Forecast. **29**, 88–99 (2013)
18. Faraway, J., Chatfield, C.: Time series forecasting with neural networks: a comparative study using the airline data. Appl. Stat. **47**, 231–250 (1998)
19. Zhang, G.P., Qi, M.: Neural network forecasting for seasonal and trend time series. Eur. J. Oper. Res. **160**, 501–514 (2005)
20. Suhartono, S., Subanar, S.: The effect of decomposition method as data preprocessing on neural networks model for forecasting trend and seasonal time series. Jurnal Keilmuan dan Aplikasi Teknik Industri 27–41 (2006)
21. Swanson, N.R., White, H.: Forecasting economic time series using flexible versus fixed specification and linear versus nonlinear econometric models. Int. J. Forecast. **13**, 439–461 (1997)
22. Suhartono: New procedures for model selection in feedforward neural networks. Jurnal Ilmu Dasa, **9**, 104–113 (2008)

The Selection Feature for Batik Motif Classification with Information Gain Value

Anita Ahmad Kasim[1,2(✉)], Retantyo Wardoyo[1], and Agus Harjoko[1]

[1] Department of Computer Science and Electronics, Faculty of Mathematics and Natural Sciences, Universitas Gadjah Mada, Yogyakarta, Indonesia
nita.kasim@gmail.com, {rw,aharjoko}@ugm.ac.id
[2] Department of Information Technology, Faculty of Engineering, Universitas Tadulako, Palu, Indonesia

Abstract. Features in the classification process have an important role. Classification of batik motif has been done by using various features such as texture features, shape features, and color features. Features can be utilized separately or combined between features. The problem in this research is how to get the potential feature to classified the motif of batik. The feature combination causes enhancement in the number of features causing dataset size changes in the classification process. In this research will be done the selection of features to the combination feature of texture and feature of shape from batik motifs. The feature selection process uses the information gain value approach. The feature selection is done by calculating the value of the information gain of each feature of texture and feature of shape. The value of the information gain will be sorted from the highest information gain value. Ten features with the highest information value will be the selected feature to be processed in the process of batik image motif classification. The classification method using in this research is an artificial neural network. The neural network consists of three layers, that is input layer, hidden layer, and the output layer. The data from selection feature processed in the artificial neural network. The result of this study shows that the accuracy of the process of batik motif classification with a combination feature of texture and feature of shape is 75%. The addition of feature selection process to batik motif classification process gives an increase of 12.5% to the yield an accuracy of 87.5%.

Keywords: Feature selection · Batik · Information gain

1 Introduction

The number of batik motifs that are owned and spread throughout Indonesia makes Batik become one of the cultural heritage of Indonesia that must be preserved. The designation of Batik into the List of Culture of Human Heritage by the United Nations (UN) make batik as an icon of Indonesian culture [1].

Batik has a variety of motif designs that are influenced by various cultures and mythology in Indonesia [2]. Classification of classical batik motif cannot be done by using the features of the shape and texture features separately so that the two methods are combined in a method of classification based on texture and shape [3]. The feature

© Springer Nature Singapore Pte Ltd. 2017
A. Mohamed et al. (Eds.): SCDS 2017, CCIS 788, pp. 106–115, 2017.
https://doi.org/10.1007/978-981-10-7242-0_9

of texture and feature of shape can be selected for use in the classification of batik motifs to obtain potential features that affect the accuracy of the classification of each type of batik. To solve the existing problems in the batik motif recognition model, we need a model that can identify the type of batik motif using the artificial neural network with combined features of texture and features of batik ornament shape.

This research is expected to improve the accuracy of the motif classification of batik so that various batik motifs can be more easily recognized. This paper will describe the feature selection process on batik motif classification includes the first part is the introduction, the second part is related works, the third part is the proposed method and the last is the conclusion.

2 Related Works

Batik is one of the oldest Indonesian art and culture. There are many batiks as a local product in all parts of Indonesia from Sabang to Merauke. Batik is a cultural heritage of Indonesia that has been recognized as a UNESCO international cultural heritage on October 2, 2009. Pattern recognition is one way to preserve the culture of batik. The research of batik pattern recognition has been done by some previous researchers [1–4].

The features used in batik pattern recognition research include texture features, shape features and color features [5–8]. The batik texture features used were obtained from sub-band image of the Co-Occurrence Matrices [9]. The image used was obtained by downloading randomly from the internet. Co-occurrence matrices method with sub-band image is done by combining Gray Level Co-occurrence Matrices and Discrete Wavelet Transform method. The results obtained are good enough to classify the image of batik. Maximum accuracy that can be achieved is 72%. The research of batik classification is also done by combining two batik features namely color (3D-Vector Quantization) and shape feature (Hu Moment) [10]. HMTSeg is used by dividing the image of batik into several areas based on similarity of features including levels of gray, texture, color, and movement [11]. The accuracy obtained by 80% can recognize the image of batik based on texture features. In addition to the use of texture features, batik imagery can also be classified by combining texture features and color stats features. Texture features extracted from Gray Level Co-Occurrence Matrices (GLCM) consisting of contrast, correlation, energy, and homogeneity. The color features are extracted with the Statistical Color Channel RGB which consists of mean, skewness, and kurtosis [12, 13]. The feature combined with the Bag of Features (BOF) method uses Scale-Invariant Feature Transform (SIFT) and Support Vector Machine (SVM) to classify batik images The results obtained are 90.6% precision, 94% recall and 94% overall accuracy [14, 15]. The acquisition of image data in one class comes from one fabric image divided into six sub-images.

The combination of texture features and shape features are also done for batik image classification. The shape features obtained feature values include compactness, eccentricity, rectangularity, and solidity [16]. The combination of features causes the number of features to be more numerous, so in this study proposed a method for selecting features of texture and features of batik shape that potential to improve the accuracy of the batik classification.

3 Proposed Method

The methods we proposed in this study consisted of the acquisition process, the pre-process, the feature extraction process, the selection feature process, and the image classification process. The classification process using the artificial neural network. The role of an artificial neural network as one of the classification method which is quite reliable due to its ability to classify the data. The artificial neural network has a high tolerance for noise-sensitive data. The neural network is able to learn from the data trained. Therefore this research analysis data classification by using ANN Backpropagation to get an accurate result.

3.1 Acquisition and Preprocess of Batik Image

The acquisition process is the process of taking the image of batik using the camera and download some images of batik through the internet. The pre-processing will process the acquired image before it is processed to obtain texture feature values and feature shapes values.

At the pre-processing process, we formed a sub-image with the image size of 100×100 pixels. Next, an image transformation to the gray scale is performed. The grayscale transformation will further facilitate the computing in obtaining the values of batik texture features.

3.2 Feature Extraction Process

Grayscale batik image produced in the process of acquisition and preprocessing into an input on feature extraction process. The texture feature extraction method used begins with the formation of a GLCM matrix from a grayscale image. The image of the normalized result is Gray Level Co-occurrence Matrices which will be used to obtain texture features values. Values obtained in GLCM are processed into textural features consist of the value of features of Angular Second Moment (ASM) or otherwise known as Energy, Contrast, Correlation, and Homogeneity [17].

The transformation from the grayscale image into a binary image by using Otsu Method. It will eliminate the noise in a binary image by using the binary morphology [18–20]. The shape features consist of the compactness, eccentricity, rectangularity, and solidity.

3.3 Feature Selection Process

The feature selection model developed in this study uses the information gain's value approach. The steps to obtain information gain are as follows:

1. Convert continuous values into discrete values by partitioning continuous values into discrete value intervals. The process of converting the continuous values into discrete values as described [21] consists of three steps as follows:
   (a) Sort the continuous values that will be made in discrete values

(b) A cut-point evaluation for dividing continuous values. The definition of cut-point refers to the real value of continuous values that divide the interval for instance into two intervals. In this researches the number of cut-points on the feature there are two cut-points, and there are three cut-points.
(c) Split the continuous values in their respective intervals into discrete values.

The process of converting the continuous values to discrete values is shown in Fig. 1.

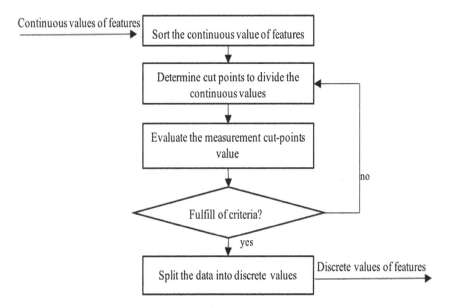

Fig. 1. Discretize process from continous data of features

2. Calculate the value of the expected information or entropy of the whole class by representing it as in Eq. (1)

$$J = (U, C \cup D) \tag{1}$$

where U denotes a set of data consisting of $\{u_1, u_2, ..., u_s\}$, C is a feature to be ranked consisting of $\{C_1, C_2, ..., C_n\}$ and D is the expected target class consisting of $\{d_1, d_2, ..., d_m\}$. If the expected target class has different values of m to distinguish each target d_i for i = 1, 2, ..., m then the amount of data included in the class. To calculate the expected value of the information or the required entropy we used Eq. (2)

$$I(s_1, ..., s_m) = -\sum_{i=1}^{m} P_i log_2 P_i \tag{2}$$

where $I(s_1,...,s_m)$ denotes the value of the expected information for the entire data. Pi is denoted the probability of sample entering in class and m is number of labels each feature on class.

3. Calculate the entropy value of each feature

   Suppose c_i is a feature to be calculated the value of the information and it has a different value v to distinguish each value in the feature, then the feature c_i can be grouped into a set S_j that consisting $\{S_1,S_2,...,S_v\}$. If S_j is the amount of data d_i in a class on a subset of sets S_j, then the feature entropy value can be obtained by Eq. (3).

$$E(c_i) = \sum_{j=1}^{v} \frac{s_{ij} + ... + s_{mj}}{s} I(s) \qquad (3)$$

where $E(c_i)$ denotes the value of feature entropy and v value in feature c_i. S_{ij} is a number of sample classes in class S_j.

4. Calculate the information gain value of each texture feature and shape feature with Eq. (4)

$$Gain(c_i) = I(s_1, ..., s_m) - E(c_i) \qquad (4)$$

5. The value of the information gain will be sorted and will be selected for feature data by taking only 10 data with the highest information gain value.

The texture-shape feature selection is done by calculating the entropy value to generate the information gain value of the texture feature and the shape features. The value of the information gain will reflect the feature's quality.

3.4   Classification Process

Motif classification model on batik image based on the feature of texture and feature of shape using neural network method using backpropagation learning method. Backpropagation is a gradient decrease method to minimize the square of output errors or algorithms that use weight adjustment patterns to achieve minimum error values [22]. Artificial neural network in the training process is the number of neurons in the input layer, the hidden layer, the output layer. The number of neurons in the input layer is determined according to the number of inputs on the artificial neural network architecture. At the beginning of the trial will be training by combining 20 features of texture and the feature of shape. For the selection, results will be selected ten features tested. The determination of the number of neurons in the hidden layer is determined using the equation described [23]. For the number of neurons in the output, the layer is tested with two outputs, i.e., three output neurons for eight classes of batik imagery and eight output neurons for eight classes of batik images. Artificial neural network architecture for the motif classification of batik image can be seen in Figs. 2 and 3.

The number of neurons in each layer used in the artificial neural network architecture in this study is shown in Table 1.

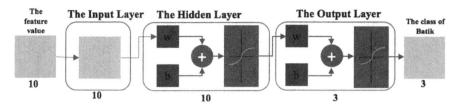

Fig. 2. Artificial neural network architecture for the motif classification of batik image with 3 outputs

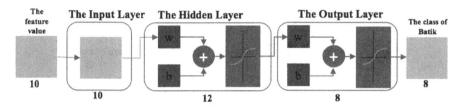

Fig. 3. Artificial neural network architecture for the motif classification of batik image with 8 outputs

Table 1. Architectural number of neurons in batik motif recognition model

Features	Input layer	Hidden layer	Output layer	NN architecture
Texture-shape	20	14	3	20:14:3
Texture-shape	20	17	8	20:17:8
Texture-shape based on feature selection	10	10	3	10:10:3
Texture-shape based on feature selection	10	12	8	10:12:8

4 Results and Discussions

The acquisition process generates images data batik with eight motif class. The motif class consists of *Ceplok* class, *Kawung* class, *Megamendung* class, parang class, *Semen* class, *Batik solo* class, *Sido asih* class and *Tambal* class. The texture's feature extraction process generates the values of ASM/Energy, the contrast value, the correlation value, and the homogeneity values of batik images. Furthermore, the values of ASM/Energy, the contrast value, the correlation value, and the homogeneity values are normalized. The minimum values and the maximum values of the normalization result from batik images can be seen in Fig. 4.

The feature extraction from the shapes generates the compactness value, the eccentricity value, the rectangularity value and the solidity value of batik images. The minimum value of the compactness value is 15.55, and the maximum value is 7296.19.

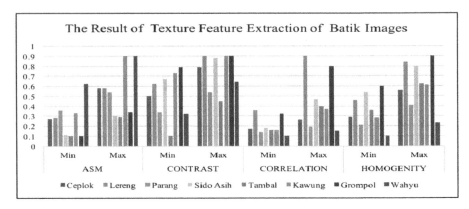

Fig. 4. Minimum value and maximum value of batik texture features

The value of eccentricity feature in the range of 1.22 to 4.08. The rectangularity values in the range of 7.62 to 1001.37 and features of solidity in the range of 0.35 to 0.86.

Figure 5 illustrates the minimum value and the maximum value of the Batik image shape feature.

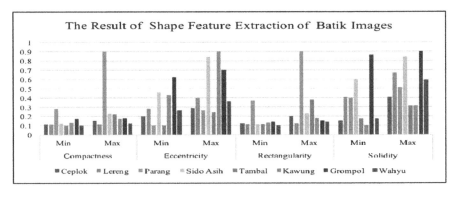

Fig. 5. Minimum value and maximum value of batik shape features

The processing of feature selection on the batik motif classification model is done by first making changes to the continuous values of the extraction result into discrete values. For each feature categorized the grade as shown in Table 2. Based on the selection model proposed in this study obtained the gain value of the texture features and the shape feature.

The resulting feature value generated ranks include the correlation 90°, eccentricity, rectangularity, correlation 135°, compactness, contrast 135°, ASM 90°, ASM 0°, contrast 90°, ASM 45°. The result of gain value ranking is shown in Table 3.

Table 2. The conversion results of continuous values into discrete values

No	Features	Discrete value		
1	Angular second moment	High	Middle	Low
2	Contrast	High	Middle	Low
3	Correlation	High	Middle	Low
4	Invers different moment	Uniform	Not uniform	
5	Compactness	Compact	Not compact	
6	Eccentricity	High	Middle	Low
7	Rectangularity	High	Low	
8	Solidity	Solid	Not Solid	

Table 3. Gain value ranking results on texture features and shape features of motif batik image

Number	Selected features	Gain value
1	correlation 90°	5.254894387
2	Eccentricity	5.157481437
3	Rectangularity	5.138537962
4	correlation 135°	4.769167528
5	Compactness	4.614079875
6	contrast 135°	3.901570398
7	ASM 90°	3.606168067
8	ASM 0°	3.581783311
9	contrast 90°	3.457986592
10	ASM 45°	3.183023418

The top ten of the rated features will be used as a combined feature of the texture feature and shape feature. Ten features will be used in the process of batik image classification using the artificial neural network.

In this study, we used two datasets. The first data set uses twenty features from the combination of texture features and shapes feature. The second data set uses the ten features of the recommended feature selection in this study. The training was conducted with two different neural network architectures. The training was performed on 64 data of batik image features. Forty data were used as training data, and twenty-four data were used as test data. The data can be classified correctly on the first test of a total of 15 data with an accuracy of 62.5%. The use of different neural network architectures yields the correct amount of data classified to 75%. The results of training and testing motif batik classification can be seen in Table 4.

The tests performed on the data using the selected selection feature with ten input neuron layer architecture, twelve hidden layer neurons and eight neurons on the output layer resulted in an accuracy of 87.5%.

There is an increase in accuracy of 12.5% of the classification results by using features from the selection feature process. The accuracy is lower than previous research. Data in the previous research was different. The amount of data in each class

Table 4. The results of training and testing motif batik classification

Dataset	Features	NN architecture	Training data	Testing data	True data	Accuracy (%)
I	Texture-shape	20:14:3	40	24	15	62.50
I	Texture-shape	20:17:8	40	24	18	75.00
II	Texture-shape based on feature selection	10:10:3	40	24	11	45.83
II	Texture-shape based on feature selection	10:12:8	40	24	21	87.50

is less, too. It is expected that in subsequent research can be used larger data so that the accuracy of classification can be increased.

5 Conclusion

The number of neurons in the neural network architecture and the number of features in the process of batik motif classification will affect the results of accuracy. There is an improvement in the accuracy of batik motif classification model with the artificial neural network from 75% (without feature selection process) to 87.5% (with feature selection process). The addition of the feature selection process on the batik motif classification model gives an increase the accuracy of 12.5%. The future works of this research develop the feature selection method with the other method. The other methods are expected to yield potential feature which will increase the accuracy of batik motif classification.

References

1. Imanudin: "Batik Identification Based On Batik Pattern And Characteristics Using Fabric Pattern Feature Extraction" (2010)
2. Kitipong, A., Rueangsirasak, W., Chaisricharoen, R.: Classification System for Traditional Textile : Case Study of the Batik. In: 13th International Symposium on Communications and Information Technologies (ISCIT) Classification, pp. 767–771 (2013)
3. Kasim, A.A., Wardoyo, R.: Batik image classification rule extraction using fuzzy decision tree. In: Information Systems International Conference (ISICO), pp. 2–4, December 2013
4. Nurhaida, I., Noviyanto, A., Manurung, R., Arymurthy, A.M.: Automatic Indonesian's batik pattern recognition using sift approach. Procedia Comput. Sci. **59**, 567–576 (2015)
5. Kasim, A.A., Wardoyo, R., Harjoko, A.: Feature extraction methods for batik pattern recognition: a review. In: AIP Conference Proceedings, vol. 70008, pp. 1–8 (2016)
6. Suciati, N., Pratomo, W.A., Purwitasari, D.: Batik motif classification using color-texture-based feature extraction and backpropagation neural network. In: IIAI 3rd International Conference on Advanced Applied Informatics, pp. 517–521 (2014)
7. Kasim, A.A., Harjoko, A.: Klasifikasi citra batik menggunakan jaringan syaraf tiruan berdasarkan gray level co-occurrence matrices (GLCM). In: Seminar Nasional Aplikasi Teknologi Informasi (SNATI), pp. 7–13 (2014)

8. Kasim, A.A., Wardoyo, R., Harjoko, A.: Fuzzy C means for image batik clustering based on spatial features. Int. J. Comput. Appl. **117**(2), 1–4 (2015)

9. Minarno, A.E., Munarko, Y., Kurniawardhani, A., Bimantoro, F., Suciati, N.: Texture feature extraction using co-occurrence matrices of sub-band image for batik image classification. In: 2nd International Conference on Information and Communication Technology (ICoICT) Texture, pp. 249–254 (2014)

10. Nugrowati, A.D., Barakbah, A.R., Ramadijanti, N., Setiowati, Y.: Batik image search system with extracted combination of color and shape features. In: International Conference on Imaging and Printing Technologies (2014)

11. Murinto, Ariwibowo, E.: Image segmentation using hidden markov tree methods in recognizing motif of batik. J. Theor. Appl. Inf. Technol. **85**(1), 27–33 (2016)

12. Aditya, C.S.K., Hani'ah, M., Bintana, R.R., Suciati, N.: Batik classification using neural network with gray level co-occurence matrix and statistical color feature extraction. In: 2015 International Conference on International, Communication Technology and System (ICTS), pp. 163–168 (2015)

13. Rao, C.N., Sastry, S.S., Mallika, K., Tiong, H.S., Mahalakshmi, K.B.: Co-occurrence matrix and its statistical features as an approach for identification of phase transitions of mesogens. Int. J. Innov. Res. Sci. Eng. Technol. **2**(9), 4531–4538 (2013)

14. Azhar, R., Tuwohingide, D., Kamudi, D., Sarimuddin, Suciati, N.: Batik image classification using sift feature extraction, bag of features and support vector machine. Procedia Comput. Sci. **72**, 24–30 (2015)

15. Setyawan, I., Timotius, I.K., Kalvin, M.: Automatic batik motifs classification using various combinations of sift features moments and k-nearest neighbor. In: 7th International Conference on Information Technology and Electrical Engineering (ICITEE), Chiang Mai, Thailand, vol. 3, pp. 269–274 (2015)

16. Kasim, A.A., Wardoyo, R., Harjoko, A.: Batik classification with artificial neural network based on texture-shape feature of main ornament. Int. J. Intell. Syst. Appl. **9**, 55–65 (2017)

17. Haralick, R., Shanmugan, K., Dinstein, I.: Textural features for image classification. IEEE Trans. Syst. Man Cybern. **3**, 610–621 (1973)

18. Bhagava, N., Kumawat, A., Bhargava, R.: Threshold and binarization for document image analysis using otsu' s Algorithm. Int. J. Comput. Trends Technol. **17**(5), 272–275 (2014)

19. Otsu, N.: A threshold selection method from gray-level histogram. IEEE Trans. Syst. Man Cybern. **20**(1), 62–66 (1979)

20. Dong, L., Yu, G., Ogunbona, P., Li, W.: An efficient iterative algorithm for image thresholding. Pattern Recognit. Lett. **29**(9), 1311–1316 (2008)

21. Liu, H., Hussain, F.,Tan, C.L., Dash, M.: Discretization: An Enabling Technique (2001)

22. Fausett, L.: Fundamental of Neural Network. Prentice Hall, New Jersey (1993)

23. Stathakis, D.: How many hidden layers and nodes? Int. J. Remote Sens. **30**(8), 2133–2147 (2009)

Eye Detection for Drowsy Driver Using Artificial Neural Network

Mohd. Razif Bin Shamsuddin[1]([⊠]),
Nur Nadhirah Binti Shamsul Sahar[1],
and Mohammad Hafidz Bin Rahmat[2]

[1] Faculty of Computer and Mathematical Sciences, Universiti Teknologi
MARA, 40450 Shah Alam, Malaysia
razif@tmsk.uitm.edu.my, nurnadhirah_ss@yahoo.com
[2] Faculty of Computer and Mathematical Sciences,
Universiti Teknologi MARA (Melaka), 77300 Merlimau, Malaysia
hafidz@tmsk.uitm.edu.my

Abstract. Driving is one of the common activities in people's everyday life and therefore improving driving skill to reduce car crashes is an important issue. Even though a lot of studies and work has been done on road and vehicle designs to improve driver's safety yet the total number of car crashes is increasing day by day. Therefore, the most factors that cause an accident is fatigue driver rather than other factors which are distraction, speeding, drinking driver, drugs and depression. To prevent car crashes that occur due to drowsy driver, it is essential to have an assistive system that monitors the vigilance level of driver and alert the driver in case of drowsy detection. This system presents a drowsy detection system based on eye detection of the driver. Vision-based approach is adopted to detect drowsy eye because other developed approaches are either intrusive (physical approach) that makes the driver uncomfortable or less sensitive (vehicle based approach). The data collected from 26 volunteers will have four (4) different type of image. Thus, the total input will be 10,800 nodes. This thesis will be classified into two (2) outputs which are drowsy eye and non-drowsy eye. The algorithm that will be used is Back-propagation Neural Network (BPNN) and will be applied in MATLAB software. The experimental result shows that this system could achieve 98.1% accuracy.

Keywords: Neural network · Eye detection · Drowsy · Drowsy detection

1 Introduction

Safety features are the first and important thing that people seek when buying a vehicle. Nowadays, many vehicles feature especially branded cars are much better and a lot of technologies are implemented in those vehicles such that airbags, black box, intelligence alarm and the latest one is the automatic braking. However, there are many vehicle accidents occur although there are many technologies have been implemented. Road accidents are the most serious issues in Malaysia and have been ranked number 20 in a list of countries with fatal road accidents by The World Health Ranking [7].

© Springer Nature Singapore Pte Ltd. 2017
A. Mohamed et al. (Eds.): SCDS 2017, CCIS 788, pp. 116–125, 2017.
https://doi.org/10.1007/978-981-10-7242-0_10

Generally, it seems that accidents occur because of the drowsy driver. A study conducted by [8], stated that a large number of a vehicle crashed is caused by fatigue driver. It also gives the estimated number of death about 1,200 people and 76,000 people with injuries. A drowsy driver is the most common factors of the road accidents especially during public holidays. People who are driving on the highways and expressways are the higher possibilities getting drowsy when driving. It has been proved by [5] that in Japan, there are many expressways and the number of traffic injuries is growing up to 0.8 million.

Nowadays, number of road vehicle accidents is getting worst and accidents occur almost every day. A study conducted by [6] stressed that drowsiness or fatigue is the highest number factors of accident causes. Drowsiness driver can be detected by the driver but they did not realize that they sleep while driving. A drowsy driver is very dangerous because it will affect the passengers and another road user. Besides that, the accidents occur because of they did not use the facilities that have been provided by the government or Projek Lebuhraya Utara Selatan Berhad (PLUS).

2   Research Motivation

A lot of vehicle safety features that have been introduced either locally or internationally and the latest is from Perodua which has been installed in Perodua Axia is Anti-lock braking system (ABS). ABS work as auto stopping the cars while in a hurry slippery road that can be very challenging to the drivers. However, drowsiness detection system is still studied by many researchers. [8] have claimed that to develop this system, it is a difficult system to be implemented in the group of avoidance system. [9] stated that there is no commercial solution or machine that can afford by all people to detect eyes.

The driver did not put this problem as the prior as other factors of the accident. This problem will become other people suffer because of self-neglectful. [9], have confirmed from The National Highway Traffic Safety Administration (NHTSA) that 100,000 reported crashed have been recorded each year. NHTSA also claimed that the reported crashed are because of the drowsy driver. It has been proved that the road accident occurs almost because of the drowsy driver.

Some of the reasons that cause most accidents is usually drunk driving, distractions and speeding. The main factor that causes the number of accident to increase is a drowsy driver [3] and based on the statistics from National Highway Traffic Safety Administration on 2008, there was 1,550 people death [4]. From the statistical of the researcher, the number of people death because of the drowsy driver is between 1,000 until 2,000 per year. This is a serious problem that must be solved and by improving the safety of the vehicle should cater the problem.

3   Relevant Research

Drowsy comes from an Old English word meaning "falling", and has evolved into falling asleep. Drowsiness or drowsy, also referred to as half asleep, sleepy, lethargy and dull is a weariness caused by exertion. Other than that, according to [2], the drowsiness is defined as feeling abnormally sleepy or tired during the day and it may lead to additional symptoms which are forgetfulness or falling asleep at wrong times.

3.1   Drowsiness Detection Technique

Drowsiness detection is one of the precise monitoring systems that can be used to increase the transportation safety. A few researchers have come out with a different technique for detecting the drowsiness of the driver. There are some of the techniques that has been studied by the researcher. Drowsiness detection can be divided into three (3) approaches which are physiological-based, vehicle-based and behavioral-based which can be referred to Fig. 1 [8]. Furthermore, each of these approaches has some techniques that is related to the approaches.

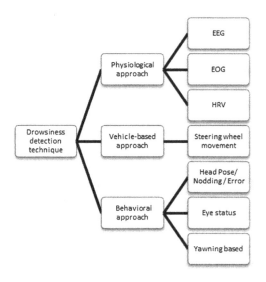

Fig. 1. Drowsiness detection approach and techniques.

Physiological approach can be defined as any of physical changes that occur in our body and usually it can be measures [1]. This approach is popular because of the accuracy of the measurement. The physical changes that can be measured in our body are the heart rate and the measurement of pulse. In this approach, there are some of the technique which is electroencephalogram (EEG), electrocardiogram (ECG), electrooculogram (EOG) and heart rate (HRV). These techniques will be explained more detail in the next subtopic.

For vehicle-based approach, it is another approach which is it can be detect the driver's level vigilance by the steering position and speed variability [1]. Usually this approach can be detected by using sensor in the specific place. So, this sensor can detect the driver's vigilance and therefore must have some alert that can inform the driver.

Besides physiological and vehicle-based approach, there is one of the common approaches that can measure the drowsiness of the driver which is behavioral approach. This approach usually can be used in detecting the behavioral of the driver by their eye status, head nodding, pose and yawning based technique [1]. Therefore, this approach will have more advantage because of this approach is nonintrusive and more reliable as well as accurate and robust [1].

3.2 Advantages and Limitation of the Approaches

From all mentioned approaches that have been discussed previously, they have their own advantage and limitations. A good approach will be chosen as a benchmark for this research. From all of the approaches, behavioral approach is used because the proposed system is based on the eye detection. This approach is chosen because of one of the advantages of this approach is non-intrusive (refer Table 1).

Table 1. Advantages and limitation of the approaches.

References	Approach	Parameters	Advantages	Limitation
Hyun et al., (2012) [11]	Physiological	EEG, EOG, HRV	Reliable	Intrusive
Charles et al., (2009) [12]	Vehicle based	Speed, acceleration and angle of steering wheel analysis	Non-intrusive	Unreliable and limited to vehicle type
Laurence et al., (2000) [13], Abtahi et al., (2011) [14]	Driving behaviour	Eye status, yawning and head position analysis	Non-intrusive, Ease of use, reliable	Lighting problem

4 Research Methodology

4.1 Data Collection

Data collection is one of the main phases in research methodology. In this project, the data of image has been collected from the volunteers from 28 February 2017 until 4 March 2017. Location of the image is different from time to time which some of the image has been captured at Terminal Sentral Kuantan, Taman Peninsular and many more. The image has been collected using handphone's camera and the image has been captured on the daylight only because of the illumination for the night mode will be different. By the time, there are 26 volunteers and each of the volunteer will have 4 different type of image. So, the total of the images collected will be 104 images. All the volunteers are in the range of age from 20 to 60 years old.

4.2 Image Processing

After all the images have been collected, the next process is image processing. Image processing is a technique to enhance raw images received from cameras or pictures taken in normal day-to-day life for various applications. The image processing mainly deals with image acquisition, image enhancement, feature extraction and image classification.

After all the image has been uploaded, the location of eye in the image must be detected. By using cascade object detector function, it will automatically detect the location of the eye. There are many type of classification model but in this project, it will use eye pair model. For the eye pair model, it is divided into two part which is eye pair big and eye pair small. Figure 2 shows the description of these two types of classification which can be generated from Matlab Toolbox.

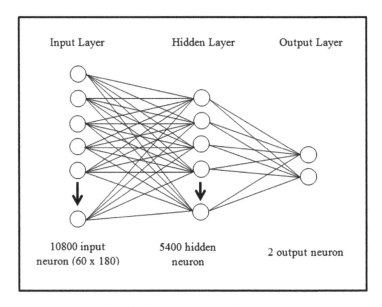

Fig. 2. Neural network architecture.

After successful in detecting the eye location, next process is crop function. It is needed for cropping the specific area of the eye for future analysis. The crop function for this process has been implemented to do the cropping automatically.

The next process is to remove noise. This is important process because this process will remove all the noise to make the image clearer to read. For this project, it will be use Gaussian type which considers the constant mean and variance of the image. This project has set up the mean is 0.6 and the variance is 0.

4.3   Design Neural Network Architecture

This activity is the most important activity for this technique which is also known as the engine of the project. In neural network architecture, there must have number of input node, number of hidden layer, number of hidden node, number of output node, weight and threshold [10]. The design of the NN architecture is shown in Fig. 2. This project will use Back Propagation Neural Network (BPNN) to classify the eyes of the driver whether it is drowsy or non-drowsy. For input layer that have about 10,800 input neuron which each input is divided by 60 rows and 180 columns. Therefore, the hidden neuron should be around 5400 neurons. For the output layer, there are 2 output neurons which are [1, 0] will indicate drowsy eyes and [0, 1] is vice versa.

5   Results and Findings

All images must be pre-processed to make computer understand and to make the image clearer to be used for training process. For this project, Image Processing Toolbox in Matlab has been used to process all the images. Pre-processing is the process that used a few of the function in the image processing toolbox which are cascade detector, cropping, resize, convert into grayscale and remove noise. Table 2 shows the step taken and result produce in image processing stage.

Table 2. Image Processing Results

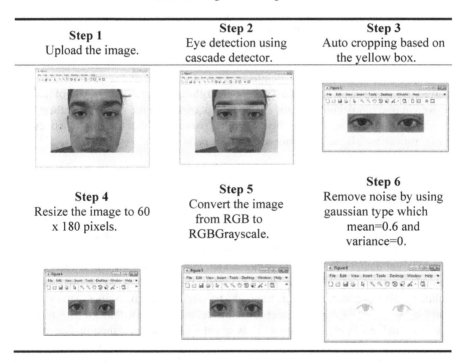

Step 1 Upload the image.	Step 2 Eye detection using cascade detector.	Step 3 Auto cropping based on the yellow box.
Step 4 Resize the image to 60 x 180 pixels.	**Step 5** Convert the image from RGB to RGBGrayscale.	**Step 6** Remove noise by using gaussian type which mean=0.6 and variance=0.

5.1   Training Results

After the images have been pre-processed, the data was trained by using BPNN algorithm. Table 3 shows the parameter involved in BPNN engine. From 107 images

Table 3. Training results

Parameter	Description
Input neuron	10,800
Hidden neuron	100
Output neuron	2
Hidden layer	1
Input value	Grayscale input
Network type	Feed-forward backpropagation
Training function	Scaled Conjugate Gradient
Data division	Random

that have been collected, about 80% was used for training data and the remaining 20% was used for validation and testing data. From 107 images that were used for training, there are 46 images classifying as non-drowsy eyes and 61 images classify as drowsy eyes. Besides that, from the pre-processing part, the image has been resized about 10,800 pixels. Therefore, the each of the input layer has 10,800 neurons. For this project, it only has 107 training data, so the hidden layer of this project will be only 1 layer but the hidden neuron of this project will be one of the parameter tuning because it will have different accuracy. Furthermore, this project will be use 2 output neurons which are [0, 1] for drowsy and [1, 0] for non-drowsy. The parameter that has been used may not suitable with this architecture because some of the problem has been occurred. By using scale conjugate gradient algorithm, the parameter tuning of learning rate, weight and momentum are not affected the accuracy of the system. So, there is no experiment on that parameter but there are other parameters that has been conducted in the experiment which are data split and hidden neuron.

5.2 Experiment on Training Function

Basically, there are some experiments that have to be conducted to get the best accuracy of the training part. One of the experiments that have been conducted is training function. This system used neural network toolbox in Matlab in order to do the learning process. First step is the dataset must be uploaded. Dataset consists of input data and target data. Next step is to decide which training function that suitable for this system. In neural network toolbox, there are many functions that can be used such that trainlm, trainbfg, trainrp, trainscg, traincgb, traincgf, traincgp, trainoss, trainoss and traindx as shown in Table 4. Trainscg is the default function that known as scaled conjugate gradient backpropagation. It is a network training function that updates weight and bias values according to the scaled conjugate gradient method. The scaled conjugate gradient algorithm is based on conjugate directions, as in traincgp, traincgf and traincgb. There were some of the conditions that occurred when the training stops which are the maximum number of epoch reached, the maximum amount of time is exceeded, performance is minimized, and performance gradient fall below min_grad and validation performance has increased more than max_fail times.

Table 4. Experiment on training function

Train function	Number of hidden neuron	Epoch	Performance (MSE)	Training Accuracy (%)	Accuracy (%)
trainrp	100	7	0.0512	62.4	81.8
trainscg	**100**	**25**	**0.00619**	**95.3**	**100**
traincbg	100	24	0.00570	96.5	90.9
traincgf	100	14	0.00291	91.8	81.8
traincgb	100	27	0.0197	92.9	81.8
traincgp	100	29	0.00884	96.5	81.8
trainoss	100	15	0.0197	90.6	81.8
traingdx	100	22	0.125	63.5	54.5

5.3   Experiment on Data Split

Several experiments have been conducted to gain the best accuracy. The data split will be produced different accuracy because it will divide the samples into different number of images. Table 5 shows the parameter tuning results for the data split. For start the number of hidden neuron must be fixed into 100 neurons. From 4 experiment that has been conduct for the data split, the data split 80:20 shows the highest accuracy and the result of performance also shows the smallest number among all the experiment. For data split 80:20, it divides the samples into 85 samples for training and 11 samples for validation and testing. From the experiment that has been conducted, the number of hidden neuron is fixed by 100 neurons and the train function that has been used is trainscg.

Table 5. Result of data split by fixed the number of hidden neuron and fixed the train function

Data Split	Number of hidden neuron	Epoch	Performance (MSE)	Training Accuracy (%)	Accuracy (%)
70:30	100	14	0.00661	100	97.2
75:25	100	28	0.000260	100	97.2
80:20	**100**	**41**	**0.00000853**	**100**	**99.1**
85:15	100	13	0.00743	100	94.4

5.4   Experiment on Hidden Neuron

While the experiment is conducted, there are some of problems that have been occurred. One of the problems that have been occurred is the program take long time to complete the system training. From the observation, the system training takes long time to complete because of the hidden neuron is too many. So, the number of hidden neuron must be reduced and experiment must be conducted to know the accuracy. From the rule of thumb, the number of hidden neuron is basically half of the number of input neuron but it is too many. So, the hidden neuron has been reduced to 3000 neurons but the results are still the same. Therefore, it must be reduced again to 1000 neurons and the result is better than the first and second experiment. After that, to gain the best accuracy, some experiment on the number of hidden neuron is conducted as shown in Table 6.

Table 6. Result of hidden neuron by fixed the data split and train function

Data Split	Number of hidden neuron	Epoch	Performance (MSE)	Training Accuracy (%)	Accuracy (%)
80:20	100	26	0.000529	100	97.2
80:20	300	16	0.00272	100	96.3
80:20	500	16	0.0120	94.1	92.5
80:20	700	23	0.000985	100	95.3
80:20	**1000**	**17**	**0.00287**	**98.8**	**98.1**

6   Conclusion

This paper proposes an artificial neural network to detect eye of the drowsy driver for learning effective features and detecting drowsiness given an input image of a driver. The previous approaches could only make binary image as for detecting driver drowsiness. Artificial neural network provides an automated set of learned features which help to classify the drowsy and non-drowsy accurately. Although the results are achieved 98.1% but the epoch is too low. If the number of the epoch is higher than this system can produce, the result may be getting better. This system also can be implementing in new technique which neuro-fuzzy and convolutional neural network to get the better result.

References

1. Khan, I.: Driver's fatique detection system based on facial features. Degree of Electrical and Electronics Engineering Thesis, Universiti Tun Hussein Onn Malaysia (2014)
2. Krucik, G.: Drowsiness_ Causes, Treatments and Prevention (2015). http://www.healthline.com/symptom/drowsiness?print=true
3. Li, K., Jin, L., Jiang, Y., Xian, H., Gao, L.: Effects of driver behavior style differences and individual differences on driver sleepiness detection. Adv. Mech. Eng. 7(4), 1–8 (2015)
4. Mbouna, R.O., Kong, S.G., Chun, M.-G.: Visual analysis of eye state and head pose for driver alertness monitoring. IEEE Trans. Intell. Transp. Syst. 14(3), 1462–1469 (2013). https://doi.org/10.1109/tits.2013.2262098
5. Miyaji, M.: Method of drowsy state detection for driver monitoring function. Int. J. Inf. Electron. Eng. 4(4), 264–268 (2014). https://doi.org/10.7763/ijiee.2014.v4.445
6. Patel, S.P., Patel, B.P., Sharma, M., Shukla, N., Patel, H.M.: Detection of drowsiness and fatigue level of driver. Int. J. Innov. Res. Sci. Technol. (IJIRST) 1(11), 133–138 (2015)
7. Road users' negligence, error main cause of fatalities (2016). Newspaper, News Straits Times Online. http://Digital.Nstp.Com.My/Nst/Includes/Shop/List_Produtestata=Nstnews&Id=Nstnews

8. Saini, V., Saini, R.: Driver drowsiness detection system and techniques a review. Int. J. Comput. Sci. Inf. Technol. (IJCSIT) **5**(3), 4245–4249 (2014)
9. Zhang, W., Cheng, B., Lin, Y.: Driver drowsiness recognition based on computer vision technology. Tsinghua Sci. Technol. **17**(3), 354–362 (2012)
10. Dreyfus, S.E.: Artificial neural networks, back propagation, and the KelleyBryson gradient procedure. J. Guid. Control Dyn. **13**(5), 926–928 (1990)
11. Hyun, J., Baek, S., Chung, K., Kwang, S.: A Smart Health Monitoring Chair for Nonintrusive Measurement of Biological Sign (2012)
12. Charles, C., Simon, G., Michael, G.: Predicting driver drowsiness using vehicle measures: recent insights and future challenges. J. Saf. Res. 239–245 (2009)
13. Laurence, H., Tim, H., Nick, M.: Review of Fatigue Detection and Prediction Technologies. Institute for Research in Safety & Transport Murdoch University Western Australia and Gerald P Krueger Krueger Ergonomics Consultants (2000)
14. Abtahi, S., Hariri, B., Shirmohammadi, S.: Driver drowsiness monitoring based on yawning detection. In: Proceedings of the Instrumentation and Measurement Technology Conference (I2MTC), pp. 1–4. IEEE (2011)

Time Series Machine Learning: Implementing ARIMA and Hybrid ARIMA-ANN for Electricity Forecasting Modeling

Wahyu Wibowo$^{(\boxtimes)}$, Sarirazty Dwijantari, and Alia Hartati

Institut Teknologi Sepuluh Nopember, Surabaya, Indonesia
wahyu_w@statistika.its.ac.id

Abstract. The aims of this paper are to develop a linear and nonlinear model in time series to forecast electricity consumption of the lowest household category in East Java, Indonesia. The installed capacity in the lowest household customer category has various power, i.e. 450 VA, 900 VA, 1300 VA, and 2200 VA. ARIMA models are family of linear model for time series analysis and forecasting for both stationary and non-stationary, seasonal and non-seasonal time series data. A nonlinear time series model is proposed by hybrid ARIMA-ANN, a Radial Basis Function using orthogonal least squares. The criteria used to choose the best forecasting model are the Mean Absolute Percentage Error and the Root Mean Square Error. The ARIMA best model are ARIMA ([1, 2], 1, 0) (0, 1, 0)^{12}, ARIMA (0, 1, 1) (0, 1, 0)^{12}, ARIMA (0, 1, 1) (0, 1, 0)^{12}, ARIMA (1, 0, 0) (0, 1, 0)^{12} respectively. The ANN architecture optimum are ANN (2, 12, 1), ANN (1, 12, 1), ANN (1, 12, 1), and ANN (1, 12, 1). The best models are ARIMA ([1, 2], 1, 0) (0, 1, 0)^{12}, ARIMA (0, 1, 1) (0, 1, 0)^{12}, ANN (1, 12, 1), and ANN (1, 12, 1) in each category respectively. Hence, the result shows that a complex model is not always better than a simpler model. Additionally, a better hybrid ANN model is relied on the choice of a weighted input constant of RBF.

Keywords: ARIMA · Hybrid ARIMA-ANN · Forecasting
Electricity consumption

1 Introduction

In the last decade, machine learning has growth rapidly as tool for making prediction. Basically, the most algorithms in machine learning are developed from classical statistical method. In the statistical perspective, machine learning frames data in the context of a hypothetical function (f) that the machine learning algorithm aims to learn. Given some input variables (Input) the function answers the question as to what is the predicted output variable (Output). For categorical data prediction, hundreds algorithm classifier can be applied in many software [1]. In the future, the feature of machine learning algorithm will be standardized, simple, specialized, composable, scalable yet cost effective in order to realize the goal of completely automated machine learning utilities, which will become an integral part of the modern software application [2].

© Springer Nature Singapore Pte Ltd. 2017
A. Mohamed et al. (Eds.): SCDS 2017, CCIS 788, pp. 126–139, 2017.
https://doi.org/10.1007/978-981-10-7242-0_11

One situation in making prediction by machine learning are dealing with the time series data. The input variables in time series data are observations from the time lag previously. There are many machine learning models for time series forecasting [3]. There is good paper to survey recent literature in the domain of machine learning techniques and artificial intelligence used to forecast time series data [4].

This paper aims to develop model for a linear and nonlinear model in time series machine learning to forecast electricity consumption of the lowest household category in East Java, Indonesia. ARIMA models are family of linear model for time series analysis and forecasting for both stationary and non-stationary, seasonal and non-seasonal time series data. A nonlinear time series model is proposed by hybrid ARIMA-ANN, a Radial Basis Function Neural Networks architecture within orthogonal least squares estimation. Radial Basis Fuction (RBF) neural networks is a class of feed forward network that use a radial basis function as its activation function [9]. The orthogonal least squares procedure choses radial basis function centers one by one in a rational way until an adequate network has been constructed [10].

The installed capacity in the lowest household customer category has various power, i.e. 450 VA, 900 VA, 1300 VA, and 2200 VA. Some papers previously have devoted to develop forecasting model for electricity. Two-level seasonal model based on hybrid ARIMA-ANFIS to forecast half-hourly electricity load in Java-Bali Indonesia [5]. The results show that two-level seasonal hybrid ARIMA-ANFIS model with Gaussian membership function produces more accurate forecast values than individual approach of ARIMA and ANFIS model for predicting half-hourly electricity load, particularly up to 2 days ahead. Meanwhile, the electricity forecasting for industrial category in East Java has been forecasted using ARIMA model [6].

2 Data and Method

The data is from PLN (Perusahaan Listrik Negara/National Electricity Company) as central holding company to serve and maintain supply and distribute electricity across Indonesia. The data is electricity consumption from the lowest household category (R1) in East Java and the time period is from January 2010–December 2016. The customer from the lowest household category is divided by various installed capacity, 450 VA, 900 VA, 1300 VA, and 2200 VA.

The forecasting method of individual series will be used by ARIMA model. ARIMA models, abbreviation of Autoregressive Integrated Moving Averaged, were popularized by George Box and Gwilym Jenkins in the early 1970s. ARIMA models are family of linear model for time series analysis and forecasting for both stationary and non-stationary, seasonal and non-seasonal time series data. There are four steps in ARIMA modelling are, model identification, parameter estimation, diagnostic checking and finally model is used in prediction purposes. There are model classification in the family of ARIMA [7].

- Model Autoregressive (AR)

  A pth-order of autoregressive or AR (p) model can be written in the form,

$$\dot{Z}_t = \varphi_1 \dot{Z}_{t-1} + \ldots + \varphi_p \dot{Z}_{t-p} + a_t \tag{1}$$

- Model Moving Averages (MA)

  A qth-order of moving averages or MA (q) model can be written in the form

$$\dot{Z}_t = a_t - \theta_1 a_{t-1} - \ldots - \theta_q a_{t-q} \tag{2}$$

- Model Autoregressive Moving Average (ARMA)

  A pth order of Autoregressive and qth order of Moving Average or ARMA (p, q) model can be written in the form,

$$\dot{Z}_t = \varphi_1 \dot{Z}_{t-1} + \ldots + \varphi_p \dot{Z}_{t-p} + a_t - \theta_1 a_{t-1} - \ldots - \theta_q a_{t-q} \tag{3}$$

- Model Autoregressive Integrated Moving Average (ARIMA)

  By taking difference to the original, a pth order Autoregressive of and qth order of Moving Average, notated by ARIMA (p, d, q), the model will be

$$\phi_p(B)(1 - B)^d \dot{Z}_t = \theta_0 + \theta_q(B)a_t \tag{4}$$

- Model Seasonal Autoregressive Integrated Moving Average (SARIMA)

  Seasonal ARIMA usually are notated by ARIMA (p, d, q) (P, D, Q)^S. The general form of Seasonal ARIMA can be written as,

$$\Phi_P(B^S)\varphi_p(B)(1 - B)^d(1 - B^S)^D \dot{Z}_t = \theta_q(B)\Theta_Q(B^S)a_t \tag{5}$$

Hybrid ARIMA-ANN is hybridisation of ARIMA and radial basis function (RBF) neural networks. RBF is a class of feed forward network that use a radial basis function as its activation function and a traditionally used for strict interpolation in multidimensional space [8], composed by three layers i.e. an input layer, a hidden layer, and an output layer. The input layer applies an orthogonal least squares procedure to choose the radial basis function centers in the input layer and a weighted input constant [9]. The hidden layer of RBF is nonlinear, whereas the output layer is linear. RBF network employs to modelling time series is $y(t)$ as the current time series value. The idea is to use the RBF network [10]

$$\widehat{y}(t) = f_r(x(t)) \tag{6}$$

as the one step-ahead predictor for $y(t)$, where the inputs to the RBF network

$$x(t) = \left[y(t-1) \ldots y(t - n_y) \right]^T \tag{7}$$

are past observation of the series. The nonlinearity $\phi(\cdot)$ within the RBF network is chosen to be the gaussian function (8) and the RBF network model writes as (9)

$$\phi(v) = \exp(-v^2/\beta^2) \tag{8}$$

$$f_r(x) = \lambda_0 + \sum_{i=1}^{n_r} \lambda_i \phi(\|\mathbf{x} - \mathbf{c}_i\|) \tag{9}$$

The weighted input constant of RBF is proposed by using measure of dispersion, it applied to build a radial basis function in the hidden layer are semi interquartile range, inter quartile range, range, and standard deviation [11].

$$Q = Q_3 - Q_1/2 \tag{10}$$

$$IQR = Q_3 - Q_1 \tag{11}$$

$$Range = Max - Min \tag{12}$$

$$s = \sqrt{\sum_{i=1}^{n} (x_i - \bar{x})^2 \bigg/ n - 1} \tag{13}$$

The criteria used to choose the best forecasting model are the Mean Absolute Percentage Error (MAPE) and the Root Mean Square Error (RMSE) [7].

$$MAPE = \frac{\sum_{t=1}^{n} \frac{|Z_t - \hat{Z}_t|}{Z_t}}{n} \times 100\% \tag{14}$$

$$RMSE = \sqrt{\frac{1}{n} \sum_{t=1}^{n} (Z_t - \hat{Z}_t)^2} \tag{15}$$

3   Results

The short description regarding the electricity consumption is presented in Table 1. The table shows that the largest contributor to total consumption is from 900 VA power capacity as well as the smallest is from 2200 VA category. This is caused by the customer number of 900 VA is the largest and the customer number of 2200 VA is the smallest. Only around 9.84% the percentage of customer is from 1300 VA and 2200 VA.

Before starting the modeling procedure, electricity consumption data is divided into two parts, namely data in-sample and out-sample. The in-sample data is used to determine the modeling, while the out-sample data is used for model selection. The in-sample data uses data from January 2010 to December 2015, while the out-sample data uses data from January 2016 through December 2016.

Table 1. Description of electricity consumption of each category

Capacity	Mean	SD	Minimum	Maximum
450 VA	298.346.202	34.628.908	183.345.937	366.022.852
900 VA	349.990.409	75.226.313	184.317.229	474.736.024
1300 VA	101.155.026	12.484.667	61.474.737	120.808.796
2200 VA	67.034.305	8.526.496	41.580.243	79.661.957

Figure 1 shows the time series plot of the electricity consumption for each category. Obviously, the pattern of each series is trend and non-stationary. From the Fig. 1, it is not clear whether there is seasonal pattern or not for each series data. So, the monthly boxplot in Fig. 2 is presented to help to identify the seasonal pattern. By observing the line between monthly means of each box plot, each series looks have the similar pattern.

Clearly that the time series plot shows that the data is non-stationary and have seasonal pattern. Therefore, it is need to make the data become stationary before constructing the forecasting model. For this purpose, it will be taken d = 1 for non-seasonal, D = 1 for seasonal and S = 12 for seasonal. It can be shown that the data become stationary as shown in Annex 1.

Model Identification. As mentioned previously, the first step in ARIMA modeling is model identification based on stationary series data. By using Autocorrelation (ACF) and Partial Autocorrelation (PACF) plot, the tentative models can be proposed, especially order the ARIMA model. The ACF and PACF of each series data set is shown in Annex 1 and the tentative best models of each series is presented in Table 2.

Model estimation. Then, based on tentative ARIMA model, the parameter of each model will be estimated. The coefficient in each model, for both seasonal and non-seasonal, must satisfy the significant criteria (Annex 2).

Model checking. This involve the residual examination by using Ljung-Box test to check whether white noise or not and checking the p-value of the coefficient, then the significant model can be determined. Table 2 shows residual of the candidate best models already meet the white noise properties.

The best model. The best model would be selected by identifying model with the minimum MAPE and RMSE. Table 2 also shows the MAPE and RMSE of each best candidate model. Therefore, it can be identified easily, which model is the best. Table 3 present the best model of each series and its model specification. All the best models are seasonal models. These results correspond to the initial hypothesis at the identification stage that the best model is likely to be seasonal.

Forecasting. Based on the best model, then it will be computed forecasting electricity consumption of each individual series for 2017 year. Table 4 presented the forecasting of each series as well as Fig. 3 shows the interval forecasting of each individual series.

The forecasting result in Table 4 corresponds to the short data description as presented Table 1. The forecasting capacity 900 VA is highest in each month as well as forecasting of 2200 VA is lowest. In addition, the forecast results also generate an

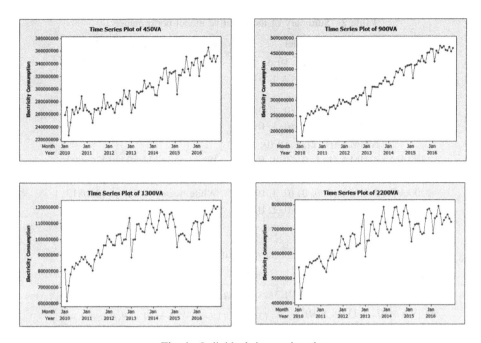

Fig. 1. Individual time series plot

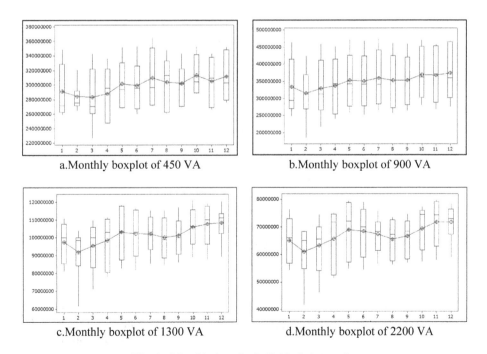

Fig. 2. Monthly boxplot individual time series

Table 2. Summary of the tentative model performance

Series	The tentative model	Residual	MAPE	RMSE
450 VA	ARIMA ([1, 2], 1, 0) (0, 1, 0)^{12}	White noise	1.78	7429122
	ARIMA (0, 1, 1) (0, 1, 0)^{12}	White noise	1.83	7586711
		White noise	3.38	22262659
900 VA	ARIMA (1, 1, 0) (0, 1, 0)^{12}	White noise	2.87	18347725
	ARIMA (0, 1, 1) (0, 1, 0)^{12}	White noise	3.24	21346976
	ARIMA ([1, 2], 1, 0) (0, 1, 0)^{12}			
1300 VA	ARIMA (1, 1, 0) (0, 1, 0)^{12}	White noise	11.69	14206993
	ARIMA (0, 1, 1) (0, 1, 0)^{12}	White noise	14.42	17186928
2200 VA	ARIMA (1, 0, 0) (0, 1, 0)^{12}	White noise	5.21	4240813
	ARIMA (1, 0, 1) (0, 1, 0)^{12}	White noise	6.13	4838444

Table 3. The best model if individual series.

Series	ARIMA best model	Model
450 VA	([1, 2], 1, 0) (0, 1, 0)^{12}	$(1 - \varphi_1 B - \varphi_2 B^2)(1 - B)^1 (1 - B^{12})^1 \dot{Z}_t = a_t$
900 VA	(0, 1, 1) (0, 1, 0)^{12}	$(1 - B)^1 (1 - B^{12})^1 \dot{Z}_t = (1 - \theta_1 B) a_t$
1300 VA	(1, 1, 0) (0, 1, 0)^{12}	$(1 - B)^1 (1 - B^{12})^1 \dot{Z}_t = (1 - \theta_1 B) a_t$
2200 VA	(1, 0, 0) (0, 1, 0)^{12}	$(1 - \varphi_1 B)(1 - B^{12})^1 \dot{Z}_t = a_t$

Table 4. Forecasting of each individual series

Month	400 VA	900 VA	1300 VA	2200 VA
January	356.437.843	487.224.399	121.158.305	72.574.265
February	325.622.443	446.809.645	110.309.109	65.479.913
March	348.720.894	481.860.192	120.259.136	72.502.790
April	342.423.412	474.746.479	121.042.694	73.766.180
May	357.343.364	497.542.040	128.142.514	78.339.198
June	359.222.035	491.795.524	125.976.680	75.561.174
July	371.806.140	495.713.077	122.120.258	71.297.066
August	353.792.327	484.458.465	125.804.791	73.248.833
September	349.300.011	482.198.764	127.441.154	74.437.918
October	358.477.706	493.836.646	131.192.699	75.684.615
November	348.381.888	479.827.444	129.456.987	74.149.104
December	357.870.168	490.326.505	130.716.836	72.879.403

increasing trend from the previous years. Then, Fig. 4 presented the time series plot of forecasting results along with the upper and lower limits to be better understand the variation from month to month, what's going down and what month is up. Interestingly, forecasting in February resulted in minimum values.

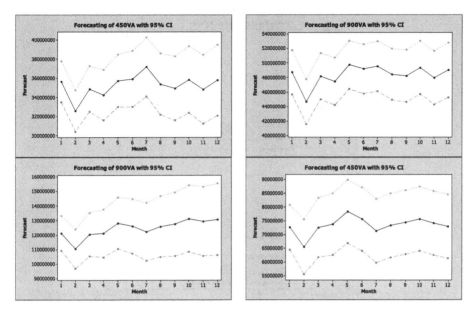

Fig. 3. Interval forecasting of each individual series

Hybrid ARIMA-ANN. The standard ARIMA model will be compared to Hybrid ARIMA-ANN. This model is built by the hybridization of ARIMA and RBF. The input is determined by the order of autocorrelation or partial autocorrelation of ARIMA model. The hidden layer is composed by the measure of dispersion constant as the width, determined by the seasonality pattern, and proceed by the orthogonal least squares. The measure of dispersion is determined by semi interquartile range (SIR), interquartile range (IQR), range, and standard deviation. The output layer is linear process as a univariate forecasting.

The architecture optimum. Based on Table 5, the architecture optimum is built by the measure of dispersion constant and proceed by the orthogonal least squares method. The architecture optimum is shown in Annex 2 (Table 6).

Forecasting. Based on the architecture optimum, it will be computed a univariate forecasting electricity consumption of each individual series in 2017. Table 7 presents the forecasting of each series as well as Fig. 4 shows the interval forecasting.

Comparation model. Based on the best model in each series, the best models are ARIMA ([1, 2], 1, 0) (0, 1, 0)^{12}, ARIMA (0, 1, 1) (0, 1, 0)^{12}, ANN (1, 12, 1), and ANN (1, 12, 1) respectively. It shows that a complex model is not always better than a simpler model. Hence, a better hybrid ANN model is relied on the choice of a weighted input constant of RBF. In addition, hybrid model forecasting results have a smaller variation compared to the ARIMA model. This can be seen in Fig. 4. Time series plot of forecasting results from month to month shows small changes. In addition, forecasting results also show a different pattern with the original data plot. If the time series of the original data in general shows an upward trend pattern, then the forecasting result of the hybrid model shows the pattern tends to go down as in forecasting 2200 VA.

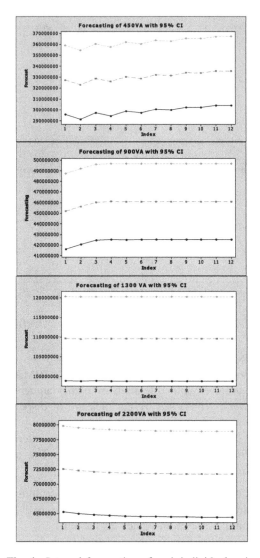

Fig. 4. Interval forecasting of each individual series

Table 5. The measure of dispersion of radial basis function

Capacity	SIR	IQR	Range	St. Deviation
450 VA	22929807.38	45859614.75	124845858	27678858.64
900 VA	53617084	107234168	281078109	64456310.27
1300 VA	9375176.38	18750352.75	56820592	11994479.18
2200 VA	6892877.25	13785754.5	38081714	8544743.80

Table 6. The tentative hybrid ARIMA-ANN architecture performance

Series	The architecture	MAPE	RMSE	Dispersion
450 VA	ANN (2, 12, 1)	7.232	28900205.8	SIR
	ANN (2, 12, 1)	10.846	47637343.31	IQR
	ANN (2, 12, 1)	7.252	30138899.10	Range
	ANN (2, 12, 1)	7.321	30605083.83	Standard deviation
900 VA	ANN (1, 12, 1)	79.75	500393532.39	SIR
	ANN (1, 12, 1)	187.58	1261928434.30	IQR
	ANN (1, 12, 1)	4.52	25172771.32	Range
	ANN (1, 12, 1)	104.89	671564249.29	Standard deviation
1300 VA	ANN (1, 12, 1)	5.13	6916929.11	SIR
	ANN (1, 12, 1)	8.73	18485362.53	IQR
	ANN (1, 12, 1)	5.99	9735898.68	Range
	ANN (1, 12, 1)	6.19	9738836.02	Standard deviation
2200 VA	ANN (1, 12, 1)	3.68	3465369.36	SIR
	ANN (1, 12, 1)	3.43	3290203.42	IQR
	ANN (1, 12, 1)	3.26	3119319.11	Range
	ANN (1, 12, 1)	3.55	3376712.69	Standard deviation

Table 7. Forecasting of each individual series

Month	450 VA	900 VA	1300 VA	2200 VA
January	327.264.333	451.794.490	109.590.586	72.538.369
February	322.862.608	456.310.202	109.485.466	72.249.617
March	328.646.077	460.195.002	109.513.222	72.055.395
April	325.904.722	461.057.594	109.506.386	71.922.917
May	330.278.944	460.815.162	109.508.103	71.831.855
June	328.665.855	460.901.946	109.507.674	71.768.973
July	332.091.946	460.872.634	109.507.781	71.725.467
August	331.347.847	460.882.490	109.507.754	71.695.299
September	333.965.777	460.879.290	109.507.761	71.674.361
October	333.826.763	460.880.314	109.507.759	71.659.827
November	335.596.047	460.879.546	109.507.760	71.649.721
December	335.709.623	460.880.186	109.507.760	71.642.705

4 Conclusion and Further Research

There are four individual series of electricity consumption, 450 VA, 900 VA, 1300 VA and 2200 VA. The time series machine learning to forecast each series is ARIMA model and the best model is ARIMA ([1, 2], 1, 0) $(0, 1, 0)^{12}$, ARIMA (0, 1, 1) $(0, 1, 0)^{12}$, ARIMA (1, 1, 0) $(0, 1, 0)^{12}$, and ARIMA (1, 0, 0) $(0, 1, 0)^{12}$ respectively. It is also noted that the largest contribution to household electricity is from 900 VA category. Meanwhile, the hybrid ARIMA-ANN forecasting generates an architecture and an architecture optimum i.e. ANN (2, 12, 1) using semi interquartile, ANN (1, 12, 1) using range, ANN (1, 12, 1) using semi interquartile, and ANN (1, 12, 1) using range respectively. The comparation model shows that there is an inconsistent model because a complex model doesn't work better in two series of electricity consumption i.e. the 450 VA series and the 900 VA series.

If the electricity consumption of each category is summed up, then it will produce the total household electricity consumption. This time series data is called hierarchical time series data. The alternative method can be used for this type of time series data is hierarchical time series model by applying the top-down, bottom up and optimal combination approach [12]. These approaches could be considered to be applied to this data in the next research.

Annex 1. ACF and PACF Plot

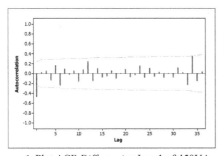

a.1. Plot ACF *Differencing* Lag 1 of 450VA

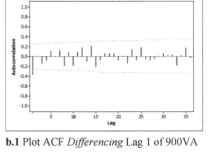

b.1 Plot ACF *Differencing* Lag 1 of 900VA

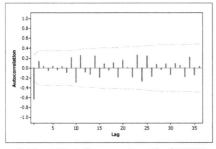

a.2. Plot ACF *Differencing* Lag 12 of 450VA

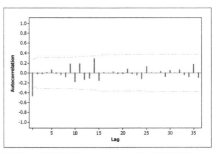

b.2. Plot ACF *Differencing* Lag 12 of 900VA

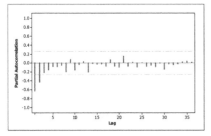

a.3. Plot PACF of 450VA

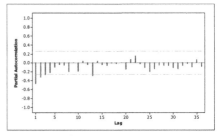

b.3. Plot PACF of 900VA

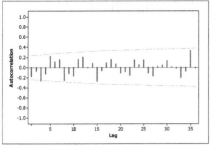

c.1. Plot ACF *Differencing* Lag 1 of 1300VA

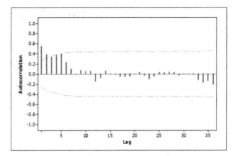

d.1. Plot ACF *Differencing* Lag 12 of 2200VA

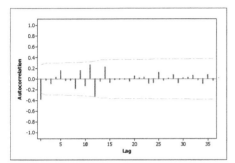

c.2. Plot ACF *Differencing* Lag 12 of 1300VA

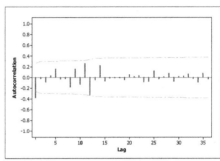

d.2. Plot ACF *Differencing* Lag 12 of 2200VA

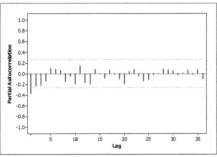

c.3. Plot PACF of 1300VA

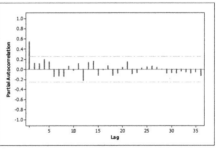

d.3. Plot PACF of 2200VA

Annex 2. The Architecture Optimum of Hybrid ARIMA-ANN

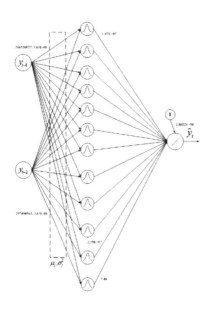

a. ANN (2,12,1) of 450 VA category

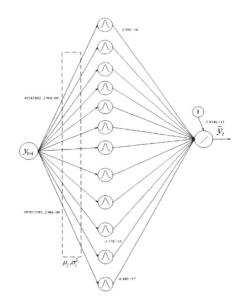

b. ANN (1,12,1) of 900VA category

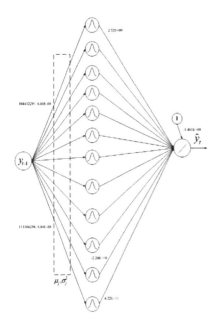

c. ANN (1,12,1) of 1300 VA category

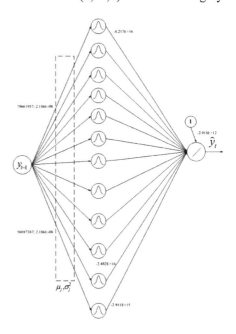

d. ANN (1,12,1) of 2200 VA category

References

1. Delgado, M.F., Cernadas, E., Barro, S., Amorim, D.: Do we need hundreds of classifiers to solve real world classification problems? J. Mach. Learn. Res. **15**, 3133–3181 (2014)
2. Cetinsoy, A., Martin, F.J., Ortega, J.A., Petersen, P.: The past, present, and future of machine learning APIs. In: JMLR: Workshop and Conference Proceedings, vol. 50, pp. 43–49 (2016)
3. Ahmed, N.K., Atiya, A.F., Gayar, N.E., Shishiny, H.E.: An empirical comparison of machine learning models for time series forecasting. Econom. Rev. **29**(5–6), 594–621 (2010)
4. Krollner, B., Vanstone, B., Finnie, G.: Financial time series forecasting with machine learning techniques: a survey. In: ESANN 2010 Proceedings, European Symposium on Artificial Neural Networks - Computational Intelligence and Machine Learning, Bruges (Belgium), 28–30 April 2010. D-Side Publishing (2010). ISBN 2-930307-10-2
5. Suhartono, Puspitasari, I., Akbar, M.S., Leem, M.H.: Two-level seasonal model based on hybrid ARIMA-ANFIS for forecasting short-term electricity load in Indonesia. In: IEEE Proceeding, International Conference on Statistics in Science, Business and Engineering (ICSSBE) (2012). https://doi.org/10.1109/ICSSBE.2012.6396642
6. Saputri, I.A.: Forecasting of Electricity in Industrial Sector at PT PLN East Java: Final Assignment, June 2016. Department Of Statistics, Faculty of Mathematics and Natural Sciences (2016)
7. Cryer, J.D., Chan, K.: Time Series Analysis With Application in R, 2nd edn. Springer Science Business Media, New York (2008). https://doi.org/10.1007/978-0-387-75959-3
8. Sivanandam, S.N., Sumathi, S., Deepa, S.N.: Introduction Neural Networks Using Matlab 6.0. Tata McGraw-Hill, New Delhi (2006)
9. Beale, M.H., Demuth, H.B., Hagan, M.T.: Neural Network Toolbox User's Guide R2011b. The MathWorks Inc., Natick (2011)
10. Chen, S., Cowan, C.F., Grant, P.M.: Orthogonal least squares learning algorithm for radial basis function networks. IEEE Trans. Neural Netw. **2**(2), 302–309 (1991)
11. Mann, P.S.: Statistics for Business and Economics. Wiley, Hoboken (1995)
12. Hyndman, R.J., Ahmed, R.A., Athanasopoulos, G., Shang, H.L.: Optimal combination forecasts for hierarchical time series. Comput. Stat. Data Anal. **55**, 2579–2589 (2011)

A Design of a Reminiscence System for Tacit Knowledge Recall and Transfer

Sabiroh Md Sabri[1](✉) [ID], Haryani Haron[2] [ID], and Nursuriati Jamil[2] [ID]

[1] Universiti Teknologi MARA Cawangan Perlis, 02600 Arau, Perlis, Malaysia
sabir707@perlis.uitm.edu.my
[2] Universiti Teknologi MARA, 40450 Shah Alam, Selangor, Malaysia
{haryani,liza}@tmsk.uitm.edu.my

Abstract. Knowledge transfer is a crucial part of a family firm's succession plan. Knowledge transfer in family firms involves transferring knowledge from a business owner to the younger generation. The process of transferring knowledge in family firms is very challenging because it involves the main source of knowledge who is business owners. The majority of the business owner who are also the founder of the family firm are elderly, whom facing difficulties in recalling their knowledge. This application was developed with the aim to assist business owners in recalling their knowledge using the cue recall technique. Apart from that, it also helps family firms to store the knowledge that has been recalled in a structured way for easy retrieval and use by the younger generation. The mobile application developed should help the business owner in recalling and store their knowledge and promote working flexibility and enhance productivity of the family firms. Successful implementation of the application would help family firms in transferring their knowledge and ensuring the firms' sustainability and competitive advantage in the future.

Keywords: Reminiscence · Family firms · Knowledge recall
Elderly · Knowledge recall system

1 Introduction

In today's world, family firms have been considered as one of the significant elements in the corporate economy. The record of family firm's performance has been proven through their long establishment since decades ago. The successful background of family firms has also been long acknowledged by scholars and practitioners. However, they agree that the reflection of the excellent performance of family firms could be assessed from various angles. Reference [1] stated that, the sustainability of family business is very much dependent on the readiness of the selected successors capabilities in running the business. This would require the successor to have a good knowledge in all aspects of the business. Although some of the knowledge and skills possessed by the successors which is acquired through formal education could be used in many business contexts [2] and enables them to identify business opportunities and implement them [3, 4], some knowledge could only be acquired through years of experience. Reference [5] stated that the successor development process is also closely linked and cannot be parted

© Springer Nature Singapore Pte Ltd. 2017
A. Mohamed et al. (Eds.): SCDS 2017, CCIS 788, pp. 140–150, 2017.
https://doi.org/10.1007/978-981-10-7242-0_12

with the experience he gained while working with family businesses. It is also crucial for the future successors to have working experience with family firm in order to increase the life expectancy of the family business and also to give a positive impact on the success of the business venture business success [6]. In addition, working experience outside the family business is considered to be a plus point to the successors. By having lots of experience working with other companies and other jobs, the successor is said to have a higher chance of being successful. Hence, it is important for the family business owner to ensure the involvement of the successor in the business operations started as early as possible. This will allow them to learn and understand how to operate the business from the ground and subsequently allow them to acquire the knowledge and experience and later on take over the business and ensure its success [5]. The process of knowledge transfer usually begins immediately after the successor involves in the family business. Apart from that, the main intention of the knowledge transfer process in a family firm is that it must ensure that the knowledge is not only received by the successor, but also the ability to apply the knowledge and skills inherited.

2   Research Background

It has been a long time ago since knowledge has been recognized as a potential source of the asset in an organization. In today's knowledge economy, knowledge has been recognized as one of the most valuable strategic assets that an organization can possess. Apart from that, knowledge also is known to be a significant source of competitive advantage, which empowers an organization to be more innovative and at the same time stay competitive in the market [7, 8]. Knowledge has been defined in many ways, ranging from a social process [9] to an active construct [10]. In addition, reference [11] captured the essence of knowledge concisely when he described it as "know-what", "know- how", "know-why", and "know-who". These types of knowledge include employee talents and skills, managerial systems and routines, experience, insights, values and norms and networks of relationships [7, 12], and accounts for 80% of a firm's valuable knowledge resources. References [15, 16] emphasize the generation of knowledge, the management of knowledge, as well as the transfer of the knowledge from the founder to the successor are important elements in ensuring the intergenerational succession of the family firm will be successful. Although in general, this knowledge could refer to both explicit and tacit forms, this research focused on the nature of knowledge which is primarily tacit.

2.1   Knowledge Transfer in Family Firms

Knowledge originates in the heads of individuals. It was built on the application of information by an individual which then is transformed and developed through personal beliefs, values, education and experience [15]. Many works of literature claimed that 90 percent of knowledge in most organizations is embedded and synthesized in peoples' heads [16, 17]. The creation of tacit knowledge in a family firm may have happened in two circumstances. Firstly, founders of family firms are viewed as entrepreneurs because they possess the ability to endure and tolerate the risks and failure that they encounter during the creation preservation of the business. This is considered as vital criteria to the accomplishment of the business success together with the capability to initiate, sees and acts on opportunities, lead, persistent, innovate and ability to live with uncertainty. The success of a family firm is often embedded in the firm's founder, who is most likely the entrepreneur who started and developed the firm [18]. Reference [19] stated that there is a substantial evidence that shows that many business decisions by the founders were made based on their own intuition, the feeling that is indescribable and intangible, sometimes referred to as a gut feeling. Reference [19] further described that, when they were asked about the reason lying behind a particular business decision that they made, entrepreneurs frequently express to how "all of the pieces seemed to be there," or "it felt right". This indicates that most of the time, the decision made was based on the entrepreneur's tacit knowledge. Reference [20] in their research stated that family firm's knowledge is based on the founder's idiosyncratic knowledge, firm-specific and often held tacitly. Therefore, the first source of tacit knowledge in a family firm is the founder himself.

Secondly, reference [21] mentioned that the creation of resources in family firms which were generated through the complex interactions between the family members, the family unit and the business itself has made family firms to be rich in tacit resources. The unique pack of these resources were created through these interactions known as the "familiness" of the firms [22]. Reference [21] stated that to ensure the family firms will survive and prosper in the future, the familiness must be preserved throughout the entire succession process and after. Apart from that, tacit knowledge may have existed and often embedded in the business operation of the firms for example in the business processes, the firm's culture, as well as the firm's structures. Thus, many researchers agree that by definition, it is easier to identify and articulate explicit knowledge, hence explicit knowledge is easier to be transferred compared to tacit knowledge [23–25]. In contrast, tacit knowledge is often embedded in the culture, processes and are more customs to the business. Tacit knowledge is not readily available for codification and generational transfer. The tacit knowledge is often detained by the founder and/or other key individuals in the business. This knowledge has significance for the development and maintenance of a firm's competitive advantage and is more important to the family firm than to non-family firms [21].

Managing tacit knowledge is a real challenge for family firms. This is because the nature of tacit knowledge which is often rooted in the mind of the founder and also in the firm's processes and culture, thus not easily identified. Despite these difficulties, due to its importance, many researchers are still encouraged to study the importance of managing and transferring tacit knowledge as part of the firm's succession plan. Reference [18] emphasized that the process of transferring knowledge from the founder to the successor in a family business is critical to the on-going success of the business. Although many family firms who undergoing the succession process may not experience stages as suggested by researchers, the provision of literature and frameworks for the study of family business succession somehow help in the understanding of succession and knowledge transfer in the family business. Based on this, it is concluded that knowledge in a family firm is a key resource in the firm; and intergenerational knowledge transfer process is one of the crucial parts in a family business succession plan.

2.2   Reminiscence and Cued Recall

Before knowledge could be transferred, it needs to be recalled from one's memory and assembled as a story to be passed on. According to [26], recall is the act of retrieving information or events from the past. In most of the time, the act of recall is carried out without having a specific cue or clue to help the person retrieve the information. When a person recalls something, he is actually reminiscing about something he has previously learned or experienced. Reminiscing involve the active use of a cognitive function in human, which tend to decline as we age, thus transferring knowledge from elderly challenges as they struggle to recall their knowledge due to their cognitive decline.

The reminiscing technique could be applied to a group as well as individual. During the reminiscing process, the intrapersonal reminiscence in turned into interpersonal memory interactions where each group members explore and gradually confront and consider specific memories. These memories create the thematic and interactional content of their experience together [26]. Reference [27] suggests that humans need to retrieve, articulate, and disseminate self-narratives from memories which lay in a form of the building blocks of these narratives. Some factor must initiate this process, a component that is called "triggers" or "cue". According to [28], cued recall is the process in which a person is given a list of items to remember and is then tested with the use of cues or guides.

When a person is provided with cues, he has a higher chances to remember the items the list that he cannot recall earlier without the cues. This can also take the form of stimulus-response recall, as when words, pictures and numbers are presented together in a pair and the resulting associations between the two items cues the recall of the second item in the pair [28]. In addition, reference [29] in their project emphasis on small group activities animated by a range of multi-sensory stimuli such as objects, images, sound and tastes to trigger memories. Stories emerging in the small groups, are then shared with the wider group and explored through dramatic improvisation, song, dance, drawing and writing as well as straightforward discussions. It was found that non-verbal exploration of memories and past skills and interest has the effect of activating and lifting the spirits of the participants and offering them ways of engaging with other, thus improve their memory.

3   Methodology

This research adopted the System Development Life Cycle (SDLC) model which consists of five phases, which are analysis, design, development, implementation and evaluation. In the analysis phase, data was collected through interview and observation to understand the knowledge transfer practice in family firms. Apart from that, observations session were also carried out to understand the reminiscence and cued recall process. In the design phase, an interview with an expert was conducted to confirm the technique used in recalling the knowledge used in the application is in accordance with the theory. The results from the expert interview were used to design the system processes for knowledge recall. The details of how the system works are explained accordingly in the following section of this paper.

4 The Application

This application was develop to provide a platform for business owners to recall and transfer their knowledge. The reminiscence system application was developed on a mobile platform to ensure the mobility and flexibility of the users in recalling their knowledge.

4.1 System Architecture

The application was developed based on the Client-Server architecture. However, due to device constraints, the client can only be any iOS devices such as iPad or iPhone. For the database, MySQL database is used because it is a freely available open source Relational Database Management System (RDBMS) that uses Structured Query Language (SQL). As for the server, the PHP web server is being used. PHP is a server-side scripting language designed for web development but also used as a general-purpose programming language. It is also a free and platform independent server. The PHP web server was found as the most suitable web server to be used because the server component offers a function or service to one or many clients, which initiate requests for any services.

When the client (iOS device) sends a simple request, technically the web server (PHP Web Server) will return a response. However, sometimes the request might contain a more complicated request, for example fetching of data that are stored in the database. When this type of request is received, the web server will access the database server as a client. Once the data in the database is accessed and return to the web server, the web server will return the result back to be read and displayed by the client (iOS device). Figure 1 below depicted the system architecture used in the application development.

4.2 The Application and Its Functions

During the application development, the researcher focus on a few main functions that was believed to be important in the intergenerational knowledge transfer. This includes (1) knowledge recall – the use of cue or trigger items to help elderly in recalling, (2) the capturing and storing of the knowledge, and (3) the retrieval of knowledge by the recipient. Figure 2 below illustrates the menu available in the application. For the purpose of prototyping, a repository that contains knowledge about herbs and its medicinal value was created. It was assumed that only one source is transferring the knowledge for one type of user. Thus, the access given to the recipient is not restricted to any type of knowledge. The details of the application are presented in the following sections.

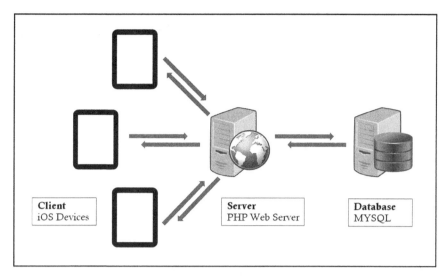

Fig. 1. System architecture

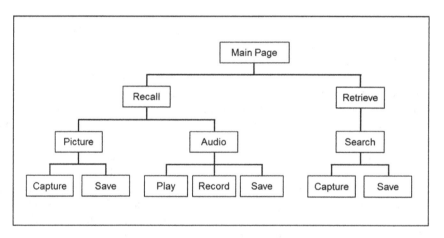

Fig. 2. The menu structure of the system

The Main Page

The application started with the main page which displays the menu where the user can select based on the function he wishes to perform. This application has two main functions, which is "Recall" and "Retrieve" the knowledge. The user can select which function he would like to perform by selecting either one button on the main page. When the user choose either one of the functions from the Main Page, he will be directed to the relevant pages that allow him to further perform the selected function. Figure 3 below shows the sequence diagram for the Main Page.

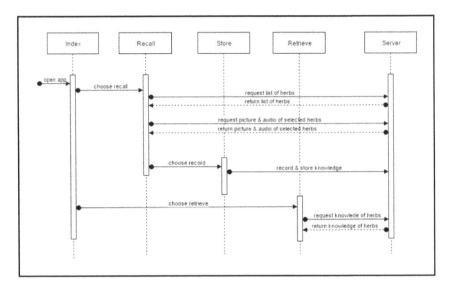

Fig. 3. Main page sequence diagram

The Recall Function

From the main page, the user can select to use the first function, which the recall function. When the user selects the "Recall" button, he will be directed to the Recall page. This page is divided into two sections, which are "Picture" and "Audio" section. The different sections refer to the use of picture and audio as the trigger or cue item to help the user to recall their knowledge. In the Picture section, first, the page will display a picture which is used to help elderly to recall. At the bottom part of the picture, there are buttons labelled as PREV and NEXT, which allow the user to navigate back and forth through the pictures of various herbs that are stored in the database. When the user views the picture, it acts as a trigger item that helps the user to recall knowledge related to the picture. Apart from that, the user also can listen to the audio related to the picture displayed. The audio also act as trigger item to help the user trigger their knowledge about the picture displayed. To listen to the audio, the user can click the button labelled PLAY located at the bottom of the Audio section. Button PREV and NEXT allow the user to navigate through the audio files stored in the database. The information at the top of the button displays the audio number that describes the picture. For one picture, there could be more than one audio recorded and submitted by a different person. This show that the user can collaborate to help each other in the collective recall.

When the user recalls the knowledge, the user presses the REC button and start to record the knowledge that he recalled. Since the users of this application are the elderly, the researcher finds that is more suitable to use voice recording rather than text input to record the knowledge. Once he finished, the user can stop the recording and press the SAVE button to store the file in the database. The use of client-server architecture using mobile technology allows the whole recalling process to be carried out simultaneously by more than one user location at different dispersed places. With this user does not need

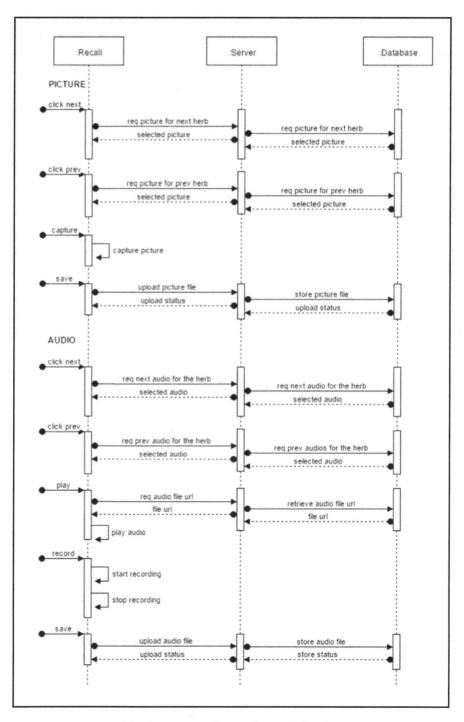

Fig. 4. Sequence diagram for recall function

to be physically together to transfer knowledge. The knowledge that has been stored in the repository can be retrieved by the recipient anytime they need to. Figure 4 below depicted the sequence diagram for the recall page that explains the whole process.

The Retrieve Function

To retrieve the knowledge, the user first selects the "Retrieve" button from the Main Page. When the user selects the "Retrieve" button, he will be directed to the Retrieve page. In this page, two buttons are provided for the user, which are HERBS and MEDICINAL VALUE. This allows users to retrieve knowledge in two perspectives, by the herbs and also by their medicinal value. When the user clicks the HERBS button, he will be directed to the search page. Using the search page, the user can search knowledge about particular herbs by typing the herb's name in the search box provided. The search results will be shown in a visualization form to allow the user to view all the information related to the herbs easily.

The other button which the MEDICINAL VALUE button works the same way to display the knowledge in terms of the medicinal value of the herbs selected. The use of repository allows the knowledge to be stored and avoid knowledge from lost especially when the elderly retire. The mobile technology used to enable the user to retrieve the knowledge anytime and anywhere. This encourages knowledge to be transferred more efficiently and effectively. Figure 5 below depicted the sequence diagram for the Retrieve page that explains the whole process.

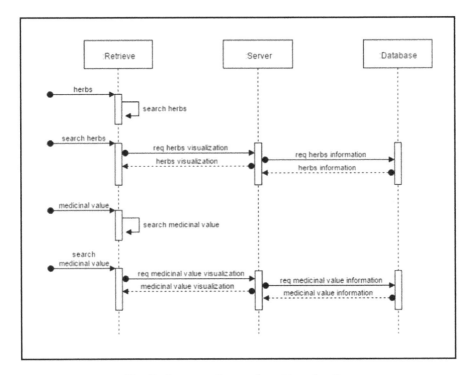

Fig. 5. Sequence diagram for retrieve function

5 Conclusion

Tacit knowledge created and owned by an individual need to be managed and transferred to others for the knowledge to be continually used. In a knowledge-centred organization like a family firm, transferring knowledge to its business successor it crucial in ensuring the continuity and sustainability of the business. The development of this mobile applications contributes to the practice of knowledge transfer in family firm as it provides a technological platform to facilitate and assist family firms in transferring their valuable knowledge and at the same time address the issues of difficulties in knowledge recall faced by the business owners. The development of the mobile application also promotes mobility and flexibility to knowledge recall and knowledge transfer in the family firm. Thus, it is hoped that the development of the reminiscence system could help the family firms to transfer their knowledge more efficiently and ensure the sustainability of their business in the future.

Acknowledgement. The authors are grateful to Universiti Teknologi MARA for their cooperation and supports towards the completion of this research.

References

1. Barbosa, S.D., Kickul, J., Smith, B.R.: The road less intended: integrating entrepreneurial cognition and risk in entrepreneurship education. J. Enterp. Cult. **16**(4), 411–439 (2008)
2. Bollinger, A.S., Smith, R.D.: Managing organizational knowledge as a strategic asset. J. Knowl. Manag. **5**(1), 1018 (2001)
3. Breton-miller, I.L., Steier, L.P.: Toward an integrative model of effective FOB succession. Entrepr. Theory Pract. **28**, 305–329 (2004)
4. Buang, N.A., Ganefri, G., Sidek, S.: Family business succession of SMEs and post-transition business performance. Asian Soc. Sci. **9**(12), 79–92 (2013)
5. Cabrera-Suarez, K., Saa-Perez, P., Garcia-Almeida, D.: The succession process from a resource- and knowledge-based view of the family firm. Fam. Bus. Rev. **14**(1), 37–48 (2001)
6. Carminatti, N., Borges, M.R.S.: Analyzing approaches to collective knowledge recall. Comput. Inform. **25**, 547–570 (2006)
7. Chirico, F.: The Accumulation process of knowledge in family firms. Electron. J. Fam. Bus. Stud. **1**(1), 62–90 (2007)
8. Chirico, F.: Knowledge accumulation in family firms: evidence from four case studies. Int. Small Bus. J. **26**(4), 433–462 (2008)
9. Dickson, P.H., Solomon, G.T., Weaver, K.M.: Entrepreneurial selection and success: does education matter? J. Small Bus. Enterp. Dev. **15**(2), 239–258 (2008)
10. Getzel, G.S.: Intergenerational reminiscence in groups of the frail elderly. J. Jew. Communal Serv. **59**(4), 318–325 (1983)
11. Grant, R.M.: Toward a knowledge-based view of the firm. Strateg. Manag. J. **17**, 109–122 (1996)
12. Habbershon, T.G., Williams, M.L.: A resource-based framework for assesing the strategic advantages of famiy firms. Fam. Bus. Rev. **12**(1), 1–22 (1999)
13. Henry, M.: Tacit knowledge transfer in family business succession. Grant MacEwan University (2012)

14. Higginson, N.: Preparing the next generation for the family business: relational factors and knowledge transfer in mother-to-daughter succession. J. Manag. Mark. Res. **4**, 1–18 (2010)
15. Lee, K.S., Lim, G.H., Lim, W.S.: Family business succession: appropriation risk and choice. Acad. Manage. Rev. **28**(4), 657–666 (2003)
16. Longenecker, J.G., Moore, C.W., Petty, J.W., Palich, L.E.: Small Business Management: An Entrepreneurial Emphasis, 13th edn. Thomson Learning/South-Western, Boston/London (2005)
17. Lynch, P.: Retrieval of Memories (2011). http://explorable.com/retrieval-of-memories
18. Malecki, E.J.: Technology and Economic Development: The Dynamics of Local, Regional and National Change, 2nd edn. Longman Scientific & Technical, New York (1997)
19. McNichols, D.: Tacit Knowledge: An Examination of Intergenerational Knowledge Transfer within An Aerospace Engineering Community. University of Phoenix (2008)
20. Nonaka, I., von Krogh, G.: Tacit knowledge and knowledge conversion: controversy and advancement in organizational knowledge creation theory. Organ. Sci. **20**(3), 635–652 (2009)
21. Robinson, P.B., Sexton, E.A.: The effect of education and experience on self-employment success. J. Bus. Ventur. **9**(2), 141–156 (1994)
22. Sambrook, S.: Exploring succession planning in small, growing firms. J. Small Bus. Enterp. Dev. **12**(4), 579–594 (2005)
23. Sardeshmukh, S.R.: Successor Development in Family Firms. Rensselaer Polytechnic Institute, Newyork (2008)
24. Sardeshmukh, S.R., Corbett, A.C.: The duality of internal and external development of successors: opportunity recognition in family firms. Fam. Bus. Rev. **24**(2), 111–125 (2011)
25. Schweitzer, P., Bruce, E.: Remembering Yesterday, Caring Today: Reminiscence in Dementia Care: A Guide to Good Practice. Jessica Kingsley Publishers, London (2008)
26. Smith, E.A.: The role of tacit and explicit knowledge in the workplace. J. Knowl. Manag. **5**(4), 311–321 (2001)
27. Thurow, L.C.: Building Wealth: the new rules for individuals, companies, and nations. Atlantic Monthly, (June), 1–12 (1999)
28. Venzin, M., von Krogh, G., Roos, J.: Future research into knowledge management. Knowing Firms **1995**, 26–66 (1998)
29. Webster, J.D., Bohlmeijer, E.T., Westerhof, G.J.: Mapping the future of reminiscence: a conceptual guide for research and practice. Res. Aging **32**(4), 527–564 (2010)

Pattern Search Based on Particle Swarm Optimization Technique for Block Matching Motion Estimation Algorithm

Siti Eshah Che Osman[1](✉) and Hamidah Jantan[2]

[1] Faculty of Computer and Mathematical Sciences, Universiti Teknologi MARA
(UiTM), 40450 Shah Alam, Selangor, Malaysia
cteshahco@gmail.com
[2] Faculty of Computer and Mathematical Sciences, Universiti Teknologi MARA
(UiTM), 21080 Kuala Terengganu, Terengganu, Malaysia
hamidahjtn@tganu.uitm.edu.my

Abstract. Block matching algorithm is a popular technique in developing video coding applications that is used to reduce the computational complexity of motion estimation (ME) algorithm. In a video encoder, efficient implementation of ME is required that affect the final result in any applications. Searching pattern is one of the factors in developing motion estimation algorithm that could provide good performance. A new enhanced algorithm using a pattern based particle swarm optimization (PSO) has been proposed for obtaining least number of computations and to give better estimation accuracy. Due to the center biased nature of the videos, the proposed algorithm approach uses an initial pattern to speed up the convergence of the algorithm. The results have proved that improvements over Hexagon base Search could achieved with 7.82%–17.57% of computations cost reduction without much value of degradation of image quality. This work could be improved by using other variant of PSO or other potential meta-heuristic algorithms to provide the better performances in both aspects.

Keywords: Motion estimation · Block matching
Particle swarm optimization and motion vector

1 Introduction

Motion estimation (ME) is one of the video compression process. ME examine the movement of object in an image sequence to try searching the best motion vector representing the estimated motion. It removes temporal redundancy exists in a video sequence and thus provides coding systems with high compression ratio [1–3]. In an effort to reduce the computational complexity of ME algorithms in video coding, several methods have been presented by many researchers such as parametric-based models, percussive techniques, optical flow and block matching algorithms [4, 5]. Block matching algorithm is chosen mainly due to its efficiency, simplicity and good compromise between motion computations and prediction quality [6].

© Springer Nature Singapore Pte Ltd. 2017
A. Mohamed et al. (Eds.): SCDS 2017, CCIS 788, pp. 151–161, 2017.
https://doi.org/10.1007/978-981-10-7242-0_13

The well-known block matching algorithm is Full Search (FS) which performs well in prediction accuracy but the foremost computation intensive algorithm. Then, various fast search algorithms are developed to reduce the search time; hence to increase the processing speed and at the same time maintaining the quality of the reconstructed signals of FS [7]. With these current approaches, there were still lack in obtaining accurate motion vector especially for large motion. In block matching algorithms, there are consist of various techniques and a pattern search techniques are widely accepted because these algorithms have less computational complexity and could improve prediction accuracy so close to FS algorithm as compared to other techniques of motion estimation [8].

Recently, some researchers have proposed the probabilistic searching method to overcome the computationally hard optimization problems in motion estimation using Genetic Algorithm (GA) [9, 10], Artificial Bee Colony (ABC) [11] and Particle Swarm Optimization (PSO) [12, 13]. These optimization algorithms are used to achieve global optima solution among all possible solutions to find the best matching block with low computational complexity which cannot be obtained by conventional block matching algorithms. Since PSO was introduced in 1995 [14], it has drawn a great attention and obtained substantial results for continues optimization problems. PSO technique which is based on Swarm Intelligence mechanism and simulates the behaviour of the bird flocking or fish schooling. It is widely accepted by the most researchers in a wide range of research areas due to its profound intelligence and simple algorithm to solve complex problems especially in motion estimation problem [7, 15–17]. The performance of the PSO basically depends on the setting of control parameters and these variations in PSO are very necessary to improve the accuracy of optimization and convergence speed as compared to original PSO [18].

In the conventional PSO, particle positions are selected at random that can cause the searching process may converge at slow rate [7]. Then, to speed up this searching process, initial position using pattern based PSO is approached. The algorithms involved were implemented and evaluated by using different video sequences. In this research work, the proposed algorithms were evaluated with aims to achieve high processing speed of motion estimation process with computation speed reduction without much degradation of image quality. There are three main factors in developing the motion estimation algorithms which are selection of search pattern, search strategy and initial center and all these factors affecting to the computational performance and PSNR [7, 19].

The organization of this paper is as follows: Sect. 2 explains the related work on block matching motion estimation and then followed by Sect. 3 that explains the experiment setup of the proposed algorithm. In Sect. 4 describes the process of the proposed algorithm. In Sect. 5 disclosed the result analysis and discussion; finally conclusion and future work in Sect. 6.

2 Related Work

2.1 Block Matching Motion Estimation Algorithm

The block matching motion estimation plays a very important role in all video coding standards such as H.261, H.264, MPEG-4 and etc. [7, 20]. Motion estimation is defined as searching the best motion vector, which is the placement of the coordinate of the best similar block in reference frame for the block in current frame [21]. Block matching (BM) is a technique used to compare images taken at two different time frames and to estimate the direction motion of two frames [22]. Block matching motion estimation algorithm can be approached as an optimization problem that will achieve fast trans-mission and reduce data storage because motion estimation based video compression helps in reducing bits by sending encoded images which have less data rather than sending original frame [23, 24]. Block matching plays a major role in image matching to improve the efficiency. The goal of image matching is to determine the similarity between the images and portions of images. The similarity measure is a key element in the matching process. In order to find out the best matching block within a search window from the previous frame, some matching criteria are considered. The micro block that results in the minimum cost is the one that matches closely to current block.

There are numerous matching criteria or distortion function that have been pro-posed such as Mean Absolute Difference (MAD), Mean Squared Error (MSE) and Peak Signals to Noise Ratio (PSNR), which are represented in (1–3) [19, 20, 22]. MSE is evaluated between original frame and reconstructed frame or between current block by the motion vectors. MAD is most commonly used as its computation cost is low and also due to its simplicity. MSE is given by (1) and MAD is given by (2).

$$\text{MSE} = \frac{1}{N^2} \sum_{i=0}^{n-1} \sum_{j=0}^{n-1} \left(C_{ij} - R_{ij}\right)^2 \qquad (1)$$

$$\text{MAD} = \frac{1}{N^2} \sum_{i=0}^{n-1} \sum_{j=0}^{n-1} \left|C_{ij} - R_{ij}\right| \qquad (2)$$

PSNR given in (3) is the most popular and it determines the motion compensated image that is created by using motion vectors and macro blocks from the reference frame. The accuracy of motion estimation can be measured using PSNR.

$$\text{PSNR} = 10 \, \text{Log}_{10} \left[\frac{255^2}{MSE}\right] \qquad (3)$$

All the above matching criteria are shown, where N is the side of the macro block, Cij and Rij are the pixels being compared in the current macro block and the reference macro block, respectively.

Fixed Search Pattern Techniques. Currently, many fast block matching algorithms (BMAs) have been developed by researchers with aim to reduce the computations of full search (FS) algorithm. All these algorithms have several techniques such as fixed pattern search, partial distortion elimination, sub-sampling techniques, variable search range and etc. [25]. Among various techniques in BMAs, fixed pattern search algo-rithms are the well-known and the most famous due to the potentially provide PSNR values that so close to FS algorithm and give less computational expenses [26]. In fixed search pattern approach, it is assumed that motion estimation matching error reduces monotonically as the search moves toward the position of the global minimum error and that the error surface is uni-modal. Therefore, the minimum matching error can be approached by following matching pattern without looking for all points in the search window. A fixed set of search pattern is used for finding the motion vectors of each block. In BMAs, at least two types of classification of pattern search technique consist of non-rectangular and rectangular search were found that could effect on the search speed and matching performance [8].

Hexagon Pattern Based Search. Among all the pattern search techniques involve in block matching algorithms, hexagon pattern is selected for further experiment in this research work. In block matching motion estimation, different shape or size of a search pattern has an impact on searching speed and distortion performance. Hexagon base search (HEXBS) algorithm is proposed by Zhu [27] is one of the block matching algorithms for fast block motion estimation that provides very good performance in estimation accuracy as well as computational complexity as compared to the diamond search (DS) algorithm and other well-known fast block matching algorithms [8]. In 2013, a diamond-shaped and square-shaped search pattern based on PSO was proposed in fast block motion estimation which have exhibited faster search speed [7]. Hexagon search pattern approach will be applied in developing fast block matching algorithm based on particle swarm optimization for further analysis performance.

2.2 Basic Particle Swarm Optimization

Particle swarm optimization (PSO) is a population based stochastic optimization technique that was originally developed by Eberhart and Kennedy in 1995 [14]. This proposed technique is developed that inspired by social behaviour from research on swarm like birds flocking and fish schooling. PSO is commonly used by researchers due to its profound intelligence background and simple algorithm structure. PSO is an evolutionary algorithm based on swarm intelligent like genetic algorithm (GA). It shares many similarities with GA but unlike GA, PSO has no evolution operators such as crossover and mutation that are exist in GA. PSO is a population based search algorithm and starts with an initial population of randomly generated solutions called particles. Each particle has a fitness value and a velocity to adjust its direction according to the best experiences of the swarm, to search for global optimum in the n-dimensional solution space. The velocity of a particle is updated based on that particle's past personal best position (pbest) and the global best position (gbest) found by all particles in the swarm. PSO reflects the swarm behavior and the particles represent points in n-dimensional space. A swarm consists of 'M' particles. A potential solution is represented by a particle that adjusts its velocity and position according to Eqs. (4) and (5)

$$V'_{id} = V_{id} + C_1 \times rand() \times (P_{id} - X_{id}) + C_2 \times rand() \times (P_{gd} - X_{id}) \qquad (4)$$

$$X'_{id} = X_{id} + V'_{id} \qquad (5)$$

where 'i' is the index of the particle i = 1, 2, ... M, 'd' is the dimension of the particle in n-dimensional space, V_{id} and V'_{id} represents the velocity of particle 'i' in time t and t + 1 respectively. X_{id} and X'_{id} is the position of it h particle in time t and t + 1 respectively. P_{id} is the position of pbest for particle 'i' and P_{gd} is the position of gbest for the entire swarm. Constants C_1 and C_2 represent the weighing terms that pull seach particle towards pbest and gbest positions known as acceleration coefficients. Generally, both are set to the value of 2. Table 1 shows the summarization of the PSO based approached in block matching motion estimation algorithm.

Table 1. Summarization o PSO based approached in block matching motion estimation.

Method approach	Parameters	Performances
PSO [28]	- Randomly initialized - Velocity equation and adding local search technique such as simplex method - MAD is used as matching criteria	Achieve better reduction in search point with little compromise in quality
PSO [29]	- Initialize in fixed pattern of diamond shape - Apply small diamond search pattern - MB size of 16 × 16 and p = 7 - MAD is used as matching criteria - Total number of iteration is limited	- In random selection, if initial position is away from global optima then the process converges at very slow rate - To speed up the performances initial position is selected in fixed pattern
PSO [7]	- Initialize in random pattern and fixed pattern of diamond shape & square pattern - MAD is used as matching criteria - Varying acceleration coefficient - Use 50 frames and MB size of 8 × 8 and p = 7 - Total number of iteration is limited	Square pattern yields high estimation accuracy and low computational cost
PSO [30]	- Randomly initialized - Total number of iteration is limited	Provide good results in both search point and PSNR
Small population based PSO [31]	- MAD is used as matching criteria - Velocity equation (3 variables) - Step size equation with small population - Iteration number is limited	Achieves better speed of search point and provides good PSNR value

3 Experiment Setup

The experiment setup is as follow. The experiment has been implemented using MATLAB. In this research work, three video sequences were used for performance comparison of different algorithms. Three video sequence standards are formatted in

Common Intermediate Format (CIF) (352 × 288). Each of the video sequences represents the various types of motion which include Football for the fast motion, Silent for less background motion (medium motion) and Container for the small motion. Fixed video parameters are used to be implemented in this experiment considering macro block = 8 × 8, search parameter, p = 7, distance = 2 and 50 frames will be used.

In performance analysis, two parameters mainly consider for analyzing block matching algorithms are estimation accuracy (PSNR) and also computational complexity (No. of Computations). The measurement of computational complexity is based on the search efficiency by counting the average number of computations. Due to less computational expensive, Mean Absolute Difference are chosen. The error function of all proposed algorithms used mean absolute different (MAD) as objective or cost function using Eq. (2). Additionally, as an alternative index, the computational complexity degradation ratio (D_{COMP}) is used in comparison as shown in (6). If the result shows the highest D_{COMP} percentage value, the proposed modified algorithms are considered as fastest algorithms.

$$D_{COMP} = -\left(\frac{comp\_complexity_{HEXBS} - comp\_comlexity_{HPSO}}{comp\_complexity_{HEXBS}}\right) \times 100\% \quad (6)$$

The estimation accuracy is characterized by the PSNR by counting the average of the PSNR values by using the Eq. (3). The higher the average values of PSNR yield the better quality of the compensated image. The PSNR degradation ratio (D_{PSNR}) is also used in the comparison as shown in (7). The least D_{PSNR} percentage values yields the better quality of the compensated image.

$$D_{PSNR} = -\left(\frac{PSNR_{HEXBS} - PSNR_{HPSO}}{PSNR_{HEXBS}}\right) \times 100\% \quad (7)$$

4 Proposed Hexagon Pattern Based PSO (HPSO) Process

In the conventional PSO, particle positions are selected at random that can cause the searching process may converge at slow rate [7]. Then, to speed up this searching process, control parameter of initialize position of the particles is modified. The hexagon pattern is selected as initial position with PSO approached. The algorithm for the proposed HPSO for motion estimation is summarized as follows:

Step 1: The particles inside the search window are initialized with the hexagon pattern. The initial particles are distributed equally in six directions to find the matching block in each direction with equal probability.
Step 2: Calculate the fitness (MAD) of each particle.
Step 3: Calculate and select the minimum among of the 7 particles as the pBest position and gBest for the entire swarm.

Step 4: Update the velocity and positions based on pBest and gBest using Eqs. (8) and (9).

$$V'_{id} = w \times V_{id} + C_1(t) \times rand() \times (P_{id} - X_{id}) + C_2(t) \times rand() \times (P_{gd} - X_{id})$$
(8)

$$X'_{id} = X_{id} + V'_{id}$$
(9)

where

$$C_1(t) = (C_{1,min} - C_{1,max}) \frac{t}{It_{max}} + C_{1,max}$$
(10)

$$C_2(t) = (C_{2,min} - C_{2,max}) \frac{t}{It_{max}} + C_{2,max}$$
(11)

where w is inertia weight set to be 0.5. The values of $C_{1,min}$, $C_{1,max}$, $C_{2,min}$ and $C_{2,max}$ can be set to 0.5, 2.5.0.5 and 2.5 respectively.

Step 5: If in conventional PSO, more number of iterations is allowed improve the results which leads to high computational complexity. In block matching algorithms, the number of iterations is limited. Stop the iteration if early termination conditions got satisfied. First condition should be maximum iteration of 3 and second if the gBest position is same for 3 iterations even maximum iterations not reached.

Step 6: The gBest position obtained gives the corresponding motion vector.

5 Results and Discussions

This paper presents experimental results to evaluate the performances of the block matching algorithms of HEXBS and HPSO from the view point of computational complexity as well as prediction accuracy. Table 2 shows the results of average number of computations and Table 3 shows the average of PSNR values with the distance, D = 2 between the current frame and the reference frame by using Football (Fast motion), Silent (Medium motion) and Container (Slow motion) video sequences respectively. It can be seen that HPSO algorithm gives least number of computations for all motion type as compared to the current value HEXBS with the average of

Table 2. Performance of block matching algorithms in terms of no. of computations.

Algorithm	Average no. of computations		
	Slow	Medium	Fast
HEXBS	11.7852	13.2144	15.9852
HPSO	10.3579	12.1814	13.1765

Table 3. Performance of block matching algorithms in terms of PSNR values.

Algorithm	Average of PSNR		
	Slow	Medium	Fast
HEXBS	34.7643	34.4513	23.2535
HPSO	34.5883	34.1083	23.2033

13.1765, 12.1814 and 10.3579 respectively. HPSO also give better PSNR values in slow motion type with the average of 34.5883. In fast motion type, the average of PSNR values comes quite close to HEXBS algorithms with the average of 23.2033. Overall, HPSO provides good performance in both computational complexity and estimation accuracy. Figures 1 and 2 demonstrate a bar graph represent the comparison results between HEXBS and HPSO in the average No. of Computations and PSNR values for all motion type.

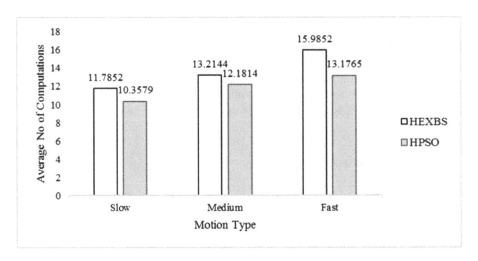

Fig. 1. The comparison the average no. of computations for all three motion type.

Table 4 demonstrates the percentage of the average speed improvement and average quality improvement for all motion type (S, M and F represent for slow, medium and fast motion type respectively). The sign $(-)$ in D_{COMP} and D_{PSNR} indicate a loss values in their performances. From the table, it can be observed that the proposed HPSO approach significantly reduce the number of computations as much as 12.11% for slow motion; 7.82% and 17.57% for medium and fast motion type respectively. In PSNR values, the proposed HPSO algorithm has provide small loss percentage considering acceptable loss in image quality for all motion type as compared to the current value.

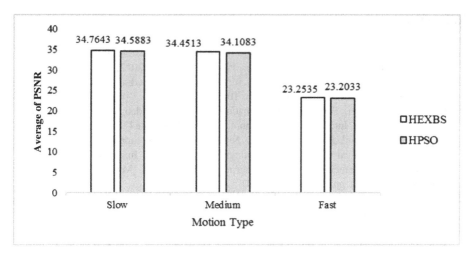

Fig. 2. The comparison the average PSNR values for all three motion type.

Table 4. D_{COMP} and D_{SNR} comparison of the HPSO over HEXBS algorithms for all motion types.

Algorithms	DCOMP (%)			DPSNR (%)		
	S	M	F	S	M	F
HPSO over HEXBS	−12.11	−7.82	−17.57	−0.51	−0.99	−0.22

6   Conclusion and Future Work

Block matching algorithm (BMA) is the most popular and efficient motion estimations in video compression. A new enhanced algorithm has been proposed to reduce the number of computations as well as to give small degradation in PSNR values. Overall, the propose HPSO has given the better solution in both issues to find the best motion vector in block matching. The conventional PSO algorithm has been modified with varying parameter which provide good solution in motion estimation problem. In PSO approach, the position of the particles is initialized in a hexagon pattern instead of selecting in random. Based on the results, the proposed approach has significantly improve the performance in term of the number of computations and in PSNR values compared to HEXBS algorithm. This work could be improved using other variant of PSO or other potential meta-heuristic algorithms to provide the better performances in both computations complexity and estimation accuracy.

Acknowledgement. This work has been supported by FRGS/1/2016/ICT02/UITM/02/2.

References

1. Acharjee, S., Chaudhuri, S.S.: Fuzzy logic based three step search algorithm for motion vector estimation. Int. J. Image Graph. Sig. Process. **2**, 37–43 (2012)
2. Babu, R.V., Tom, M., Wadekar, P.: A survey on compressed domain video analysis techniques. Multimedia Tools Appl. **75**(2), 1043–1078 (2014)
3. Pinninti, K., Sridevi, P.V.: Motion estimation in MPEG-4 video sequence using block matching algorithm. Int. J. Eng. Sci. Technol. (IJEST) **3**, 8466–8472 (2011)
4. Philip, J.T., Samuvel, B., Pradeesh, K., Nimmi, N.K.: A comparative study of block matching and optical flow motion estimation algorithms. In: Annual International Conference on Emerging Research Areas: Magnetics, Machines and Drives (AICERA/iCMMD), pp. 1–6. IEEE (2014)
5. Cuevas, E., Zaldívar, D., Pérez-Cisneros, M., Oliva, D.: Block matching algorithm based on differential evolution for motion estimation. Eng. Appl. Artif. Intell. **26**, 488–498 (2013)
6. Hadi, I., Sabah, M.: A novel block matching algorithm based on cat swarm optimization for efficient motion estimation. Int. J. Digit. Content Technol. Appl. (JDCTA) **8**, 33–44 (2014)
7. Pandian, S.I.A., JoseminBala, G., Anitha, J.: A pattern based PSO approach for block matching in motion estimation. Eng. Appl. Artif. Intell. **26**, 1811–1817 (2013)
8. Baraskar, T., Mankar, V.R., Jain, R.: Survey on block based pattern search technique for motion estimation. In: International Conference on Applied and Theoretical Computing and Communication Technology, pp. 513–518. IEEE (2015)
9. Li, S., Xu, W.-P., Wang, H., Zheng, N.-N.: A novel fast motion estimation method based on genetic algorithm. In: 1999 International Conference on Image Processing, pp. 66–69. IEEE (1999)
10. Ho, L.T., Kim, J.-M.: Direction integrated genetic algorithm for motion estimation in H.264/AVC. In: Huang, D.-S., Zhang, X., Reyes García, C.A., Zhang, L. (eds.) ICIC 2010. LNCS, vol. 6216, pp. 279–286. Springer, Heidelberg (2010). https://doi.org/10.1007/978-3-642-14932-0_35
11. Karaboga, D.: An Idea Based on Honey Bee Swarm for Numerical Optimization. Erciyes University, Engineering Faculty (2005)
12. Yuan, X., Shen, X.: Block matching algorithm based on particle swarm optimization for motion estimation. In: International Conference on Embedded Software and Systems, ICESS 2008, pp. 191–195. IEEE (2008)
13. Cai, J., Pan, W.D.: On fast and accurate block-based motion estimation algorithms using particle swarm optimization. Inf. Sci. **197**, 53–64 (2012)
14. Kennedy, J., Eberhart, R.: Particle swarm optimization. In: IEEE International Conference on Neural Networks, pp. 1942–1948. IEEE (1995)
15. Jalloul, M.K., Al-Alaoui, M.A.: A novel parallel motion estimation algorithm based on particle swarm optimization. In: International Symposium on Signals, Circuits and Systems (ISSCS), pp. 1–4. IEEE (2013)
16. Jalloul, M.K., Al-Alaoui, M.A.: A novel cooperative motion estimation algorithm based on particle swarm optimization and its multicore implementation. Sig. Process.: Image Commun. **39**, 121–140 (2015)
17. Priyadarshini, K., Moni, D.J.: Analysis of block matching algorithm based on particle swarm optimization and differential evolution. Int. J. Appl. Eng. Res. **11**, 2055–2058 (2016)
18. Chavan, S.D., Adgokar, N.P.: An overview on particle swarm optimization: basic concepts and modified variants. Int. J. Sci. Res. **4**, 255–260 (2015)

19. Yaakob, R., Aryanfar, A., Halin, A.A., Sulaiman, N.: A comparison of different block matching algorithms for motion estimation. In: The 4th International Conference on Electrical Engineering and Informatics (ICEEI 2013), vol. 11, pp. 199–205 (2013)
20. Hadi, I., Sabah, M.: Enhanced hybrid cat swarm optimization based on fitness approximation method for efficient motion estimation. Int. J. Hybrid Inf. Technol. **7**, 345–364 (2014)
21. Manjunatha, D.V., Sainarayanan: Comparison and implementation of fast block matching motion estimation algorithms for video compression. Int. J. Eng. Sci. Technol. (IJEST) **3**, 7608–7613 (2011)
22. Sorkunlu, N., Sahin, U., Sahin, F.: Block matching with particle swarm optimization for motion estimation. In: IEEE International Conference on Systems, Man, and Cybernetics, pp. 1306–1311. IEEE (2013)
23. George, N.P., Anitha, J.: Motion estimation in video compression based on artificial bee colony. In: 2015 2nd International Conference on Electronics and Communication Systems (ICECS), pp. 730–733. IEEE (2015)
24. Pal, M.: An optimized block matching algorithm for motion estimation using logical image. In: International Conference on Computing, Communication and Automation (ICCCA2015), pp. 1138–1142. IEEE (2015)
25. Madhuvappan, C.A., Ramesh, J.: Video compression motion estimation algorithms-a survey. Int. J. Sci. Eng. Res. **5**, 1048–1054 (2014)
26. Arora, S.M.: Fast motion estimation algorithms. Int. J. Multi. Res. Dev. **3**, 450–456 (2016)
27. Zhu, C., Lin, X., Chau, L.-P.: Hexagon-based search pattern for fast block motion estimation. IEEE Trans. Circ. Syst. Video Technol. **12**, 349–355 (2002)
28. Damodharan, K., Muthusamy, T.: Analysis of particle swarm optimization in block matching algorithms for video coding. Sci. J. Circ. Syst. Sig. Process. **3**, 17–23 (2014)
29. Britto, J.D.J., Chandran, K.R.S.: A predictive and pattern based PSO approach for motion estimation in video coding. In: International Conference on Communications and Signal Processing (ICCSP), pp. 1572–1576. IEEE (2014)
30. Jacob, A.E., Pandian, I.A.: An efficient motion estimation algorithm based on particle swarm optimization. Int. J. Electron. Sig. Syst. **3**, 26–30 (2013)
31. Bakwad, K.M., Pattnaik, S.S., Sohi, B.S., Devi, S., Gollapudi, S.V.R.S., Sagar, C.V., Patra, P.K.: Small population based modified parallel particle swarm optimization for motion estimation. In: 16th International Conference on Advanced Computing and Communications, pp. 367–373. IEEE (2008)

Enhancing Parallel Self-organizing Map on Heterogeneous System Architecture

Muhammad Firdaus Mustapha$^{(\boxtimes)}$, Noor Elaiza Abd Khalid$^{(\boxtimes)}$, Azlan Ismail, and Mazani Manaf

Universiti Teknologi MARA, 40450 Shah Alam, Selangor, Malaysia
firdaus19@gmail.com, azlanismail08@gmail.com,
{elaiza,mazani}@tmsk.uitm.edu.my

Abstract. Self-organizing Map (SOM) is a very popular algorithm that has been used as clustering algorithm and data exploration. SOM consists of complex calculations where the calculation of complexity depending on the circumstances. Many researchers have managed to improve online SOM processing speed using discrete Graphic Processing Units (GPU). In spite of excellent performance using GPU, there is a situation that causes computer hardware underutilized when executing online SOM variant on GPU architecture. In details, the situation occurs when number of cores is larger than the number of neurons on map. Moreover, the complexities of SOM steps also increase the usage of high memory capacity which leads to high rate memory transfer. Recently, Heterogeneous System Architecture (HSA), that integrated Central Processing Unit (CPU) and GPU together on a single chip are rapidly attractive the design paradigm for recent platform because of their remarkable parallel processing abilities. Therefore, the main goal of this study is to reduce computation time of SOM training through adapting HSA platform and combining two SOM training processes. This study attempts to enhance the processing of SOM algorithm using multiple stimuli approach. The data used in this study are benchmark datasets from UCI Machine Learning repository. As a result, the enhanced parallel SOM algorithm that executed on HSA platform is able to score a promising speed up for different parameter size compared to standard parallel SOM on HSA platform.

Keywords: Parallel self-organizing map · GPU computing
Heterogeneous system architecture

1 Introduction

Self-organizing Map (SOM) is an unsupervised neural network that has been used as data analysis method. It is being widely used and applied to solve clustering and data exploration problems in various domain areas [1]. There were many researches have been found in the literature that used SOM to solve clustering problem [2, 3]. Despite its excellent performance, there are problems related to slow processing when visualizing large map size [4]. This imposed heavy workload on the processor especially

© Springer Nature Singapore Pte Ltd. 2017
A. Mohamed et al. (Eds.): SCDS 2017, CCIS 788, pp. 162–174, 2017.
https://doi.org/10.1007/978-981-10-7242-0_14

when dealing with winner-search and updating weightage of neurons on the map [1]. On the other hand, the datasets dimension also have high influence in SOM processing [5].

This situation attracts much interest among researchers to improve SOM processing by parallelizing the algorithm. Among the common ways to parallelize SOM are network or map partitioning [6, 7] and data or example partitioning [8, 9]. However, there also efforts to parallelize SOM algorithm through combining both network and data partitioning [10, 11] with the interest to gain advantages of both parallelism. In the meantime, most of research works on improving SOM are aimed to achieve efficiency and scalability in their proposed works. In details, proposed parallel SOM that efficient should be faster in term of processing than the previous version [12, 13]. Meanwhile, some research works are attempting to increase the utilization of processing elements in executing the SOM algorithm [7, 14]. Furthermore, several research works pursue to lower the power consumption [15], and solve computer cluster problems [16, 17].

On the other hand, two computer architectures mostly used by researchers in improving SOM algorithm are; Single Instruction Stream Multiple Data Stream (SIMD) and Multiple Instruction Streams Multiple Data Streams (MIMD). Some of the SIMD architectures used Graphic Processing Unit (GPU) computing [18, 19], Field Programmable Gate Array (FPGA) [6, 20], and specialized hardware architecture for Artificial Neural Network (ANN) [21, 22]. Meanwhile, MIMD architectures are employed by researchers to parallelize SOM consists of different types of computer clusters. Among several computer architectures, GPU computing offers an efficient solution at lower cost compared to others.

GPU or widely known as General Purpose Graphic Processing unit (GPGPU) is a many core processor consisting hundreds or even thousands of compute cores [23]. It has been used to speed up applications of scientific computing and simulations. GPU computing has been proven to have high throughput in processing large data floating point operations in graphic applications [24]. Since the introduction of GPU programming frameworks such as of Compute Unified Device Architecture (CUDA) in 2007 and Open Computing Language (OpenCL) in 2009 [24], the GPUs have become popular in designing parallel algorithm in the quest for higher speed. In the meantime, the evolution of hardware technology, has made it possible to design high performance scientific computing software [25]. Essentially, GPU is an accelerator to Central Processing Unit (CPU) that has been mainly used for graphic purposes before applying to process scientific data. Combination of CPU and GPU that work closely together creates a paradigm known as Heterogeneous Computing (HC) [26].

On top of that, many researchers have attempted to capitalize HC to execute SOM algorithm in parallel manner. However Hasan et al. [5] found that when parallelizing SOM with larger map size and high attribute dimension, it will significantly slow down the processing even with both CPU and GPU. Many researchers agreed that executing SOM on GPU shows significantly increase processing speed for large data compared to executing on CPU only [11, 20, 27]. Moreover, due to the restrictions imposed by past GPU architectures, most of these frameworks treated the GPU as an accelerator which can only work under close control of the CPU. Consequently, the communication protocol between CPU and GPU is a source of high latency which causes bottleneck. This drawback is enhanced when using distributed memory of HC where the memory management has to be manually configured by the programmer to manage data movement between CPU and GPU which includes data transfer between host code and device code [28].

A more recent technology hardware design of heterogeneous systems is a single integrated unify CPUs and GPUs circuit chip which is known as Heterogeneous System Architecture (HSA) [28]. This technology provides a unified programming platform which eliminates programmability barrier and reduces CPU and GPU communication latency. The introduction of OpenCL 2.0 that supports HSA specification in July 2013 is able to improve communication by allowing the GPU to manage their own resources as well as access some of the CPU's resources. It also introduces Shared Virtual Memory (SVM) which allows the host and the device to share a common virtual address range [29]. This reduces overhead by eliminating deep copies during host-to-device and device-to-host data transfers. Deep copies involve complete duplicating objects in the memory thus reduce redundancies [28].

In view of the above discussion, this study proposed an enhanced parallel SOM which executes on HSA compliant processor to solve clustering problem. The data used are benchmark datasets that were acquired from UCI Machine Learning Repository [30]. The performance of the algorithm is evaluated in terms of efficiency, scalability and accuracy. Efficiency and scalability are based on processing time while accuracy is based on number of class generated.

This paper is organized as follows. Section 2 explains about the proposed work on parallel SOM using HSA. Section 3 provides explanation on experimental setup while Sect. 4 discusses the experimental result and discussion. Lastly, Sect. 5 provides conclusion and future research directions.

2 Proposed Work

The proposed work consists of enhanced parallel SOM architecture and enhanced parallel SOM algorithm which will be explained in Sects. 2.1 and 2.2 respectively.

2.1 Enhanced Parallel SOM Architecture

Previous studies that highlighted parallel SOM have been successfully executed on GPU. Almost all the researchers in the literature apply parallelism at calculate distance and find Best Matching Unit (BMU) steps. There are many of them apply parallelism at update weight step. Consequent of that, this study proposed to parallelize these three steps into the new enhanced parallel SOM architecture. Meanwhile, heterogeneous system compromises a promising solution for reducing latency in communication between CPU and GPU. In order to gain these advantages, the proposed architecture is utilizing OpenCL 2.0 platform which specifically SVM feature. The implementation of this work is based on fined-grained SVM buffers. The fined-grained SVM buffers are synchronized during the implementation of SVM buffer which could reduce communication latency between CPU and GPU. The design of the proposed architecture is extended from the previous work [31] where it is introduced with two parallel kernels for distance calculation and find BMU as depicted in Fig. 1. The idea of parallel

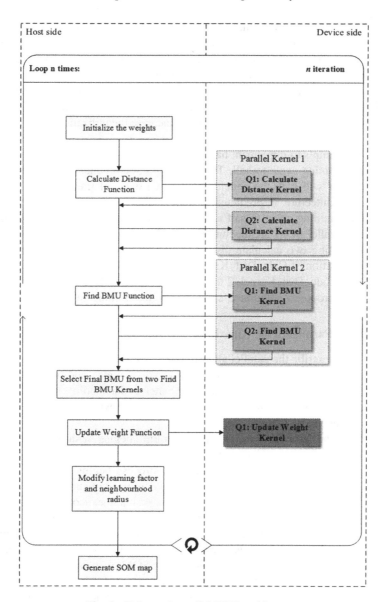

Fig. 1. Enhanced parallel SOM architecture

kernels is based on the analogy of biological neural networks in which neurons respond to the stimuli in parallel [32]. However this work is differ from [32] in term of SOM training types and hardware architecture. The main reason of duplicating the kernels is to increase utilization of work units in GPU.

The design of parallel kernels is realized by implementing multiple stimuli into the architecture. The main reason of parallelized the kernels are to increase utilization of work units in GPU. This work is supported by OpenCL where OpenCL allows a programmer to create more than one queue for execution. The queue will process based on out-of-order execution [29]. In out-of-order execution mode there is no guarantee that the enqueued commands will finish in the order they were queued. The first parallel kernel is Parallel Kernel 1 that consists of Calc_Dist_Kernel_1 and Calc_Dist_Kernel_2. Meanwhile, Parallel Kernel 2 consists of Find_BMU_Kernel_1 and Find_BMU_Kernel_2. The execution of kernels in each parallel kernel might overlap each other. This situation will create a chance to increase the utilization of work units in GPU side.

This study attempts to implement batch learning process through duo stimuli which leads to reduce training cycle to half. This solution that combines both online training and batch training will gain the benefits of the two types of training. The enhanced parallel SOM architecture is predicted to reduce the computation time due to the reduction of the training cycle to half and maintain the final results as online training. The enhanced parallel SOM architecture is essentially implement batch learning processing for executing two calculate distance kernels and two find BMU kernels. For instance, the execution of Calc_Dist_Kernel_1 is considered as executing one task and the execution of Calc_Dist_Kernel_2 also considered as another task. The rest of the algorithm of the proposed solution is similar to online training SOM algorithm.

2.2 Enhanced Parallel SOM Algorithm

This section will describe the parallel algorithm that applies the enhanced parallel SOM architecture. The enhanced parallel SOM algorithm includes additional three steps compared to the original SOM algorithm. For better understanding to the reader, the enhanced parallel SOM algorithm in this study is labeled with e-FGPSOM. Figure 2 illustrates pseudocode of e-FGPSOM.

In depth of the proposed works, the algorithm begins with initializing SOM parameters such as learning factor and weights at the host side. The input data is retrieved and stored into an array. The training process begins with selecting duo stimuli randomly. Each input will be assigned to a command queue. In order to realize duo stimuli method, e-FGPSOM requires two sets of parameter of two calculate distance kernels and two update weight kernels due to the algorithm employs separate command queue for each kernel execution. Each kernel requires a map array. Both map arrays should contain similar values of neurons' weights before the kernels processing is started. Right after the map has initialized using SOM_map_array_1, the values of SOM_map_array_1 will be copied into SOM_map_array_2. The SOM map array is very important because it will be employed within all kernel processing. At the step 2, the algorithm obtains duo stimuli from the dataset randomly and then at the step 3, the host broadcast the two inputs to device side, specifically the host assigns to two calculate distance kernels; Calc_Dist_Kernel_1 and Calc_Dist_Kernel_2, and two find BMU kernels; Find_BMU_Kernel_1 and Find_BMU_Kernel_2. Once all information has been broadcasted, the kernels are ready for the execution especially for calculate distance kernel at step 4. Each kernel at the GPU side is invoked by function respectively. The functions also provide setting, initializing parameters, and call the kernels. For example, the calculate distance function is used to call Calculate Distance kernel and it is done the same way with the other two kernels.

```
<BEGIN>

1.Randomly initialize the neuron' weights W [w_{j1}, w_{j2}, ... w_{jm}]

2.The HOST side ramdomly selects duo stimuli, x_i and x_{i+1}

from  X [x_1, x_2, ... x_m] according to sequence.
3.The HOST side broadcasts the two input at once

X [(x_i, x_{i+1}), ..., (x_{m-1}, x_m)] to DEVICE side.
4.Execute calculate distance kernels, Calc_Dist_Kernel_1
and Calc_Dist_Kernel_2 based on out-of-order execution
mode.
5.Execute find BMU kernels, Find_BMU_Kernel_1 and
Find_BMU_Kernel_2 based on out-of-order execution mode.
6.The HOST side select the winner of the winners from
both find BMU kernels.
7.The HOST side braoadcasts the final BMU to DEVICE
side.
8.Execute update weights kernel.
9.Repeat step 2 to 8 until certain termination
condition.
10.Generate the map using visualization algorithm.
<END>
```

Fig. 2. Enhanced parallel SOM algorithm

Essentially, the original flow of SOM algorithm is maintained unless with the additional two parallel stimuli execution that has applied for execution of two calculate distance kernels and two find BMU kernels in step 4 and step 5 respectively. The executions of two calculate distance kernels and two find BMU kernels are based on out-of-order execution mode. There is no guarantee that the enqueued commands will finish execution in the order because the execution of kernel is based on when the clEnqueueNDRangeKernel calls are made within a command-queue.

Calculate Distance Kernel. The calculate distance kernel is used to calculate the distance between neurons and current input vector. The amount of work units is employed to parallelize the calculation distance step is mapped by amount of work-items on GPU where the amount of work-items is equal to the number of neurons in the SOM map. Specifically, each work-item of the kernel is responsible for finding the distance between a single neuron and the current input vector. This study applies Manhattan distance calculation.

Find BMU Kernel. The Find BMU kernel applies two stages reduction method. The kernel utilizes work items the same amount of neurons on SOM map. The first stage of reduction method is to find the minimum distance for each local work group. The values of minimum distances of each work group will be stored into local array. The second stage is to find the minimum distance for each Compute Unit (CU). The minimum values of each CU then stored into global array and the host will determine the winning neurons. After the execution of both Find BMU 1 and Find BMU 2, at the step 6, the winner among the winners or final BMU from the both kernels will be selected with the minimum value at the host side. With the selected of the final BMU means the input vector which has the final BMU will use to update the map meanwhile the loser input will be eliminated. The final BMU will be broadcasted to device side for execution of third kernel, the update weight kernel.

Update Weight Kernel. This kernel is the third kernel in the proposed architecture that updates the weight of neurons based on learning rate and neighborhood function. The learning rate defines how much a neuron's vector is altered through an update with referring to how far the distance of the neuron from the BMU on the map. The BMU and its close neighbors will be altered the most, while the neurons on the outer edges of the neighborhood are changed the least. The enhanced version includes array copying process at host side right after the update weight kernel completed the execution. The array copying process is to copy the updated map array into another map array which is not selected as final BMU. The array copying codes are simply assign one array to another array with the same size through looping. Immediately after executing the three kernels, the learning factor and neighborhood radius are updated with the new values. All of the steps included in the loop block will repeat until certain number iterations or epochs before the SOM map is generated.

3 Experimental Setup

Two experiments have been conducted in this study which is result evaluation and result validation. These experiments are performed with the interest to evaluate the proposed work. Initially, this study employs four benchmark datasets from UCI Machine Learning Repository [30] for results validations. The details of the datasets are shown in Table 1. These datasets have been selected because information on the number of classes is known in advance and ease for the validation process. The information of number of classes is as provided by Fränti [33]. These datasets are processed by e-FGPSOM where each experiment is performed to validate each dataset.

Table 1. Benchmark datasets for result validations

Benchmark dataset	Number of training data	Number of attributes of pattern	Number of classes
Iris	150	4	3
Glass	214	9	6
Wine	178	13	3
Yeast	1484	8	10

On the other hand, for evaluations purpose, this study employs Bank Marketing benchmark dataset from UCI Machine Learning Repository [34]. Initially, data pre-processing takes place before the experiment is conducted. Data pre-processing is one of the important steps in data mining to obtain final datasets that can be considered as correct and useful for further data mining algorithm [35]. There are four methods in data pre-processing which are data cleaning, data integration, data transformation and data reduction [36]. Thus, this study uses method of data reduction by using discretization technique to convert values of the selected attributes [37]. This study converts all the values that appear in categorical into numeric values for easy processing in SOM.

Firstly, the experiment starts with result validation. The objective of result validation is to make sure the proposed work is capable to generate correct results by comparing the results from e-FGPSOM with results from other researchers. Four dimension sizes have been used: 4, 9, 13 and 8 parameters according to specific benchmark datasets as shown in Table 2. The algorithm is tested on the same map size, 40 × 40 by using 250 iterations.

Table 2. Experimental design for enhanced parallel SOM evaluation

Experiment series	Datasets parameter		SOM parameter		Performance measurement
	No. of samples	No. of parameter	No. of iteration	Map size	
Result validation	Iris, Glass, Wine, Yeast	4, 9, 13, 8	250	40 × 40	Number of classes
Result evaluation	10000, 15000	3, 5, 8	30	50 × 50	Time, s and speed up

On the other note, result evaluation is concerning on evaluation of the proposed work on different dimension size or number of parameter of dataset. Three dimension sizes have been used: 3, 5 and 8 parameters. The algorithm is tested on the same map size, 50 × 50 by using 30 iterations. Result evaluation applies time comparison and speed up [38] for performance measurements.

Moreover, for analysis purpose, the result of this study is compared to the result of our previous work [31] referred as FGPSOM. FGPSOM is based on standard parallel SOM on HSA platform. Both FGPSOM and e-FGPSOM implemented on OpenCL 2.0. These experiments are conducted on a laptop that equipped with Intel Skylake i7-6700HQ processor and built in Intel® HD Graphics 530.

4   Experimental Result and Discussion

4.1   Result Validations

Firstly, the experimental result starts with result validation. As mentioned in Sect. 3, result validation for e-FGPSOM is based on four benchmark datasets which are Iris, Glass, Wine, and Yeast dataset. Figure 3 illustrates the generated results in map visualization by e-FGPSOM. This figure shows that Iris dataset and wine dataset clearly produce three classes. This results can be validated with Fränti [33] that the proposed work is capable to generate the similar results. Meanwhile, the results of glass

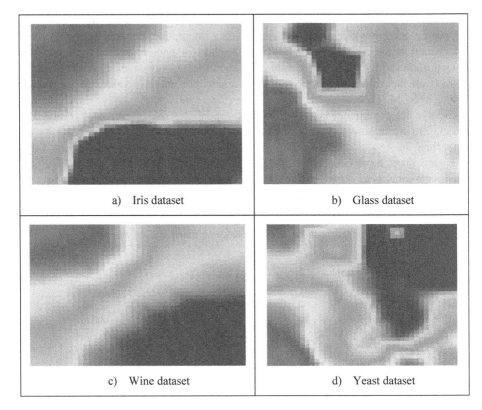

Fig. 3. Result produced by using algorithm of the enhanced parallel SOM architecture

and yeast datasets are quite subjective to be decided. However, one could be identified the number of classes via observation that the proposed work also capable to generate equivalent results [33].

4.2   Results Evaluation

For result evaluation purpose, the proposed algorithm tested on the same dataset size and map size. However, this study is focusing on experimenting the proposed algorithm on different dimension size of dataset; 3, 5, and 8 parameters. Figure 4 demonstrates the comparison result between FGPSOM and e-FGPSOM. From this figure, the results of e-FGPSOM perform better than results of FGPSOM for both datasets. The result of e-FGPSOM shows the improvement compared to FGPSOM where all the results of e-FGPSOM capable to reduce the processing time. The experiment exposes that e-FGPSOM successfully improved the processing speed compared to FGPSOM when experimenting on larger dimension size of dataset.

For the details, according to 10000 dataset, e-FGPSOM achieves 1.24x, 1.13x, and 1.10x of speed up for 3, 5, and 8 parameters respectively. Meanwhile, for 15000 dataset, e-FGPSOM scores 1.16x, 1.11x, and 1.10x of speed up for 3, 5, and 8 parameters

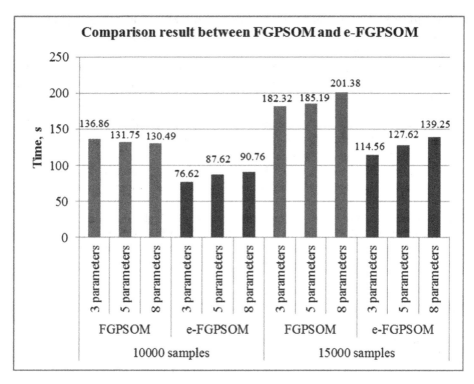

Fig. 4. Result produced by using algorithm of the enhanced parallel SOM architecture

respectively. The proposed algorithm is capable to score speed up over FGPSOM due to it applies the proposed architecture which consists of parallel kernels. The execution of parallel kernels has triggered multiple stimuli processing. The processing of multiple stimuli capable to increase the utilization of GPU cores compared to FGPSOM. However, due to complexity of processing larger dataset and larger parameters size, the speeds up scores are reducing over the increasing the dataset size and parameter size.

5 Conclusion

In this study, we proposed an enhanced parallel SOM that based on heterogeneous system architecture. The proposed architecture is extended from parallel SOM researches that consist of three kernels: calculate distance kernel, find BMU kernel, and update weight kernel. The proposed architecture is included with two calculate distance kernels and two find BMU kernels. The proposed architecture is designed with the aim to increase the utilization of processing element on GPU.

The overall results indicate that the enhanced parallel SOM or e-FGPSOM is able to improve in terms of efficiency and scalability performance. This is due to e-FGPSOM that optimizes the usage of cores in the GPU. The enhanced parallel SOM or e-FGPSOM also demonstrates some advantages from various perspectives as below:

- The proposed work is capable to generate comparable results (number of classes) compared to other researches by using benchmark datasets. Thus, the proposed work maintains the accuracy of the result.
- The proposed work more scalable in terms of GPU cores utilization due to its imposition with multiple stimuli method. The implementation of multiple stimuli method also improves the proposed work to more efficient.

The proposed work has a limitation where the synchronization point at find BMU step could burden the processing of the find BMU kernel. Based on the limitation, there seems to be an opportunity for improvement. In the future, the consumption time of synchronization point could be reduced through eliminating the access of BMUs values at the host side. The final BMU values should be identified at kernel processing.

Acknowledgement. This study was funded by Ministry of Higher Education (MOHE) of Malaysia, under the FRGS, grant no. FRGS/1/2015/ICT02/UITM/02/6 and Academic Staff Bumiputera Training Scheme (SLAB). The authors also would like to thank the Universiti Teknologi MARA for supporting this study.

References

1. Kohonen, T.: Essentials of the self-organizing map. Neural Netw. **37**, 52–65 (2013)
2. Llanos, J., Morales, R., Núñez, A., Sáez, D., Lacalle, M., Marín, L.G., Hernández, R., Lanas, F.: Load estimation for microgrid planning based on a self-organizing map methodology. Appl. Soft Comput. **53**, 323–335 (2017)
3. Matic, F., Kovac, Z., Vilibic, I., Mihanovic, H., Morovic, M., Grbec, B., Leder, N., Dzoic, T.: Oscillating adriatic temperature and salinity regimes mapped using the self-organizing maps method. Cont. Shelf Res. **132**, 11–18 (2017)
4. McConnell, S., Sturgeon, R., Henry, G., Mayne, A., Hurley, R.: Scalability of self-organizing maps on a GPU cluster using OpenCL and CUDA. J. Phys: Conf. Ser. **341**, 12018 (2012)
5. Hasan, S., Shamsuddin, S.M., Lopes, N.: Machine learning big data framework and analytics for big data problems. Int. J. Adv. Soft Comput. Appl. **6**, 1–17 (2014)
6. Kurdthongmee, W.: A novel Kohonen SOM-based image compression architecture suitable for moderate density {FPGAs}. Image Vis. Comput. **26**, 1094–1105 (2008)
7. Kurdthongmee, W.: A low latency minimum distance searching unit of the SOM based hardware quantizer. Microprocess. Microsyst. **39**, 135–143 (2015)
8. Moraes, F.C., Botelho, S.C., Filho, N.D., Gaya, J.F.O.: Parallel high dimensional self organizing maps using CUDA. In: 2012 Brazilian Robotics Symposium and Latin American Robotics Symposium, pp. 302–306 (2012)
9. Sul, S.J., Tovchigrechko, A.: Parallelizing BLAST and SOM Algorithms with MapReduce-MPI library. In: 2011 IEEE International Symposium on Parallel and Distributed Processing Workshops and Phd Forum, pp. 481–489 (2011)
10. Mojarab, M., Memarian, H., Zare, M., Hossein Morshedy, A., Hossein Pishahang, M.: Modeling of the seismotectonic provinces of Iran using the self-organizing map algorithm. Comput. Geosci. **67**, 150–162 (2014)
11. Richardson, T., Winer, E.: Extending parallelization of the self-organizing map by combining data and network partitioned methods. Adv. Eng. Softw. **88**, 1–7 (2015)

12. Garcia, C., Prieto, M., Pascual-Montano, A.: A speculative parallel algorithm for self-organizing maps. In: Proceedings of Parallel Computing 2005 (ParCo 2005), vol. 33, pp. 615–622 (2005)
13. MacLean, D., Valova, I.: Parallel growing SOM monitored by genetic algorithm. In: 2007 International Joint Conference on Neural Networks, pp. 1697–1702 (2007)
14. Dlugosz, R., Kolasa, M., Pedrycz, W., Szulc, M.: Parallel programmable asynchronous neighborhood mechanism for kohonen SOM implemented in CMOS technology. IEEE Trans. Neural Netw. **22**, 2091–2104 (2011)
15. Khalifa, K.B., Girau, B., Alexandre, F., Bedoui, M.H.: Parallel FPGA implementation of self-organizing maps. In: Proceedings of the 16th International Conference on Microelectronics, ICM 2004, pp. 709–712 (2004)
16. Yang, M.-H., Ahuja, N.: A data partition method for parallel self-organizing map. In: International Joint Conference on Neural Networks, IJCNN 1999, vol. 3, pp. 1929–1933 (1999)
17. Schabauer, H., Schikuta, E., Weishäupl, T.: Solving very large traveling salesman problems by SOM parallelization on cluster architectures. In: Proceedings of Parallel and Distributed Computing, Applications and Technologies, PDCAT, pp. 954–958 (2005)
18. Gajdos, P., Platos, J.: GPU based parallelism for self-organizing map. In: Kudělka, M., Pokorný, J., Snášel, V., Abraham, A. (eds.) Intelligent Human Computer Interaction. Advances in Intelligent Systems and Computing, vol. 179, pp. 3–12. Springer, Heidelberg (2013). https://doi.org/10.1007/978-3-642-31603-6_20
19. Nguyen, V.T., Hagenbuchner, M., Tsoi, A.C.: High resolution self-organizing maps. In: Kang, B.H., Bai, Q. (eds.) AI 2016. LNCS, vol. 9992, pp. 441–454. Springer, Cham (2016). https://doi.org/10.1007/978-3-319-50127-7_38
20. Lachmair, J., Merényi, E., Porrmann, M., Rückert, U.: A reconfigurable neuroprocessor for self-organizing feature maps. Neurocomputing **112**, 189–199 (2013)
21. Asanović, K.: A fast Kohonen net implementation for spert-II. In: Mira, J., Moreno-Díaz, R., Cabestany, J. (eds.) IWANN 1997. LNCS, vol. 1240, pp. 792–800. Springer, Heidelberg (1997). https://doi.org/10.1007/BFb0032538
22. Porrmann, M., Witkowski, U., Ruckert, U.: A massively parallel architecture for self-organizing feature maps. IEEE Trans. Neural Netw. **14**, 1110–1121 (2003)
23. Perelygin, K., Lam, S., Wu, X.: Graphics processing units and open computing language for parallel computing. Comput. Electr. Eng. **40**, 241–251 (2014)
24. Kirk, D.B., Hwu, W.W.: Programming Massively Parallel Processors. Elsevier, Amsterdam (2013)
25. Rauber, T., Rünger, G.: Parallel Programming: For Multicore and Cluster Systems. Springer, Heidelberg (2010). https://doi.org/10.1007/978-3-642-04818-0
26. Mittal, S., Vetter, J.S.: A survey of CPU-GPU heterogeneous computing techniques. ACM Comput. Surv. **47**, 69:1–69:35 (2015)
27. De, A., Zhang, Y., Guo, C.: A parallel image segmentation method based on SOM and GPU with application to MRI image processing. Neurocomputing **198**, 180–189 (2016)
28. Mukherjee, S., Sun, Y., Blinzer, P., Ziabari, A.K., Kaeli, D.: A comprehensive performance analysis of HSA and OpenCL 2.0. In: 2016 IEEE International Symposium on Performance Analysis of Systems and Software (2016)
29. Khronos OpenCL: OpenCL Specification (2014)
30. Lichman, M.: UCI Machine Learning Repository. http://archive.ics.uci.edu/ml
31. Mustapha, M.F., Abd Khalid, N.E., Ismail, A.: Evaluation of parallel self-organizing map using heterogeneous system platform. J. Appl. Sci. **17**, 204–211 (2017)

32. Yasunaga, M., Tominaga, K., Kim, J.H.: Parallel self-organization map using multiple stimuli. In: International Joint Conference on Neural Networks, IJCNN 1999 (Cat. No. 99CH36339), vol. 2, pp. 1127–1130 (1999)

33. Fränti, P., et al.: Clustering datasets. http://cs.uef.fi/sipu/datasets/

34. Moro, S., Cortez, P., Rita, P.: A data-driven approach to predict the success of bank telemarketing. Decis. Support Syst. **62**, 22–31 (2014)

35. Berkhin, P.: A survey of clustering data mining techniques. Group. Multidimens. Data **25**, 71 (2006)

36. Han, J., Kamber, M., Pei, J.: Data preprocessing. In: Data Mining Concept and Techniques, pp. 83–134 (2012)

37. Nawi, N.M., Atomi, W.H., Rehman, M.Z.: The effect of data pre-processing on optimized training of artificial neural networks. Procedia Technol. **11**, 32–39 (2013)

38. Hennessy, J.L., Patterson, D.A.: Computer Architecture, Fourth Edition: A Quantitative Approach. Morgan Kaufmann Publishers Inc., San Francisco (2006)

Data Visualization

Geovisualization Using Hexagonal Tessellation for Spatiotemporal Earthquake Data Analysis in Indonesia

Ridho Dwi Dharmawan$^{(\boxtimes)}$, Suharyadi, and Nur Muhammad Farda

Department of Geography Information Science, Gadjah Mada University,
Bulaksumur, Sleman, D.I. Yogyakarta 55281, Indonesia
ridho.dwi.d@mail.ugm.ac.id

Abstract. Geovisualization process is the key in map design to extract information from geospatial data set especially on big data. The use of point feature in the geovisualization process of earthquake spatial data in Indonesia caused some problems such as overlapping between symbols, complications due to the large amount of data, and uneven spatial distribution. This research aims to create geovisualization of earthquake spatial data in Indonesia using hexagonal tessellation method, analyze earthquake map in Indonesia based on geovisualization using hexagonal tessellation and interpret earthquake map in Indonesia based on geovisualization using hexagonal tessellation spatiotemporally. The method used in this research is geovisualization of earthquake spatial data in Indonesia using hexagonal tessellation. This research use earthquake epicenter density analysis to discover and illustrate the spatial phenomena pattern of the earthquake epicenter into more easily understood information. The density analysis includes distance matrix analysis to examine the visualization result and proximity analysis to know the proximity of earthquake density represented by centroid hexagon point with the tectonic plate fault line. The result of this research is earthquake map in Indonesia based on geovisualization using hexagonal tessellation in Indonesia 2010 to 2015. The result of this research shows that the map design of earthquake geospatial information using hexagonal tessellation geovisualization method can show the density distribution of earthquake point spatiotemporally. The earthquake epicenter density analysis in Indonesia based on geovisualization using hexagonal tessellation showed that the hexagon centroid point with high density attribute earthquake data of all magnitude classes tended to have a closer distance to the tectonic plate fault lines spatially.

Keywords: Geovisualization · Tessellation · Hexagonal · Earthquake

1 Introduction

The spatial data is related to the spatial analysis aimed to obtain new information that can derive from big data. The important aspect of the spatial analysis is the representation or visualization of data [6]. Visualization techniques are needed to produce an effective and easy-to-read map. Visualization of the spatial data in large quantities,

A. Mohamed et al. (Eds.): SCDS 2017, CCIS 788, pp. 177–187, 2017.
https://doi.org/10.1007/978-981-10-7242-0_15

especially for a point feature has its own problems in the process of visualization such as overlapping among symbols, many complications because of the amount of data, and the uneven spatial distribution [14].

One of the methods to visualize point-based spatial data is using tessellation. Tessellation method can visualize spatial data points using basic geometric shape. Geometries that can be used is square, hexagon, triangle, or other geometry shapes. Hexagonal shape in cartography simplifies the process of changing the cell size when the scale of the map was changed without reducing the level of readability on the map [14]. Meanwhile, using a square shape causes the reader uncomfortable and difficult to explaining the spatial patterns [4]. Hexagon has a higher representational accuracy and when analyzed statistically, the hexagon has a direct border with six neighbors, instead of four neighbors like the square shape [13]. Cell size variation that can be adjustable in a hexagon shape allows customization of cell sizes appropriate for different scale levels and have the potential to resolve Modifiable Areal Unit Problem [10].

Examples of point-based spatial data that has a large amount and problems in the representation, as mentioned above, is the earthquake spatial data in Indonesia issued by the Meteorological, Climatology and Geophysics Agency showed on Fig. 1. Indonesia is located on the active tectonic plates fault, annually earthquake-ravaged from small scale to large scale. Spatial data generated would continue to increase so that the visualization process should be appropriate in order to avoid a decrease in the level of readability when presented in map form. Visualization of earthquake data in Indonesia still has problems that commonly occur in the representation of spatial data point.

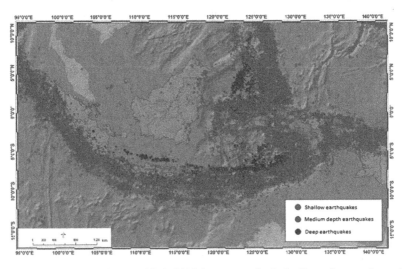

Fig. 1. Indonesia earthquake map 1973–2010 by meteorological, climatology and geophysics agency (BMKG)

The problem of earthquake spatial data visualization in Indonesia causes the difficulty of epicenter spatial distribution analysis. The map which exists today represented the earthquakes in the point-based spatial data inflicts many points of earthquake events overlap one another. Its effect on the information of earthquake events distribution cannot be analyzed temporally well. In addition, seismic data attributes such as earthquake magnitude and depth are not represented their distribution on the map with the result is difficult to analyze spatially and temporally.

The occurrence of earthquakes is closely related to the movement of tectonic plates. Earthquakes often occur in regions that are close with tectonic faults called earthquake zone [7]. However, the visualization in the form of a point cannot describe the spatial and temporal how the actual relationship between the earthquake epicenters point with a fault line. Therefore, this study aimed to examine the earthquake spatial data visualization in Indonesia using hexagonal tessellation method and then used as a model of the earthquake epicenter dot density analysis in spatial-temporal distance to determine how the relationship with the fault lines of tectonic plates.

2   Collecting and Processing Earthquake Spatial Data

Earthquake spatial data in Indonesia is obtained from the Agency of Meteorology and Geophysics through the official website. The earthquake epicenter data is downloaded based on the time range from 2010 to 2015. The territorial boundaries studied are 6° LU–11° LS and 94° BT–142° BT. The downloaded data is in text format so it is necessary to convert into a table format using Micosoft Excel numeric software. Table data contains coordinates, earthquake depth, magnitude earthquake, and earthquake event time attributes. This research only uses coordinates attributes and magnitude of the earthquake.

Data processing begins by converting table data into point spatial data using Quantum GIS software. Based on the coordinate information in the attribute table, the earthquake epicenter point obtained and then stored in shapefile spatial data format. The earthquake spatial data are classified according to their magnitudes. The classification of earthquake magnitude used is less than 4.0 RS, 4.0–5.0 RS, and more than 5.0 RS.

Figure 2 showed that the earthquake spatial data presented using the point feature encountered many problems in reading, analysis, and interpretation. The number of

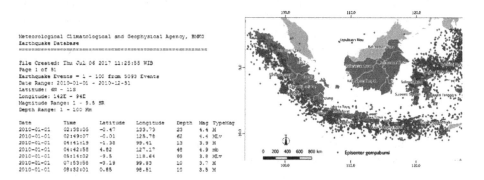

Fig. 2. Converting earthquake table data to earthquake spatial data

earthquake epicenter points with an average of 5181 epicenter annually from 2010 to 2015 in the territory of Indonesia resulted visualization using points caused covering each other at the frequent earthquake epicenter. It reduces the level of information understanding that can be extracted from the data so that its usage is not maximal and cannot be analyzed further.

3 Geovisualization Using Hexagonal Tessellation

Tessellation is a process of preparing a data representation by creating partitions using one or more geometric shapes that complement each other without any overlap and gap on each side [15]. Based on the variation of the shape and size, tessellation is divided into two regular tessellation and irregular tessellation [11]. Regular tessellation is a tessellation that uses a uniform shape and size of a geometry cell such as triangles, square, and hexagon to present data. Irregular tessellation is a more complex tessellation which the shape and size of the cell may vary according to its data attributes. Difference between regular and irregular tessellation shown in the Fig. 3.

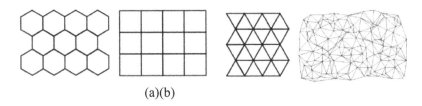

(a)(b)

Fig. 3. Regular tessellation (a) and irregular tessellation (b)

The usage of hexagon geometry reduces ambiguity in determining neighbour cells [2]. Hexagon geometry has a shorter perimeter than square with the same area where it can reduce bias due to edge effects [2]. Tessellation for observation has been widely applied to modeling, simulating, and even studying ecosystems. This is because the geometry with the same pattern is an efficient way of surveying, sampling, and experimenting [9]. Hexagonal tessellation can be used to analyze population density in an area for example citizen population in a country [5]. The size of the cells resolution used for hexagonal tessellation mapping on a national scale is 65 km while on the urban scale is 2 km so the map reader can still distinguish the larger pattern without ignoring minor differences [14]. The study area of this research is entire territory of Indonesia that according to Shelton is categorized as national scale so the cells resolution used is 65 km.

3.1 Creating Hexagonal Tessellation Cell

The geovisualization process of point spatial data using the hexagonal tessellation method can be executed after the coordinate system used is similar or equivalent.

Therefore, hexagon cells is created using Web Mercator projection same as the earthquake spatial data. The coordinate boundaries used to generate hexagon cells are 7.875° N–12.194° LS and 93.28° E–141.387° east. The benefit of the coordinate boundary is to determine the area to be mapped using hexagonal tessellation where the hexagon cell will cover the entire region as in Fig. 4. Determination of the limit refers to the study area that is all regions of Indonesia that have a history of earthquake occurred from 2010 to 2015.

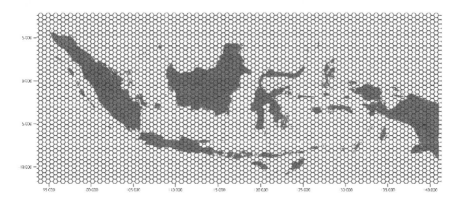

Fig. 4. Creating hexagonal tessellation cell

3.2 Hexagonal Tessellation

Geovisualization of hexagonal tessellation executed by covering the epicenter earth-quake epicenter data with hexagon cells in each year and each earthquakes classifi-cation. The number of points that fall on each hexagon is calculated and stored in the polygon data attribute using analysis tool in QGIS that is counts points in polygon. This research obtained tessellation for each class of earthquake magnitude annually so that there are 18 layers produced and three cumulative maps of the year 2010-2015 in each class of magnitude.

Figure 5 shows the density of earthquakes stored in hexagon cell attributes is classified to symbolized by their density levels. The density information of the

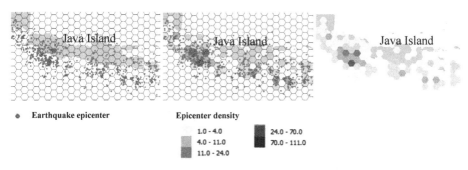

Fig. 5. Hexagonal tessellation process

earthquake is used to assess the relation density of earthquakes with major tectonic plate fault which each hexagon is represented by one centroid point.

4   Analysis

Spatial analysis techniques is various, such as visual observation and mathematics or applied statistics [12]. The advantages of spatial analysis is can be used to explain how the location relationship between the occurrence of a phenomenon to the surrounding phenomenon [8]. The density of earthquakes geospatial data of in Indonesia that have been visualized using hexagonal tessellation method can be analyzed spatially and temporally. The analysis that has been done in this research includes analysis to test the result of hexagonal tessellation and interpretation to know the density distribution or frequency of earthquake occurrence spatiotemporally.

```
d_lyr = QgsVectorLayer('LineString?field=distance:float', 'dist', 'memory')
QgsMapLayerRegistry.instance().addMapLayer(d_lyr)

p_lyr = QgsMapLayerRegistry.instance().mapLayersByName('Centroid Hex Mag kecil 2010')[0]
l_lyr = QgsMapLayerRegistry.instance().mapLayersByName('Fault line')[0]
d_lyr = QgsMapLayerRegistry.instance().mapLayersByName('dist')[0]
prov = d_lyr.dataProvider()
feats = []
for p in p_lyr.getFeatures():
    minDistPoint = min([l.geometry().closestSegmentWithContext(QgsPoint(p.geometry().asPoint())) for l in l_lyr.getFeatures()])[1]
    feat = QgsFeature()
    feat.setGeometry(QgsGeometry.fromPolyline([QgsPoint(p.geometry().asPoint()), QgsPoint(minDistPoint[0], minDistPoint[1])]))
    feat.setAttributes([feat.geometry().length()])
    feats.append(feat)

prov.addFeatures(feats)
d_lyr.updateExtents()
d_lyr.triggerRepaint()
```

Fig. 6. NNjoin code to compute distance between centroid and break line

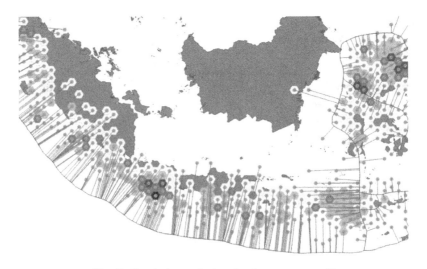

Fig. 7. Proximity analysis using hexagon centroid

4.1   Proximity Analysis

Proximity analysis is one of spatial analysis by considering the distance of an object with other objects to know their relationship [3]. Proximity analysis was used in this study to assess the density of occurrence of earthquake with major plate fault line. The proximity analysis utilizes an additional NNjoin plugin on QGIS software that enables the calculation of distances between different features which in this study is the hexagon centroid point that contains the attributes of earthquake density with plate fault lines in Indonesia. NNjoin code showed on Fig. 6.

The results of the analysis are then presented in the form of a two-variable scatter plot that is the distance from the fault line and the density level of earthquake

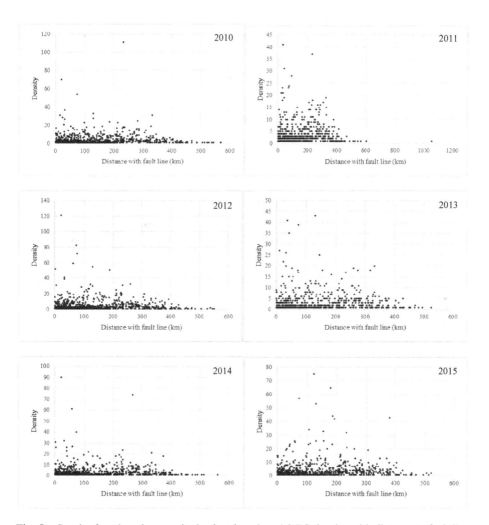

Fig. 8. Graph of earthquake magnitude class less than 4.0 RS density with distance to fault line in Indonesia year 2010–2015

occurrence for each class of magnitude and the year of occurrence. The result of proximity analysis is showed on Fig. 7.

The purpose of the proximity analysis is to see how exactly the distance relationship between the two spatial objects is. The distance analysis was carried out using QGIS with additional NNjoin plugins that allowed the calculation of the distance between different features which in this study was the level of earthquake density with plate fault lines. The results of the analysis are then presented in the form of a two-variable scatter plot that is the distance from the fault line and the density level of earthquake occurrence for each magnitude class as well as the year of occurrence.

The distance measuring result of the hexagon centroid with plate fault line is presented on scatter plot which shows the distribution of density of earthquake. Each class of earthquake magnitude has six years of temporal time in from 2010 to 2015.

Scatter plot in Fig. 8 shows the results of spatial data proximity analysis of earthquake data from 2010 to 2015 in Indonesia. Distance between the centroid with fault line is on x axis and density of epicenter is on y axis. The analysis results show spatially the centroid that has high density attribute is closer to plate fault line. This may indicate that the existence of a plate fault line affects the occurrence of earthquakes where the density of the occurrence is higher at closer distance to the fault line (Figs. 9, 10 and 11).

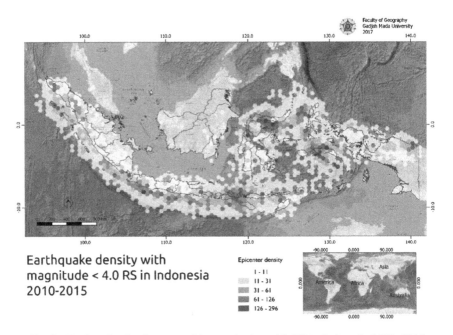

Fig. 9. Earthquake density map with magnitude < 4.0 RS in Indonesia 2010–2015

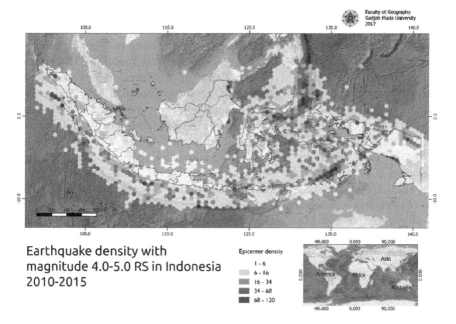

Earthquake density with
magnitude 4.0-5.0 RS in Indonesia
2010-2015

Epicenter density
1 - 6
6 - 16
16 - 34
34 - 68
68 - 120

Fig. 10. Earthquake density map with magnitude 4.0–5.0 RS in Indonesia 2010–2015

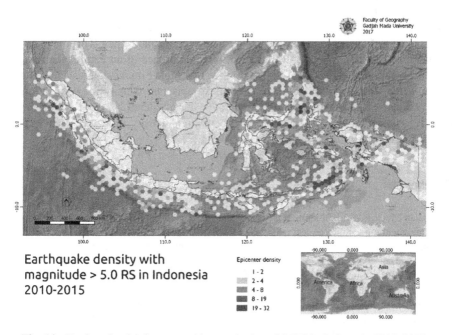

Earthquake density with
magnitude > 5.0 RS in Indonesia
2010-2015

Epicenter density
1 - 2
2 - 4
4 - 8
8 - 19
19 - 32

Fig. 11. Earthquake density map with magnitude > 5.0 RS in Indonesia 2010–2015

5   Interpretation

Earthquake epicenter shows the existence of tectonic activity in a region because of tectonic earthquake closely related to the existence of plate fault line. The pattern of density level of the earthquake may indicate tectonic activity in an area where in high epicenter density area tectonic activity in the region is classified as intensive. Areas with low episenter densities indicate that tectonic activity in the region is less intensive.

The results of the interpretation of earthquake maps using hexagonal tessellation geovisualization considering the density of earthquake epicenter shows some areas with intensive tectonic activity such as North Maluku, Sulawesi, southern Java region, southern region of Sumatra, and Papua. One example of significant tectonic earthquake activity in 2010 was in the Mentawai Islands that killed 286 people. In 2015 tectonic activity causes more destructive earthquakes in eastern Indonesia. There are seven recorded events, namely Manggarai Earthquake, Banggai Earthquake, Membramo Raya Earthquake, Sorong Earthquake, Alor Earthquake, West Halmahera Earthquake, and Ambon Earthquake.

6   Conclusion

Hexagonal tessellation geovisualization method can show the density distribution of earthquake epicenter point in Indonesia spatiotemporally that may indicate tectonic activity in an area. The proximity analysis shows that spatial centroid hexagon having high density attributes for earthquake data all magnitude classes tend to have a closer distance to major fault lines. Conversely, hexagon centroids that have low density earthquake attributes tend to be farther away from major fault lines. This may indicate that the existence of a plate fault line affects the occurrence of an earthquake where if spatially observed the density of the occurrence of a higher quake at a distance close to the fault line. Interpretation result of earthquake episenter density maps in Indonesia using hexagonal tessellation indicates that the density level of the earthquake can indicate tectonic activity in an area where in high density epicenter area tectonic activity in the region is classified as intensive.

References

1. Badan Meteorologi Klimatologi dan Geofisika. Gempabumi (2016). http://inatews.bmkg.go.id/new/tentang_eq.php. Accessed 30 Sep 2016
2. Birch, C.P., Oom, S.P., Beecham, J.A.: Rectangular and hexagonal grids used for observation, experiment and simulation in ecology. Ecol. Model. **206**, 347–359 (2007)
3. Blinn, C.R., Queen, L.P., Maki, L.W.: Geographic Information Systems: A Glossary. Universitas Minnesota, Minnesota (1993)
4. Carr, D.B., Olsen, A.R., White, D.: Hexagon mosaic maps for display of univariate and bivariate geographical data. Cartogr. Geog. Inf. Syst. **19**(4), 228–236 (1992)
5. Dewi, R.S.: A GIS-Based Approach to The Selection of Evacuation Shelter Buildings and Routes for Tsunami Risk Reduction. Universitas Gadjah Mada, Yogyakarta (2010)

6. Dittus, M.: Visualising Structured Data on Geographic Maps Evaluating a Hexagonal Map Visualisation. Centre for Advanced Spatial Analysis, University College London, London (2011)
7. Kious, W.J., Tilling, R.I.: This Dynamic Earth: The Story of Plate Tectonics, illustrated edn. DIANE Publishing, Washington, DC (1996)
8. Longley, P.A., Goodchild, M.F., Maguire, D.J., Rhind, D.W.: Geographic Information Science and Systems, 4th edn. Wiley, New Jersey (2016)
9. Olea, R.: Sampling design optimization for spatial functions. Math. Geol. **16**(4), 369–392 (1984)
10. Poorthuis, A.: Getting Rid of Consumers of Furry Pornography, or How to Find Small Stories With Big Data. North American Cartographic Information Society, Greenville (2013)
11. Rolf, A., Knippers, R.A., Sun, Y., Kraak, M.: Principles of Geographic Information Systems. ITC Educational Textbook, Enschede (2000)
12. Sadahiro, Y.: Advanced Urban Analysis: Spatial Analysis using GIS. Associate Professor of the Department of Urban, Tokyo (2006)
13. Scott, D.W.: Averaged shifted histograms: effective nonparametric density estimators in several dimensions. Ann. Stat. **13**(3), 1024–1040 (1985)
14. Shelton, T., Poorthuis, A., Graham, M., Zook, M.: Mapping the data shadows of hurricane sandy: uncovering the sociospatial dimensions of 'big data'. Geoforum **52**, 167–179 (2014)
15. Wang, T.: Adaptive tessellation mapping (ATM) for spatial data mining. Int. J. Mach. Learn. Comput. **4**, 479–482 (2014)

Incremental Filtering Visualization of JobStreet Malaysia ICT Jobs

M. Bakri[1], Siti Z. Z. Abidin[2(✉)], and Amal Shargabi[2]

[1] Fakulti Sains Komputer dan Matematik,
UiTM Melaka (Kampus Jasin), 77300 Jasin, Melaka, Malaysia
bakri@tmsk.uitm.edu.my
[2] Fakulti Sains Komputer dan Matematik, UiTM Selangor
(Kampus Shah Alam), 40450 Shah Alam, Selangor, Malaysia
zaleha@tmsk.uitm.edu.my, amalphd83@gmail.com

Abstract. In Malaysia, JobStreet is one of the most popular job search engines for job seekers. There are thousands of available positions posted on the website by numerous companies. Due to many choices available for the job seekers, ability to quickly find the jobs that met their criteria such as location, industry, job role and specialization will be a help. Currently, JobStreet only allows job filtering by specialization and state. To get more details for the job, they need to spend significant amount of time reading the descriptions. Thus, in order to assist the users in their job seeking process, this project proposes a system which will use interactive visualization with incremental filtering that consists of heat map and barcharts. The developed system allows users to filter and get an overview of all posted jobs that met their criteria in less time.

1 Introduction

Finding a job in today's market is a major challenge. A common way to look for a job is to use job search engines. Rather than taking the time to search newspapers, company websites, and other traditional job postings, a job search engine can do it all with the click of a button. A job search engine facilitates the matching and communication of job opportunities between jobseekers and employers.

In Malaysia, job seekers use numerous job search engines. One of the leading job search engines among Malaysian job seekers is JobStreet. It was founded in 1997 and it is now Southeast Asia's largest online employment company in Malaysia, Philippines, Singapore, Indonesia and Vietnam. Recently, JobStreet became part of the Australian Stock Exchange-listed SEEK Limited – the world's largest online marketplace by market capitalization. As part of the SEEK family, JobStreet leverages world-class products to match talented job seekers with reputable employers across the region.

When choosing a job, the location of the job has the potential to significantly affect an individual's lifestyle. Thus, location is one of the most important factors to be considered when looking for a new job. People often tend to focus their job search on a particular area. As such, a job search engine worth using need to has the ability to search for suitable jobs based on location. Another attributes that usually be considered when searching for a job is the job role, specialization and the company's industry.

© Springer Nature Singapore Pte Ltd. 2017
A. Mohamed et al. (Eds.): SCDS 2017, CCIS 788, pp. 188–196, 2017.
https://doi.org/10.1007/978-981-10-7242-0_16

In JobStreet search engine, it only allows users to filter the jobs by specialization and state. All the information displayed is in textual form. In order to know the specific location, role and company's industry for each job, users need to spend significant amount of time opening and reading the job details.

We therefore propose a visual analytic system that allows users to incrementally filter data in the visualization during the analysis process. The system comprises of two main modules, data extraction and data visualization. It is designed to help users for analyzing large amount of data with multiple attributes. It also will allow users to select and filter any attributes to help them during analysis process.

2 Literature Review

Nowadays data has been generated at such high volumes in a very fast rate. Most of them are available in digital form. Even simple transactions such as using the telephone and buying groceries in supermarket are typically recorded by computers. Usually, many parameters are recorded, resulting in multivariate data. The data are recorded because it is believed that a potential source of valuable information can be obtained which will give competitive advantage. However, finding the valuable information in them is a difficult task.

With today's data management systems, it is only possible to view small portion of the data at a time. Having no tools to explore the large amount of data collected makes the data become useless. Visualization can help to get overview of the whole data. However, when there is too much information displayed on the visualization, it will hamper the effective retrieval of relevant information. Thus, effectively handling and comprehending such data becomes key to gaining value and insight from them.

2.1 Multivariate Data

Every job information posted in the JobStreet consists of multiple attributes such as position, salary, company, location, etc. This type of data is called multivariate data. Multivariate data is defined as ubiquitous data type which employ the attribute values of homogenous sets of items to describe them specifically. It is often used to record complex real-world data and widely used in all kind of fields [1]. Therefore, analysing the collected data from various kinds of sources is now very important to ensure the continuous advancement of technology.

There are several techniques that can be used to analyse multivariate data. Among the most widely used are cluster analysis, regression analysis and visualization technique [2]. Since the purpose of the project is to allow users to search and filter list of jobs based on their chosen criteria, visualization seems to be the most suitable technique to be used. Cluster analysis and regression analysis provide users with only static results, whereas visualization allow them to interact with the output and dynamically change the output based on their preferences.

2.2   Data Visualization

The fundamental idea of data visualization is to provide data in some visual form that would let the human understand them, gain insight into the data, draw conclusions, and directly influences a further process of decision-making [3]. It capitalizes the inherent and remarkable ability of human's visual perception to identify trends, patterns, and unusual occurrences in datasets [4].

Basic idea of data visualization is to present the data in a visual manner that will reduces the cognitive works required in order to understand it. By leveraging the process of gaining insight of the data through the detection of patterns in visual representation decrease the gap between the data and the user's mental model of it [5].

By using the basis provided by graphic design and advanced statistics, visualization techniques had been growing in strength steadily throughout the years. One of the technique, which is information visualization, had been gaining prominence in recent years. The technique transforms a data sets into interactive interfaces which make it easier for the users to decipher and focus on the information in the data [6].

Visualization of multivariate data are usually generated based on the values of the attributes [2]. As the size and dimensionality of the data sets increases, it will make it easier to comprehend the relationships and patterns of the data if all the attributes are simultaneously projected [7]. It can be achieved by using the proper and suitable visualization techniques. In addition, visualization technique and interactive technologies could make the process of studying the multivariate data to be more efficient by promoting analytical reasoning and easing the understanding of the data [8].

Interactivity has now become an increasingly fundamental part of visualization. When the volume of the multivariate data increases, analysis will become harder. There might me information overload which results in inefficiency during the process of extracting information. Therefore, the interaction with the visualization result will be crucial in order for users to thoroughly analyse the data [9].

2.3   Related Work

There are a few job listing websites that integrate data visualization to show the job availability on a map. The most popular is JobExplorer by Glassdoor Inc. It creates a customized visual map of job opportunities for each category as shown in Fig. 1. From the map, users can click on any state to reveals relevant job openings in that area for the particular category. There is no further information provided on the map.

Another job listing websites that also implement data visualization is Indeed Jobs as in Fig. 2. However, the maps can only be generated for one company's job listing at a time. Users are not able to visualize jobs by category or field. They are only able to know the specific location for the jobs offered by the company.

The interaction with the visualization offered is very limited. Users still have to browse and read the job details manually. All the job details need to be included inside the visualization so that a user doesn't have to leave the visualization when searching for a job. However, placing too much information inside visualization may reduce the effectiveness of it.

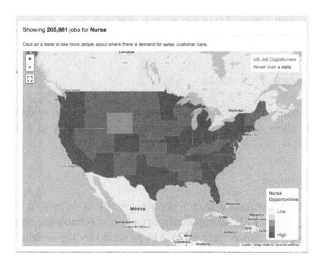

Fig. 1. Glassdoor JobExplorer

Fig. 2. Indeed interactive job map

2.4   Interactive Filtering

Decision making during analysis of data visualization may be hampered if the data representations if ineffective or simply too much raw data [12]. When there is too much data being displayed in a visualization, our visual cognition is constantly utilized in unconsciously processing unnecessary and nonessential visual information [13]. It may overwhelm users' perceptual and cognitive capacities. Reducing this data through sampling or filtering can elide interesting structures or outliers [10].

The main purpose of interaction in data visualization is to decide what data that one wants to see next based on what one has learned in the previous step. Most typically, one may want to see just a subset of the previous data [10]. By doing this, they can focus more on the relevant data during the analysis process. The interactive filtering concept will be implemented inside the system that will allow users to filter the jobs based on their chosen criteria.

3   System Design

This system is designed to help users for analyzing large amount of data with multiple attributes. It will allow users to select and filter any attributes to help them during analysis process.

3.1   Data Preparation

For a case study, data is extracted from a popular job listing website in Malaysia, www. jobstreet.com.my. Jobs with Computer/Information Technology specialization are selected to be extracted. Since there is no API provided by the website to retrieve the data, a crawler and scraper is developed in order to retrieve all the required information. For each listed job in the selected specialization, information about data posted, company address, state, role, industry, and specialization are scraped.

To get a better view on the job distribution in Malaysia, the company address must be translated into GPS coordinate so that it can be plotted into a map. To achieve that, address information that are scraped from the website will be queried using google geocoding API to get the exact location for each job. All these information is stored in JSON format using MongoDB database system.

3.2   System Architecture

The system architecture shows the core part of the system. The architecture is depicted in Fig. 3. There are three main components, data retrieval module, visualization module, and filtering module.

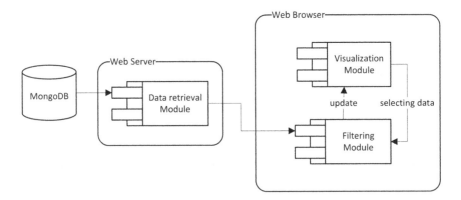

Fig. 3. System architecture

Data retrieval module's purpose is to handle connection with the MongoDB database for retrieving the data. Retrieved data is in JSON format, which will be passed to the filtering module for initialization. The module is developed using python language and placed at the server side.

To assist users in the analysis process, filtering module is used. This module is located in the client side to make sure that there is no overhead in the filtering process. This module allows users to do incremental filtering on the data using the time series and the bar chart. Selecting a specific time frame in time series chart or selecting specific category in bar chart will trigger the filtering module. The module will filter all other chart reflecting the data that has been selected. Clicking at any other place inside the chart will reset the current filtering. During initialization process, the filtering module will index all data to improve the speed of the filtering process. All data are set to be visible initially.

Visualization module will also be located at the client side, and developed with Javascript using D3 visualization library. The module will prepare each attribute of the data to be visualized inside the web browser. Data to be visualized is received by the filtering module. The type of visualization used depends on the type of the attribute. Currently, the system only handles three types of data which shown in the table below (Table 1):

Table 1. Data type and type of visualization used.

Data type	Type of visualization
Data/Time	Time series chart
GPS coordinate	Heat map
Category/Others	Bar chart

4 Exploring ICT Jobs in JobStreet Malaysia

Using the developed crawler and scraper, 953 jobs information are scraped. The data was stored inside the database and will be retrieved by the web system for visualization purpose. Address of the companies were translated into GPS coordinates using Google Geolocation API.

4.1 Main Dashboard

The screenshot of the system using the data is shown in Fig. 4. A chart will represent each attribute in the data. There are six different charts that represent each of the attribute. Date posted will be visualized using time series chart, gps location will be visualized using heat map and other attributes will be visualized using bar chart to represent the count.

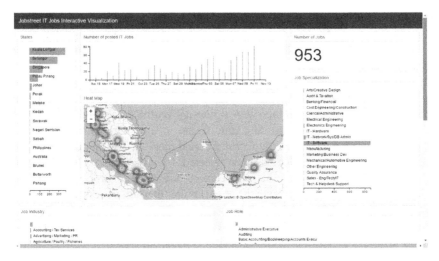

Fig. 4. Main page of the system

4.2   Incremental Filtering

Users can select a specific time frame to show the job posted during that time. As shown in Fig. 5, jobs that are posted between 7 and 15 November 2016 are selected. As shown in the figure, there are 323 jobs that are posted during that time frame.

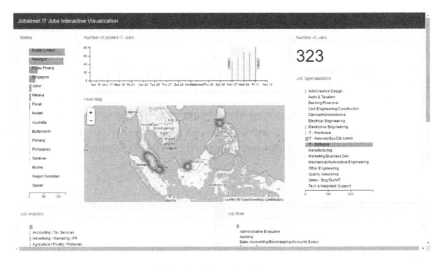

Fig. 5. Filter time frame

Users can also do incremental filtering by selecting another data to be filtered. For example, jobs that are posted between a 7 and 15 November 2016 and located in Pulau Pinang can be viewed by filtering these two attributes. As shown in Fig. 6, there are 29 jobs that met the conditions and those jobs are in Electronic Engineering and IT-Software Specialization. Users can further filter the data using any other attributes available.

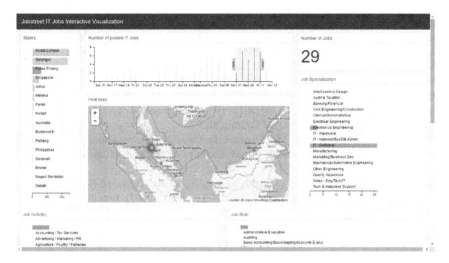

Fig. 6. Filter time frame and state

5 Conclusion

The interactive visualization developed for the JobStreet data successfully integrate all jobs information in one visualization without hampering the ability of users in analyzing them. The interactive filtering module provided allows users to focus only on the jobs that they are interested in. At the same time allow them to further filter the jobs with other attributes and reading the detail job description in one place

Acknowledgment. The authors are grateful to the Research Management Centre (RMC) UiTM for providing and supporting the research (600-RMI/DANA 5/3/REI (16/2015)).

References

1. Jäckle, D., Fischer, F., Schreck, T., Keim, D.A.: Temporal MDS plots for analysis of multivariate data. IEEE Trans. Vis. Comput. Graph. **22**(1), 141–150 (2016)
2. Claessen, J.H., Van Wijk, J.J.: Flexible linked axes for multivariate data visualization. IEEE Trans. Vis. Comput. Graph. **17**(12), 2310–2316 (2011)
3. Metev, S.M., Veiko, V.P.: Laser Assisted Microtechnology, 2nd edn. Springer, Berlin (1998). https://doi.org/10.1007/978-3-642-87271-6. Osgood Jr., R.M. (Ed.)
4. Breckling, J. (ed.): The Analysis of Directional Time Series: Applications to Wind Speed and Direction. Lecture Notes in Statistics, vol. 61. Springer, Berlin (1989). https://doi.org/10.1007/978-1-4612-3688-7

5. Zhang, S., Zhu, C., Sin, J.K.O., Mok, P.K.T.: A novel ultrathin elevated channel low-temperature poly-Si TFT. IEEE Electron Device Lett. **20**, 569–571 (1999)
6. Khan, W., Hussain, F., Prakash, E.C.: Web 2.0 mash-up system for real time data visualisation and analysis using OSS. In: Handbook of Research on Trends and Future Directions in Big Data and Web Intelligence, pp. 208–231. IGI Global (2015)
7. Livingston, M.A., Decker, J.W., Ai, Z.: Evaluation of multivariate visualization on a multivariate task. IEEE Trans. Vis. Comput. Graph. **18**(12), 2114–2121 (2012)
8. Kobayashi, H., Furukawa, T., Misue, K.: Parallel box: visually comparable representation for multivariate data analysis. In: 2014 18th International Conference on Information Visualisation (IV), pp. 183–188. IEEE, July 2014
9. Sjöbergh, J., Li, X., Goebel, R., Tanaka, Y.: A visualization-analytics-interaction workflow framework for exploratory and explanatory search on geo-located search data using the meme media digital dashboard. In: 2015 19th International Conference on Information Visualisation (iV), pp. 300–309. IEEE, July 2015
10. Wegmuller, M., von der Weid, J.P., Oberson, P., Gisin, N.: High resolution fiber distributed measurements with coherent OFDR. In: Proceedings of ECOC 2000, paper 11.3.4, p. 109 (2000)
11. Sorace, R.E., Reinhardt, V.S., Vaughn, S.A.: High-speed digital-to-RF converter. U.S. Patent 5 668 842, 16 September 1997. The IEEE website (2007). http://www.ieee.org/
12. Shell, M.: IEEEtran webpage on CTAN (2007). Signal Processor (MC68175/D), Motorola (1996). http://www.ctan.org/tex-archive/macros/latex/contrib/IEEEtran/FLEXChip
13. "PDCA12–70 data sheet," Opto Speed SA, Mezzovico, Switzerland. Karnik, A.: Performance of TCP congestion control with rate feedback: TCP/ABR and rate adaptive TCP/IP. M. Eng. thesis, Indian Institute of Science, Bangalore, India, January 1999

Software Application for Analyzing Photovoltaic Module Panel Temperature in Relation to Climate Factors

Zainura Idrus[1]($\boxtimes$), N. A. S. Abdullah[1], H. Zainuddin[2], and A. D. M. Ja'afar[1]

[1] Faculty of Computer and Mathematical Sciences,
Universiti Teknologi MARA, 40450 Shah Alam, Selangor, Malaysia
{zainura,atiqah}@tmsk.uitm.edu.my,
daijaafar@gmail.com
[2] Faculty of Applied Sciences, Universiti Teknologi MARA, 40450 Shah Alam,
Selangor, Malaysia
zainuddinhedzlin@gmail.com

Abstract. Discovering factors and relationships that affect the temperature of photovoltaic module panel is vital to solar energy based research in an attempt to generate green energy effectively. Despite the availability of visualization tools that assist in the discovery, they are not dedicated. They require some efforts, time and certain level of skills to gain benefits out of them. Thus, this research has designed and developed a dedicated application system called RAID Explorer to assist the researchers in their attempts to explore various factors that contribute to photovoltaic module temperature. With the system, discoveries are through observation that stimulates human cognitive ability. The system strength's resides in its three main components, which are data entry, visualization and visual analytics. The components are flexible as such, visual interaction is intuitive and users are able to manage it with minimum instruction. They also enable personal customization as such judicious judgment can be made. For performance analysis, the system was implemented and utilized by Green Energy Research Centre (GERC), Universiti Teknologi MARA (UiTM), Malaysia to determine its functionality and usability. The findings show that RAID capabilities are beyond merely to discover the factors contributed to photovoltaic module temperature. In fact, it has successfully supported the discovering of various relationships in the photovoltaic dataset. Thus, RAID Explorer supports scientists' need for flexible, speed and interactive exploration of huge photovoltaic dataset in relation to relationship discovery.

Keywords: Visual analytics · Data visualization · Parallel coordinate graph
Photovoltaic module temperature · Solar energy

1 Introduction

The operations of producing solar energy are dependent on factors such as site, weather and system. Nice bright daylight weather that commonly found in equator region led to better performance. However, the system's performance is periodically changing

© Springer Nature Singapore Pte Ltd. 2017
A. Mohamed et al. (Eds.): SCDS 2017, CCIS 788, pp. 197–208, 2017.
https://doi.org/10.1007/978-981-10-7242-0_17

because of shifting weather conditions [1]. Other aspects that contribute to system's performance are photovoltaic module temperature, shading, cables quality, charge controller, photovoltaic cells, inverter and battery. Higher photovoltaic module temperature (PMT) means more energy is lost due to the ineffectiveness of module cells to convert photon to electric. There are many elements that are devoted to the PMT. These elements are solar radiation, wind speed, gust speed, wind direction, ambient temperature and relative humidity [1]. Even though system located within equator region usually get more photon from sun light, the solar radiation that come within the package unintentionally gave rise to PMT, which increase the power lost. Thus, one of the main concerns of the scientists is to discover on how these factors (independent parameters) affect the temperature of the module (dependent parameter).

The discovering process is very consuming in term of time and manpower. There are ranges of visual tools that are meant to assist scientists in the discovery through observation which speed the process of stimulating human cognitive abilities. However, these tool supports generalized problems thus requires certain level of skill to gain benefit out of it. Even though some tools support simplicity in the usage, they are lacking in the processing ability thus limit the data to a day range at best and up to five parameters for three days' range at most [2]. Moreover, some of the systems support visualization at the static level with minimum interaction ability. Even commonly used tools such as spreadsheet requires time and work in organizing data, plotting and labeling the graphs [3]. Moreover, any manipulation to the graph such as adding new parameter is performed manually. Thus, there is a need to design and develop a system that can assist scientists in the discovering process of PMT in a flexible, speed and interactive manner. To fulfill the need, this research has designed and developed a system named RAID Explorer. Hence, the following sections discuss the issues on the development of the system. It starts with Sect. 2 highlights the background study of visual analytic technology in photovoltaic area of research. Section 3 presents the design and development of RAID Explorer where its main components are discussed. Section 4 discusses the performance analysis of the system before the conclusion remark in Sect. 5.

2 Background Study

Visual analytic is a process of mapping data into its visual form for speed understanding and analysis. It assists in decision making, planning and prediction [4]. The activities of analyzing visual data involved both visual analytic tools and human visual sense. The tools give insights; reveal pattern and relationship toward massive data. Then, human with their cognitive ability, interpret the patterns and relationships [5] which are useful for various purposes in supporting various domains [8–12]. Through this technique, it makes complex data more accessible, understandable and usable. The technique has long time gained its interest in the area of photovoltaic related area of research.

Photovoltaic technology gave much impact to society within recent few decades [13]. The word "photovoltaic" is a combination of two words of "photon" and "voltage" [2]. Photovoltaic technology is a technology that processes the conversion of photon (substance of light) into electric, where it is stored in battery cells or sold to power station. The current issue of the energy is on the installation technology where the margin cost is huge as compared to other energy technology. Nevertheless, the process of designing and installing is straightforward [14, 15]. Photovoltaic technology is also free pollutant product and low maintenance [16]. Thus, contributes to green environment.

2.1 Existing Photovoltaic Systems in Visual Analytic Approach

Normally, photovoltaic system runs as a stand-alone system. But, it is often combined with electricity power grid. Having back-up by other power source, enhance the reliability of the system [2].

In Bogota Columbia, the research developed a system to monitor the energy production by photovoltaic system has already started. The system is user-friendly and provides the visualization of real-time data which are acquired by sensors. The sensors measure temperature, voltage, AC of the inverter, solar panel DC, solar radiation, power that generated by photovoltaic modules and the amount of delivered power [17]. However, the system functionality on monitoring module temperature is too general, thus, not applicable for critical study. It became costly to analyze PMT through this system.

Another system named Geographic Information System (GIS) is used in Punjab to determine the applicable location for photovoltaic system installation to maximize the output [18]. The system is integrated with Google Earth to prepare for the predictive function of spatial data to analyze the geographic environment. This system is available in internet browser, mobile and desktop which prove to have great flexibility in supporting different type of users. However, the system strength is on its ability to analyze geographical need for effective photovoltaic output and not photovoltaic monitoring functions.

Other researchers focus on troubleshooting based research since there are many factors need to be considered in order to ensure the effectiveness of photovoltaic system in long period term. It is not rare for a photovoltaic system to not functioning as it was supposed to while troubleshooting consume extra costs and time. Hence, detecting any components failure from photovoltaic arrays, batteries and to utility connection in real time would be beneficial [19]. However, the core of this system is more to maintenance and troubleshooting purposes and not for data exploration purposes.

Researcher in the area of automated analysis on the other hand focuses on efficiency in energy extraction from monitoring photovoltaic systems which connected to power grid. For instance a system called S-lar is dedicated to manage data monitoring. First, the data that has been integrated from all devices such as meters, inverters and environmental sensors are stored in Relational Database Management System (RDBMS). Then, through its numerical and graphical visualization tool, the system automatically access all the information and represent them in a series of time scales [20]. The representations automatically determine the standard measurements that are related to power losses and performance of photovoltaic system. Unfortunately, the system required large amount of training data that appeared to be expensive or difficult to obtain.

PQMS, on the other hand, provide high resolution in long term period of the data on momentary appearance that could occur in the inverter's DC and AC conditions. The system continuously monitor data measured every 200 ms which are stored in local server. The PQMS effectively used to analyze and visualize these data. The system helps local distribution companies (LDCs) to improve asses to power which has been generated by photovoltaic system. LDCs also used PQMS to authenticate the claims of compliances subjected to inverter manufacturers [21]. Nevertheless, the system's main objective is to analyze inverter to produce solar energy effectively.

In short, the available systems and tools are inadequate to support the discovering process of PMT and its relationships. Thus, there is a need to have a visual analytics system to support the discovering process. The system assist scientist in understanding out climate, thus, actions can be taken for the benefits of our society and world as a whole.

3   Method

This section starts by defining the dataset used. Then the discussions are on compo-nents that make up RAID Explorer.

3.1   Definition

In order to find the need in the data for RAID design and development, a set of data has been collected. The data is from a center called Green Energy Research Centre (GERC) located at high educational institution MARA University of Technology, Malaysia [2]. It is a center dedicated for photovoltaic energy base research mainly to effectively produce electric from solar energy.

Most of the GERC projects are funded by government entities such as Ministry of Higher Education (MOHE), Technology & Innovation (MOSTI), Ministry of Science, and Ministry of Energy, Green Technology and Water. The data are acquired through various logs sensors of photovoltaic system. The collection processes are initiated every five minute, every day, for the whole year starting from the first day of the project. Thus, the amounts of data are growing rapidly over time.

There are a total of 111,151 dataset collected for the period of 39 days and they are stored in Excel format. Let the data be denoted as D. The data are divided into rows and column. Let row be denoted as R as such

$$D \rightarrow R \qquad (1)$$

On the other hand, column C stores nine parameters of each dataset $r \in R$ and C = {*Solar Radiation (W/m^2), Wind Speed (m/s), Gust Speed (m/s), Wind Direction (ø), Ambient temp (°C), Relative Humidity (%), date, time,* $c_x, c_{x+1,} c_{x+2} ... c_n$}. Thus

$$D \rightarrow R \, x \, C \qquad (2)$$

In addition, C can either be in one of the two available positions as such $P = \{inde\text{-}pendent, dependent\}$ as such

$$D \to R \, x \, C \, x \, P. \tag{3}$$

Moreover, C also varying in format which can be denoted as F, thus

$$D \to R \, x \, C \, x \, P \, x \, F \tag{4}$$

3.2   Design and Development

There are three main components, which have been designed and developed for RAID Explorer. There are flexible data entry, visualization, and visual analytics component.

3.2.1   Data Entry Component

To support flexibility, RAID allowed scientist to upload their data into the system with minimum requirement as shown in Fig. 1. Users just have to click on the 'Browse' button, locate the file and confirm the selection. The component accepts two types of files which are CSV and Excel. The Excel file will be converted to CSV prior to processing. The parameters $c \in C$ does not have to be in any particular order. However, if date is a combination of date and time in a single parameter, the component is responsible to detect them and parse them into two separate parameters which are date and time. The separation enabled independent axes in the graph where particular range of data can be explored.

Fig. 1. Upload function

Thus, data entry component is a function responsible to input data and convert them into suitable format and can be represent formally as

$$DE : D \to C \, x \, F$$
$$DE(c_a, f_b) = \{\langle c, f, DE(c, f) \rangle | \, c \in C, f \in F\} \tag{5}$$

3.2.2   Visualization Component

Photovoltaic data not only huge but its parameters are also varies in types and formats. Thus, this multivariate data suited best with parallel coordinate as a mean for visualization. Each of its parameter $c \in C$ is represented by an axis and it is depicted in its own measurement and scaling factor regardless others parameters' data format. In situation where big data are used, these parameters axis can handle data representation as to reduce clutter of information. In term of time taken to analyze big data, this method is well ahead compared to other methods [22]. Moreover, there are many techniques that can be applied to maximize parallel coordinate analysis. Hence, visualization component is a function that transform $D \rightarrow D'$ where D' are data in its visual format.

To develop the graph, there are nine steps which begun with creating an empty canvas. The canvas is the medium for displaying the data. In second step, data $r \in R$ is read and parameter's quantity $c \in C$ reside within the data are determined. The quantity is used again in third step where axes created are depended on the quantity. The maximum and minimum of data $r_{max}, r_{min} \in R$ in each parameter $c \in C$ are extracted in fourth step to visualization algorithm since they are needed in fifth to give scale to axis created in third step. In sixth step, each cell of data is parsed to their own format and mapped on the axis. The formats are crucial to ensure user readability of the data in the graph. In seventh step, the mapped data are connected by drawing line according to their row. These lines are colored in eighth to provide connection view of the data where these connections are vital for analysis. In the last step, each parameter's name is read and displayed at the top of its axis.

Thus, visualization component is a function that convert data D into its visual form D' as such

$$DC(d) = \{\langle d, DC(d) \rangle | d \in D\} \tag{6}$$

3.2.3   Visual Analytics Component

The third component is responsible to provide interactive analysis where scientists can transform the graph into its different versions hence can be viewed from different perspectives. Five visual analytics methods have been found suitable for discovering relationship between the independent and dependent parameters of photovoltaic dataset. They are brushing, filtering, ranging, coloring and sorting.

(a) *Brushing*

Region in multiple parameters are selected using brushing technique. These selected regions and others whose lines are associated with it are highlighted.

The function is applied on the axes where users can have one brush on one axis and can be extended to all axes. The lines that user brushed are highlighted and non-brushed are shadowed. User can still see the shadowed lines but it is not in the focus. The purpose of the showed is to create awareness of the selected data as relative to the whole data [23, 24]. Figure 2(a) is the design of brushing technique. The orange arrow on the left in Fig. 2(b) shows the shadowed lines and the blue arrow on the right, shows the highlighted lines after two brushes are applied on *SR* and *time* axes.

(b) *Filtering and Ranging*

The parameters created in parallel coordinates can be removed using filtering technique to reduce clutter. While brushing highlighted selected region, ranging technique remove unwanted range of data.

(a)

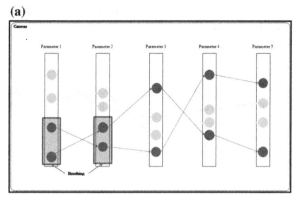

(b)

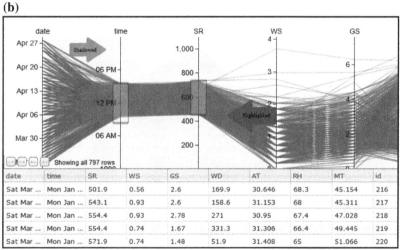

Fig. 2. (a) Brushing design. (b) Brushing function – two brushes.

For example, parameter 4 in Fig. 3 is removed from the graph after check box of parameter 4 in table is unchecked. The process can be set by default or by users.

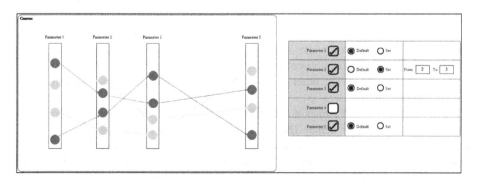

Fig. 3. Filtering and ranging function.

Once the graph is loaded, the default algorithm is called and all checkboxes input are checked. The checked values mean that the axes are visible. The axes are hidden if its checkboxes values are unchecked.

The ranging process starts when a user clicks *Set* option, enter new value of the parameter and click *Update* button. The changes before and after the process are shown in Fig. 4. The scale of the module temperature is updated to a new scale by a controller. The data which are outside the range of the new scale are removed. This enables more detail exploration.

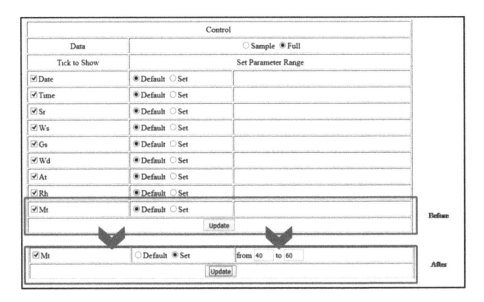

Fig. 4. Ranging function – control.

(c) *Coloring*

The purpose of line color is to create obvious relationship between dependent and independent parameters. In this case, since module temperature is the dependent parameter, the coloring is anchor to its axis.

The green arrow in Fig. 5(a) indicates temperature. Higher data position in axis means hotter temperature while lower position means cooler temperature. To show hotter temperature, the lines are colored with more red as it goes higher while lower are colored with more blue. The other data that are related to PMT parameter also followed the coloring technique.

The coloring is performed automatically according to the last parameter in the dataset regardless of its name. The color representation is according to the level of hotness of each data. The lines of data sets that make up the high temperature are in red color, while in opposite side they are lean toward blue color. Referring to Fig. 5(b), most of date sets that make up high module temperature come from a particular range

(a)

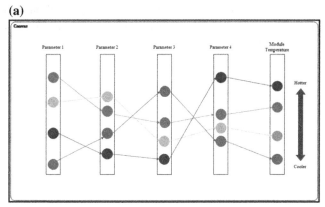

(b)

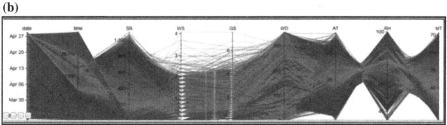

Fig. 5. (a) The design of coloring technique. (b) Coloring function on the data. (Color figure online)

of time. Let *VA* be the function to perform manipulation on data rows, columns and parameters:

$$VA : D' \rightarrow R \, x \, C \, x \, P$$
$$VA(r_a c_b, p_c) = \{\langle r, c, p, VA(r, c, p)\rangle | r \in R, c \in C, p \in P\} \tag{7}$$

RAID Explorer, *RE*, therefore is a functions that support discovery process through visual exploration as such:

$$RE = (DE(c, f), \, DC(d), (VA(r, c, p)) \tag{8}$$

4   Performance Analysis

The graph is designed to cater for CSV file format and each of its data rows and columns are mapped onto their respective axes. Moreover, the graph can further be transformed onto its various versions to assist in the discovering process.

In order to access the system performance, it has undergone two types of testing, which are functionality and usability. Functionality testing is performed in two phases, which are unit testing and integration testing. For unit testing, there are eight different units testing that have been undertaken. They are parsing, upload, mapping, brushing, filtering, ranging, color, and sorting. Their purposes and expected outcomes are summarized in Table 1. Next, integration testing is performed to ensure all the functionalities work collaboratively as a whole.

Table 1. Functionality testing performed on RAID explorer.

Component	Functions	Processes	Expected Outcome
One	Parse	Split the parameter *Date Time* to *Date* and *Time* and convert the original data in excel format to CSV format	The original parameter is spited and new CSV format is formed
	Upload	Locate, select and upload CSV file	Selected CSV file is loaded
Two	Map	Parallel coordinates graph is created	Present data in parallel coordinates graph
Three	Sort	Select axes by double click the axes to activate sort function	The axes are sorted ascending or descending
	Brush	Mouse arrow is clicked in axis and drag to upper or lower as necessary	Highlight selected particular range of lines on axes while the other lines are shadowed
	Filter	Trigger function to show or hide axes when user checked or unchecked checkboxes	Hide axes whose checkboxes are unchecked
	Range	Select the range of data to be activated	Hide data that are not selected.
	Color	Coloring the lines from blue to red according to their respective of module temperature values	The lines are colored with blue to red according to values of module temperature

Finally, usability testing is conducted in GERC where five scientists are asked to input their own dataset and identified factors that contribute to PMT (dependent parameter). Then, they are to select other dependent parameters and identify factors that lead to the production of the dependent parameters.

The result proves that RAID Explorer's with its three components support flexible, speed understanding and ease in discovering relationship among various independents and dependent parameters of photovoltaic dataset.

5   Conclusion

The result proves that RAID Explore is beyond merely to discover the factors that contribute to photovoltaic module temperature. In fact, it has successfully supported the discovering of various relationships in the photovoltaic dataset. Besides, it is useful to discover other relationships among photovoltaic dataset, thus, make the system flexible. Therefore, RAID Explorer can support the scientists' need for flexible, speed, and interactive exploration of huge photovoltaic dataset in relation to relationship discovery.

Acknowledgments. The authors would like to thank Universiti Teknologi MARA, Malaysia and Ministry of Higher Education Malaysia for the facilities and financial support under the national grant 600-IRMI/FRGS 5/3 (022/2017), FRGS/1/2017/1CT04/UITM/02/2.

References

1. Ramawan, M.K., Othman, Z., Sulaiman, S.I., Musirin, I., Othman, N.: A hybrid bat algorithm artificial neural network for grid-connected photovoltaic system output prediction. In: Proceedings of the 2014 IEEE 8th International Power Engineering and Optimization Conference, PEOCO 2014, pp. 619–623 (2014)
2. Hussain, T.N., Sulaiman, S.I., Musirin, I., Shaari, S., Zainuddin, H.: A hybrid artificial neural network for grid-connected photovoltaic system output prediction, pp. 108–111 (2014)
3. Imazawa, A., Naoe, N., Hanai, H., Nakajima, S.: Software application to help make discoveries based on air temperature data, pp. 276–281 (2016)
4. Sklar, E.: TR-2009005: Visual Analytics : A Multi-Faceted Overview (2009)
5. Chen, M., Ebert, D., Hagen, H., Laramee, R.S., Van Liere, R., Ma, K.-L., Ribarsky, W., Scheuermann, G., Silver, D.: Data, information, and knowledge in visualization. Comput. Graph. Appl. IEEE **29**(1), 12–19 (2009)
6. Chen, C.: Visual analytics. In: Chen, C. (ed.) Mapping Scientific Frontiers, pp. 321–339. Springer, London (2013). https://doi.org/10.1007/978-1-4471-5128-9_9
7. Keim, D., Andrienko, G., Fekete, J.-D., Görg, C., Kohlhammer, J., Melançon, G.: Visual analytics: definition, process, and challenges. In: Kerren, A., Stasko, J.T., Fekete, J.-D., North, C. (eds.) Information Visualization. LNCS, vol. 4950, pp. 154–175. Springer, Heidelberg (2008). https://doi.org/10.1007/978-3-540-70956-5_7
8. Yafooz, W.M., Abidin, S.Z., Omar, N., Idrus, Z.: Managing unstructured data in relational databases. In: Conference on Systems, Process & Control (ICSPC), pp. 198–203. IEEE (2013)
9. Kamaruddin, N., Wahab, A., Quek, C.: Cultural dependency analysis for understanding speech emotion. Expert Syst. Appl. **39**(5), 5115–5133 (2012)
10. Aliman, S., Yahya, S., Aljunid, S.A.: Innovation using rasch model in measuring influence strategies of spiritual blogs. In: 5th International Conference on Information and Communication Technology for the Muslim World (ICT4M), pp. 1–5. IEEE (2013)
11. Abdullah, N.A.S., Rusli, N.I.A., Ibrahim, M.F.: A case study in COSMIC functional size measurement: angry bird mobile application. In: IEEE Conference on Open Systems (ICOS), pp. 139–144 (2013)
12. Idrus, Z., Abidin, S.Z.Z., Omar, N., Idrus, Z., Sofee, N.S.A.M.: Geovisualization of non-resident students' tabulation using line clustering. In: Regional Conference on Sciences, Technology and Social Sciences (RCSTSS) (2016)
13. Bagnall, D.M., Boreland, M.: Photovoltaic technologies. Energy Policy **36**(12), 4390–4396 (2008)
14. Yu, Z.J., Fisher, K.C., Wheelwright, B.M., Angel, R.P., Holman, Z.C.: PVMirror: a new concept for tandem solar cells and hybrid solar converters. IEEE J. Photovolt. **5**(6), 1791–1799 (2015)
15. Parida, B., Iniyan, S., Goic, R.: A review of solar photovoltaic technologies. Renew. Sustain. Energy Rev. **15**(3), 1625–1636 (2011)

16. Hamakawa, Y.: Thin-Film Solar Cells: Next Generation Photovoltaics and Its Applications. Springer Science & Business Media, Heidelberg (2013). https://doi.org/10.1007/978-3-662-10549-8
17. Diaz, R.R., Jutinico Alarcón, A.L., Moreno, R.J.: Monitoring system for global solar radiation, temperature, current and power for a photovoltaic system interconnected with the electricity distribution network in Bogota. In: 2013 IEEE 56th International Midwest Symposium on Circuits and Systems (MWSCAS), pp. 485–488 (2013)
18. Luqman, M., Ahmad, S.R., Khan, S., Akmal, F., Ahmad, U., Raza, A., Nawaz, M., Javed, A., Ali, H.: GIS based management system for photovoltaic panels. J. Geogr. Inf. Syst. 7(4), 392–401 (2015)
19. Zhao, Y., Ball, R., Mosesian, J., de Palma, J.-F., Lehman, B.: Graph-based semi-supervised learning for fault detection and classification in solar photovoltaic arrays. IEEE Trans. Power Electron. 30(5), 2848–2858 (2015)
20. Trillo-Montero, D., Santiago, I., Luna-Rodriguez, J.J., Real-Calvo, R.: Development of a software application to evaluate the performance and energy losses of grid-connected photovoltaic systems. Energy Convers. Manag. 81, 144–159 (2014)
21. Kandil, S., Farag, H.E., Hilaire, L.S., Janssen, E.: A power quality monitor system for quantifying the effects of photovoltaic penetration on the grid, pp. 237–241 (2015)
22. Raidou, R.G., Eisemann, M., Breeuwer, M., Eisemann, E., Vilanova, A.: Orientation-enhanced parallel coordinate plots. IEEE Trans. Vis. Comput. Graph. 22(1), 589–598 (2016)
23. Idrus, Z., Abidin, S.Z.Z., Hashim, R., Omar, N.: Social awareness: the power of digital elements in collaborative environment. J. WSEAS Trans. Comput. 9(6), 644–653 (2010)
24. Idrus, Z., Abidin, S.Z.Z., Hashim, R., Omar, N.: Awareness in networked collaborative environment: a comparative study on the usage of digital elements. In: Proceedings of 9th WSEAS International Conference on Applications of Computer Engineering, pp. 236–241 (2010)

Budget Visual: Malaysia Budget Visualization

Nur Atiqah Sia Abdullah[(⊠)], Nur Wahidah Ab Wahid,
and Zainura Idrus

Faculty of Computer and Mathematical Sciences, Universiti Teknologi MARA,
40450 Shah Alam, Selangor, Malaysia
atiqah@tmsk.uitm.edu.my

Abstract. Data visualization is an efficient method to analyze complex data. It eases the reader to identify data patterns. Based on the previous studies, countries such as United States, China, Russian, and Philippines have used data visualization method to present their budgets. However, Malaysia Budget is still reported in speech text, which is difficult for Malaysian citizen to understand the overall distribution. Therefore, this paper proposed treemap technique to visualize the Malaysia Budget using dynamic approach. This technique is chosen as it allows the users to explore the data interactively, which covers various ministries. The result shows that treemap technique works well and presents the whole budget effectively. This study can be further improved by performing the comparison between previous and the current budget.

Keywords: Visualization · Budget data visualization · Treemap
Malaysia Budget

1 Introduction

Data visualization is a field of information that can represent data in graphic form to ease the viewers in identifying data patterns. Data visualization is very significant because human brain processes information more efficiently in visual forms. Human brain recognizes shapes and patterns of large and complex data instantly. Moreover, data visualization is easier to be understood than data in text form [1].

Data visualization is a quick and fast method to analyze concepts in a first glance. It can also identify areas that need attention and improvement. Data visualization allows the users to execute process interactively. It can reflect the information pattern, data relation, data change trend, thus help scientific researchers and government officers to view and analyze data in the style of direct graphs to find the principles hiding behind the data [2].

Data visualization has changed the way analysts work with data. In order to respond to the issues more rapidly, the data analysts usually need to mine the data for more insights by looking at data from different perspective and more imaginatively. Data visualization promotes data exploration. For example, by using data visualization, businesses are able to analyze large amounts of data and devive significant conclusions from the data pattern. Since it is considerably faster to analyze information in graphical format, therefore it can address problems or answer questions in a timelier manner [1].

© Springer Nature Singapore Pte Ltd. 2017
A. Mohamed et al. (Eds.): SCDS 2017, CCIS 788, pp. 209–218, 2017.
https://doi.org/10.1007/978-981-10-7242-0_18

Most of the government information is presented using texts in articles or graphical information. For example, in Malaysia Budget (BAJET 2017), the citizen has to understand the information through speech texts provided in the official website [3]. They need time to read the budget in the speech article. Some of the local newspapers tried to simplify the budget information using information graphic (info graphic) approach [4, 5]. Info graphic is one of the visual representations of information, data and knowledge [1]. However, these info graphics are not comprehensive to represent the whole distribution of the budget.

This study aims to visualize the Malaysia Budget in a more dynamic approach. It displays the whole distribution of Malaysia Budget 2017 based on the ministries. This study helps the citizen to view the information at the first glance. It is also imperative for them to understand the overall budget through data visualization.

2 Reviews on Budget Visualization

Data visualization techniques are usually used to represent data, which are useful for various purposes in supporting various domains [6, 7]. Normally these techniques can be divided into Maps, Temporal, Multidimensional, Hierarchical/Tree and Network. Since the study focuses on the budget visualization, the literature reviews are closely related to the budget or distribution. This study reviewed previous studies in budget distribution, which include Obama's 2016 budget [8], Philippines budget [9], Russian 2015 budget [10], US Federal budget [11] and China government budget [12].

Obama's 2016 budget [8], as shown in Fig. 1, uses treemap data visualization technique. Treemap is suitable for interactive budget, because it allows pop up box to display details when a viewer hovers over the smallest rectangle. In order to use a treemap, all the data sets should be in positive values [13]. In this case, Obama budget uses rectangles to represent name of category for each distribution. Therefore, when hovering over on each rectangle, it shows a pop up box with total amount of budget. The amount of budget distribution depends on the size of the rectangle. It also uses colors to differentiate the departments. It is easy to see the portion among the distribution based on the sizing. Normally, when using treemap, it is a challenge to view the labels for the smaller rectangles in the first glance (see Fig. 1). However, treemap allows the viewers to read the details using pop up box. Moreover, this treemap visualization is not suitable for budget comparison.

Philippines use a dynamic column chart to show the comparison of the economic indicators from 2014 until 2017 [9], as shown in Fig. 2. Bar chart can be the best alternative for relative information [14]. It is more well-known technique because it is easy to understand and generally it is useful to show information that is arranged into ordinal classification and comparison between categories of data [14]. As in Fig. 2, time series is plotted on bar charts. There are two main goals in time series analysis, which is identifying the nature of the phenomenon represented by the sequence of observations, and forecasting such as predicting future values of the time series variable [15]. Based on both visualization techniques, the bar chart represents the changes (in percentage) for the economic indicators, while the time series indicates the total amount (in million pesos). Besides, when hovering over the chart, it shows

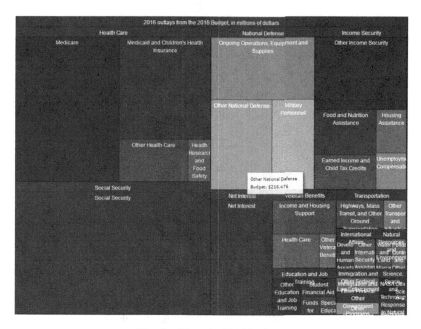

Fig. 1. Obama's 2016 budget.

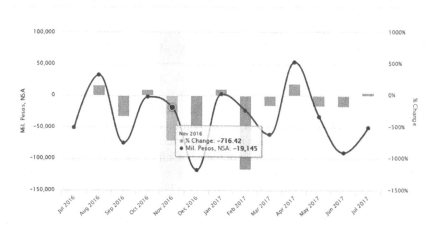

Fig. 2. Philippines government budget balance.

the detail of each indicator and its respective amount. However, it does not show the overall budget for each year.

Figure 3 shows Russian 2015 budget [10]. This budget uses various data visualization techniques to represent the distribution of budget. There are many choices for the user to view the budget, which include circle packing, bar chart, pie chart, table, scatter plot, and time series.

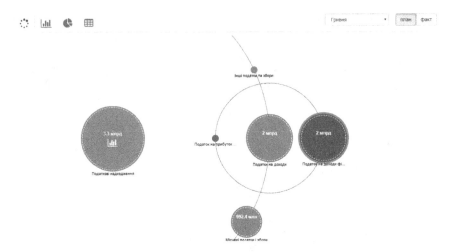

Fig. 3. Russian budget 2015.

Circle packing is a variation of treemap that uses circles instead of rectangles [16]. Every branch of the tree is represented as a circle and its sub branches are represented as surrounding circles. The area of each circle can be used to represent an additional random value, such as quantity or file size. Color may also be used to assign categories or to represent another variable via different shades [16]. In Russian 2015 budget, each circle represents different categories and the size of the circle depends on the amount of the budget. Besides, it can zoom in to the circle to show the details of specific department when the user mouse over the circle. Besides, it is more interactive using circle packing. However, by using circle packing, it cannot directly display all the distributions of budget as in treemap.

US Federal 2013 Budget [11] uses collapsible tree layout, which is one of the hierarchical or tree techniques. Collapsible tree layout is more dynamic than treemap because it is equipped with zoom in and out functionalities for the distribution. Any parent node can be clicked on to collapse the portion of the tree (child node) below it. Conversely, it can be clicked on parent node to return the original state. Besides, the size of the node depends on the amount of budget. Collapsible tree layout is suitable for the situation where there is a comparison among categories. In this case, it shows federal, state and local categories of US budget (see Fig. 4). However, the viewer has to click on each node to expand the child nodes. It is not suitable to view the overall budget at once. Nevertheless, the fully collapsed tree will be too crowded if the data involves too many categories.

China government budget from 2006 until 2016 [12] uses column chart to show the comparison of the government budget throughout the years (see Fig. 5). Column chart would be a good alternative because it is easy to understand and generally useful for showing comparison between categories [14]. In this budget website, it provides other options to view the budget, include line graph, area chart, candlestick, histogram, and etc. Instead of increment and decrement of the budget, there is a comparison between previous budgets and the current budget. However, this visualization does not provide

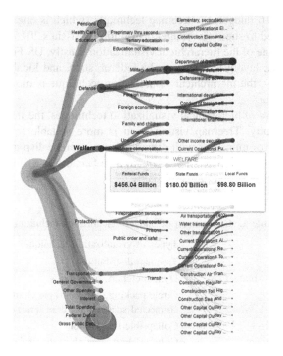

Fig. 4. US federal 2013 budget.

any details about the budget. Thus, this kind of chart is only suitable to show the overall budget distribution throughout a series of times.

From the literature, Table 1 shows the summary of the data visualization techniques used in each budget.

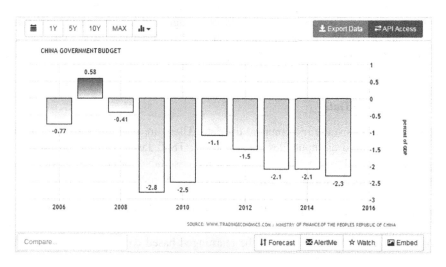

Fig. 5. China government budget 2006–2016.

In Obama's 2016 budget, the treemap technique (which is one of the hierarchical visualization) is used to show the budget distribution. Russia's 2015 budget uses circle packing, which is one of the hierarchical visualization. Lastly, US Federal 2013 budget uses collapsible tree layout to visualize the federal, state, and local budget. Based on the literature review, the hierarchical visualization technique is more suitable to visualize the budget data.

From the reviews of hierarchical visualization techniques, the treemap technique is selected for this study. Treemap visualization is more suitable for Malaysia budget because the number of ministries with different sectors can be displayed at once. Circle packing and collapsible tree layout are not suitable because it will be difficult to view the overall distribution of the budget directly.

Table 1. Comparison of budget visualization techniques.

Country	Types of visualization techniques
Obama 2016 budget	Treemap distribution
Philippine 2010–2013 budget	Dynamic chart, time series
Russia 2015 budget	Circle packing, bar chart, pie chart, table, connected scatter plot, time series
US federal 2013 budget	Collapsible tree layout
China 2006–2016 budget	Column chart, area chart, line graph, histogram

3 Treemap Representation for Budget

After the reviews, the treemap visualization technique is chosen for visualizing the Malaysia Budget. Treemap is generated using a straightforward algorithm known as slice-and-dice [17]. It is area-based visualization, either sliced horizontally or vertically, to create smaller rectangles with area dependent upon the value of a particular weighting attribute. To prepare the visualization page, this study used D3.js features. D3.js is data driven documents approach that use JavaScript library, HTML, Scalable Vector Graphics (SVG) and Cascading Style Sheet (CSS) [18] for manipulating document based on data.

3.1 Prepare the Data

The first step in this study is to prepare the data. There are many formats of input data file that can be used as input files such as CSV, TSV, Jason, and etc. In this study, Jason file is selected as input file type. Jason file is suitable for collection of data with many clusters, which allows sub sections for each category.

In this step, data cleaning is very important. In Malaysia Budget speech text, the distribution is based on priorities. Each priority involves many ministries. Therefore, the distribution for each ministry cannot be defined clearly if the visualization is based on priorities. Thus, the data needs to be rearranged based on ministries.

After data cleaning, the data have to be clustered to ease the visualization. Therefore, the data are rearranged in specific format for the visualization. In order to use a treemap, a hierarchical tree structure has to be defined. The variables are created to store the information, such as "name", "size", "details" and etc. In this case, "name" is used to store the ministry name; "size" stores the size of each rectangle that needs to be visualized. The size is based on the amount of distribution. There are parents and children in the Jason file. Parent consists of ministry name, which is known as header. Children are known as the slices or nodes. Each ministry is represented with different colors.

3.2 Create the Layout

Next is to create a canvas in CSS style. It consists of canvas of the rectangle, which has SVG element. Then, it continues with setting treemap diagram properties, which starts with creating treemap layout. It needs several stages to produce the treemap layout. The treemap layout consists of the combination of height and width, which represented the entire rectangle. The process continues by creating a container to map the treemap layout. Scale function is used to set the size of the rectangles. This function takes an interval and transforms it into a new interval. Then, data is assigned to the nodes before the development of filtering function. Data filtering is needed to differentiate children and parent nodes. Data is known as "children", which it has no child. For the children, the screen area is sliced to create smaller rectangles with area dependent upon the value of a particular weighting attribute. Data with child is known as "parent" or "header". After data filtering, parents and children nodes need to be created for clustering data. Besides, the tooltip class for the parents and children also need to be created.

3.3 Set the Transitions and Tooltip

The following step is to complete the transition parts, include enter, update, exit, and zoom transition. Each transition needs parents and children. "Enter transition" is needed to call the zoom function. "Update transition" is used to set up the attribute of treemap after zooming. "Exit transition" is used to stop the transition. "Zoom transition" focuses on zoom in and out. "Zoom in" is used to display the specific data. "Zoom out" is called when the user wants to see the whole treemap layout.

Next step is to create the tooltip box. Tooltip box will appear when the user hovers over the header (parents) and rectangle (children). Tooltip is useful to show the details for some small rectangles. The tooltip will disappear when the user move away from the rectangle and header. After the tooltip is completed, text wrapping is created to display the data on the rectangle with readable format. Foreign object is used to allow the text wrapping.

3.4 Set the Color

Lastly, color function is created in order to differentiate the data clusters. In this study, data are clustered based on ministry. There are 23 different colors used in this study. The color function is created with the intention to differentiate the ministries. Finally, the Jason file is integrated in D3.js. It is needed to visualize the treemap diagram. The d3.json function is used to load a file and returns its contents as Jason data object. Figure 6 shows the process of setting the treemap algorithm.

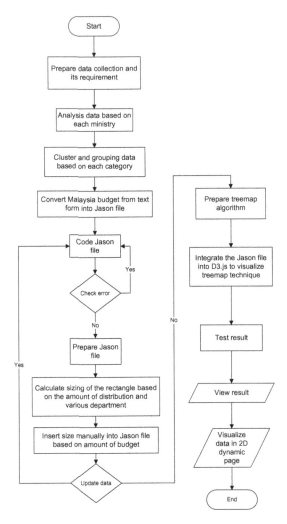

Fig. 6. Flow chart for setting treemap webpage.

4   Malaysia Budget Visualization

This study has created data visualization for Malaysia Budget 2017. In the treemap, the size of the rectangle is based on the amount of budget allocation and different colors are used to represent the 23 ministries. The headers of the rectangles display the ministries' names and the respective distributions. The total budget allocation is based on the budget stated in Malaysia Budget, which is in speech text.

When the user hovers over the header of the rectangle, the ministry name and its budget are displayed on the tooltip box. The tooltip box is very useful since some headers cannot be displayed clearly as in Fig. 7.

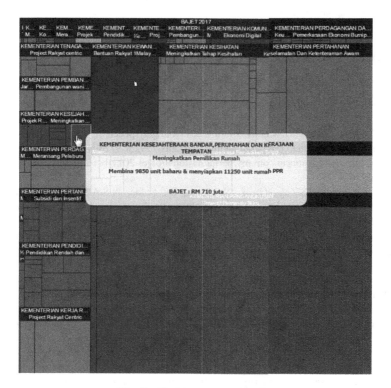

Fig. 7. Tooltip to show details.

When the user double clicks on the webpage, the zoom in function will display the second layer of information that contains the specific distribution of the selected ministry. In the second layer, the tooltip box is displayed when the user hovers over the rectangle. The detail of the budget is displayed in tooltip. It can ease the reader to understand all the information on the portion.

5   Conclusion

This study has applied the data visualization technique to help visualize the Malaysia Budget. The data visualization technique shows the potential to be used in any of the domain that contain large amount of data. It helps to promote creative data exploration by significantly reflect the information pattern, data relation, and data change trend. This study can be further improved by enhancing the visualization techniques. The visualization can be improved by adding more layers to help the user to have more details about the budget. Furthermore, the visualization of Malaysia budget also can be improved by comparing the previous and the current budget.

Acknowledgments. The authors would like to thank Faculty of Computer and Mathematical Sciences, Universiti Teknologi MARA and Internal Grant (600-IRMI/DANA 5/3/LESTARI (0125/2016) for sponsoring this paper.

References

1. Gohil, A.: R Data Visualization Cookbook. Packt Publishing, Birmingham (2015)
2. Kirk, A.: Data Visualization: A Successful Design Process. Packt Publishing, Birmingham (2012)
3. Bajet Malaysia. http://www.bajet.com.my/2017-budget-speech/
4. Bernama. https://twitter.com/bernamadotcom/status/799238372694704129
5. Utusan. http://www.utusan.com.my/berita/nasional/bajet-2017-br1m-dinaikkan-kepada-rm1-200-1.397664
6. Idrus, Z., Abidin, S.Z.Z., Omar, N., Idrus, Z., Sofee, N.S.A.M.: Geovisualization of non-resident students tabulation using line clustering. In: Regional Conference on Sciences, Technology and Social Sciences, Malaysia, pp. 251–261 (2016)
7. Abdullah, N.A.S., Shaari, N.I., Abd Rahman, A.R.: Review on sentiment analysis approaches for social media data. J. Eng. Appl. Sci. **12**(3), 462–467 (2017). https://doi.org/10.3923/jeasci.2017.462.467
8. Office of Management and Budget. https://obamawhitehouse.archives.gov/interactive-budget
9. Philippines Government Expenditures. https://www.economy.com/philippines/government-expenditures
10. Russian Budget Visualization. http://openbudget.in.ua/public/budget_files/55a818d06b617309df652500?locale=uk
11. US Federal Budget 2013. http://www.brightpointinc.com/2013-federal-budget/
12. China Government Budget. http://www.tradingeconomics.com/china/government-budget
13. Burch, M.: Interactive similarity links in treemap visualizations. In: 18th International Conference on Information Visualisation IV, Paris, pp. 34–39. IEEE (2014). https://doi.org/10.1109/IV.2014.35
14. Keim, D.A., Hao, M.C., Dayal, U.: Hierarchical pixel bar charts. IEEE Trans. Vis. Comput. Graph. **8**(3), 255–269 (2002). https://doi.org/10.1109/TVCG.2002.1021578
15. Haroz, S., Kosara, R., Franconeri, S.: The connected scatterplot for presenting paired time series. IEEE Trans. Vis. Comput. Graph. **22**(9), 2174–2186 (2015). https://doi.org/10.1109/TVCG.2015.2502587
16. Carrabs, F., Cerrone, C., Cerulli, R.: A tabu search approach for the circle packing problem. In: 17th International Conference on Network-Based Information Systems, Italy, pp. 165–171. IEEE (2014). https://doi.org/10.1109/NBiS.2014.28
17. Asahi, T., Turo, D., Shneiderman, B.: Visual decision-making: using treemaps for the analytic hierarchy process. In: Companion on Human Factors in Computing Systems, pp. 405–406. ACM, New York (2003). doi:10.1145/223355.223747
18. D3 Data-Driven Documents. https://d3js.org/

Fuzzy Logic

Fuzzy Arithmetical Modeling of Pressurizer in a Nuclear Power Plant

Wan Munirah Wan Mohamad[1], Tahir Ahmad[1,2(✉)],
Niki Anis Ab Rahman[1,2], and Azmirul Ashaari[1]

[1] Department of Mathematical Science, Faculty of Science,
Universiti Teknologi Malaysia, 81310 UTM Skudai, Johor, Malaysia
tahir@ibnusina.utm.my
[2] Centre for Sustainable Nanomaterials,
Ibnu Sina Institute for Scientific and Industrial Research,
Universiti Teknologi Malaysia, 81310 UTM Skudai, Johor, Malaysia

Abstract. A pressurizer is to control pressure and temperature in a nuclear power plant. In this paper, a model using fuzzy arithmetic which is based on the Transformation Method (TM) is presented. The analytical solution is used to evaluate the state space model. The TM is then used to quantify the influence of each parameter, and the estimation of relative and absolute measures of uncertainty. To ensure the operating system is maintained and efficient, the method is applied to a pressurizer for simulation and analysis. Simulation and analysis of TM's efficiency are presented in this paper.

Keywords: Fuzzy arithmetic · Measure of influence and pressurizer

1 Introduction

Nuclear power plant generates electricity. There are many types of nuclear power plant in the worldwide. The most popular is the pressurized water reactor (PWR) which is studied in this work. A nuclear power plant or nuclear power station is a thermal power station which produces heat. The heat is used to generate steam which drives steam turbine to produces electricity [1]. There are two types of PWR, namely, primary and secondary systems. The primary system consists of five components namely pressurizer, reactor vessel, steam generator, main turbine, and main condenser. However, only modelling of a pressurizer will be presented in this paper.

The function of a pressurizer is to ensure the temperature of moderator and pressure of PWR is maintained. The pressurizer has two main functions in primary system, namely to maintain the pressure within the specified limits at 123 *bar* and to control the coolant mass flow in the primary circuit [2]. The pressure is monitored either by increasing the amount of bubble in pressurizer through heating or increasing the steam condensation through water spray [3, 4] (*see* Fig. 1).

In this paper, a reduced transformation method is introduced to model a pressurizer in a nuclear power plant. The transformation method is described in Sect. 2. Section 3 presents the application of the pressurizer in a nuclear power plant. The conclusion for the modeling is given in the last section of this paper.

© Springer Nature Singapore Pte Ltd. 2017
A. Mohamed et al. (Eds.): SCDS 2017, CCIS 788, pp. 221–229, 2017.
https://doi.org/10.1007/978-981-10-7242-0_19

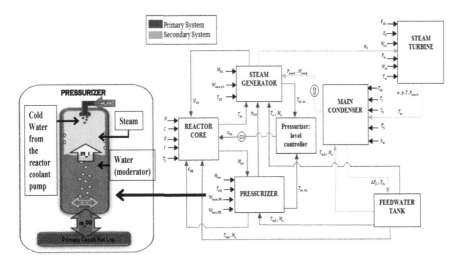

Fig. 1. Pressurizer of PWR [5].

2 The Transformation Method

They are two types of transformation method (TM): general and reduced forms. This method can be used to evaluate static or dynamic systems with fuzzy valued parameters. In this paper, the reduced TM is preferred since the characteristic of input parameters show the monotonic behavior. The method was proposed in [6–11].

The interval [0, 1] of the μ-axis is subdivided into m intervals of length $\Delta\mu = \frac{1}{m}$ and the discrete μ_j of the (m + 1) levels of membership are given by

$$\mu_j = \frac{j}{m}, \quad j = 0, 1, 2, \ldots, m. \tag{1}$$

The input fuzzy numbers $\tilde{p}_i$ with $i = 1, 2, \ldots, n$, are decomposed into α-cuts, where each set P_i consists of the (m + 1) intervals.

$$P_i = \left\{ X_i^{(0)}, X_i^{(1)}, \ldots, X_i^{(m)} \right\} \tag{2}$$

$$X_i^{(j)} = \left[a_i^{(j)}, b_i^{(j)} \right], \quad a_i^{(j)} \le b_i^{(j)}, \tag{3}$$

$$i = 1, 2, \ldots, n, \quad j = 0, 1, \ldots, m.$$

The intervals are transformed into arrays $\hat{X}_i^{(j)}$ in the form

$$\hat{X}_i^{(j)} = \left(\overbrace{\alpha_i^{(j)}, \beta_i^{(j)}, \alpha_i^{(j)}, \beta_i^{(j)}, \ldots, \alpha_i^{(j)}, \beta_i^{(j)}}^{2^{i-1}\,pairs} \right) \tag{4}$$

$$\text{with} \quad a_i^{(j)} = \underbrace{\left(a_i^{(j)}, \ldots, a_i^{(j)}\right)}_{2^{n-i}\,elements}, \beta_i^{(j)} = \underbrace{\left(b_i^{(j)}, \ldots, b_i^{(j)}\right)}_{2^{n-i}\,elements} \tag{5}$$

The evaluation is then carried out at each of 2^n positions of the combination using the conventional arithmetic for crisp numbers. It can be its decomposed and transformed by as combination $\hat{Z}^{(j)}$, $j = 0, 1, \ldots, m$, with k^{th} element of the array $\hat{Z}^{(j)}$ is given as

$$^k\hat{Z}^{(j)} = F(^k\hat{x}_1^{(j)}, {}^k\hat{x}_2^{(j)}, \ldots, {}^k\hat{x}_n^{(j)}) \atop k = 1, 2, \ldots, 2^n, \tag{6}$$

Finally, the fuzzy-valued result $\tilde{q}$ of the expression can be determined in its decomposed form

$$Z^{(j)} = [a^{(j)}, b^{(j)}], \quad a_i^{(j)} \le b_i^{(j)}, \quad j = 0, 1, \ldots, m \tag{7}$$

The corrected are obtained by the recursive formulas

$$a^{(j)} = \min_k\left(a^{(j+1)}, {}^k\hat{z}^{(j)}\right), \tag{8}$$

$$b^{(j)} = \max_k\left(b^{(j+1)}, {}^k\hat{z}^{(j)}\right), \quad j = 0, 1, \ldots, m - 1 \tag{9}$$

$$a^{(m)} = \min_k(^k\hat{z}^{(m)}) = \max_k(^k\hat{z}^{(m)}) = b^{(m)} \tag{10}$$

Then the reduced transformation method is used for the analysis. The coefficients $\eta_i^{(j)}$, $i = 1, 2, \ldots, n$, $j = 0, 1, \ldots, m - 1$, are determined

$$\eta_i^{(j)} = \frac{1}{2^{n-1}\left(b_i^{(j)} - a_i^{(j)}\right)} \sum_{k=1}^{2^{n-i}} \sum_{l=1}^{2^{i-1}} ({}^{s_2}\hat{z}^{(j)} - {}^{s_1}\hat{z}^{(j)}) \tag{11}$$

$$s_1(k, l) = k + (l - 1)2^{n-i+1} \tag{12}$$

$$s_2(k, l) = k + (2l - 1)2^{n-i} \tag{13}$$

The values $a_i^{(j)}$ and $b_i^{(j)}$ denote the lower and upper bound of the interval $X_i^{(j)}$, and $^k\hat{z}^{(j)}$ respectively. The coefficients $\eta_i^{(j)}$ are gain factors. Furthermore mean gain factors $\eta_i^{(j)}$ is given as

$$\bar{\eta}_i = \frac{\sum_{j=1}^{m-1} \mu_j \eta_i^{(j)}}{\sum_{j=1}^{m-1} \mu_j} \tag{14}$$

Finally, the degree of influence ρ_i is determined for $i = 1, 2, \ldots, n$, using

$$\rho_i = \frac{\sum_{j=1}^{m-1} \mu_j \left| \eta_i^{(j)} \left(a_i^{(j)} + b_i^{(j)} \right) \right|}{\sum_{q=1}^{n} \sum_{j=1}^{m-1} \mu_j \left| \eta_q^{(j)} \left(a_q^{(j)} + b_q^{(j)} \right) \right|} \tag{15}$$

$$\sum_{i=1}^{n} \rho_i = 1 \tag{16}$$

The reduction of fuzzy arithmetic to multiple crisp-number operations indicates that the transformation method can be implemented into a suitable software environment [12, 13]. Hanss [14] successfully developed a software called FAMOUS, the decomposition and transformation can be handle by the software. Figure 2 shows the structure of FAMOUS.

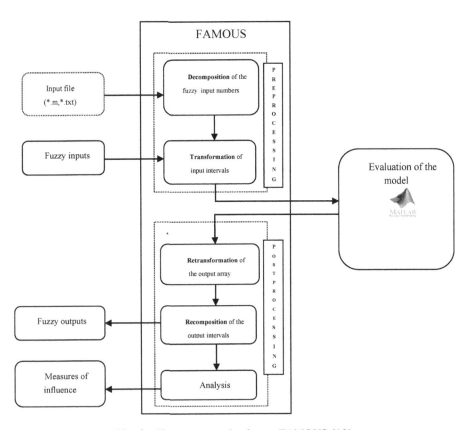

Fig. 2. The structure of software FAMOUS [10].

3   Application to a Pressurizer in a Nuclear Power Plant

The model by Fazekas in [15] is used in this study. The state space of pressurizer by [1] as below:

$$
\begin{pmatrix} dM_{PR}/dt \\ dM_{PC}/dt \\ dT_{PR}/dt \end{pmatrix} = \begin{pmatrix} -\dfrac{V_{pC}^0 c_{\varphi,1}\left(m_{in}T_{pc,l}-m_{out}\right)}{M_{pc}M_{pR}} & \dfrac{(m_{in}-m_{out})}{M_{pc}} & -\dfrac{V_{pC}^0 c_{\varphi,1}\left(c\psi+6K_{T,SG}T_{SG}-W_{loss,pc}\right)}{c_{p,pc}M_{pc}T_{pR}} \\ \dfrac{\varphi\left(T_{pC}\right)V_{pC}^0}{M_{PR}} & \dfrac{(m_{in}-m_{out})}{M_{PC}}-1 & 0 \\ 0 & 0 & (X_{m_{PR}>0}-1) \end{pmatrix} \begin{pmatrix} M_{PR} \\ M_{PC} \\ T_{PR} \end{pmatrix}
$$

$$
+ \begin{pmatrix} a & 0 & 0 \\ 0 & 0 & 0 \\ \dfrac{X_{m_{PR}>0}c_{p,PC}}{c_{p,PR}} & \dfrac{-1}{c_{p,PR}M_{PR}} & \dfrac{1}{c_{p,PR}M_{PR}} \end{pmatrix} \begin{pmatrix} T_{PC} \\ W_{loss,PR} \\ W_{heaty,PR} \end{pmatrix}
$$

$$\tag{17}$$

$$
a = \left(\dfrac{V_{pC}^0 c_{\varphi,1}\left(c_{p,pc}m_{in}+6K_{T,SG}\right)}{c_{p,pc}M_{pc}} - \dfrac{2V_{pC}^0 c_{\varphi,2}\left(m_{in}T_{pc,l}-m_{out}\right)}{M_{pc}} - \dfrac{2V_{pC}^0 c_{\varphi,2}\left(c\psi+6K_{T,SG}T_{SG}-W_{loss,pc}\right)}{c_{p,pc}M_{pc}} + \dfrac{2V_{pC}^0 c_{\varphi,2}\left(c_{p,pc}m_{in}+6K_{T,SG}\right)T_{PC}}{c_{p,pc}M_{pc}} \right)
$$

The output is

$$
\begin{pmatrix} l_{PR} \\ p_{PR} \end{pmatrix} = \begin{pmatrix} \dfrac{-V_{pC}^0}{M_{PR}A_{PR}} & \dfrac{1}{\varphi(T_{PC})A_{PR}} & 0 \\ 0 & 0 & p_*^T \end{pmatrix} \begin{pmatrix} M_{PR} \\ M_{PC} \\ T_{PR} \end{pmatrix}
$$

$$\tag{18}$$

In order to reduce computing time, a model reduction approach is applied by using integrating factor. The output is given by Eq. (18). Such that

$$
p_{PR} = p_*^T T_{PR}
$$

$$\tag{19}$$

From the Eq. (17), the equation of T_{PR} is calculated.

$$
dT_{PR}/dt = (X_{m_{PR}>0}-1)T_{PR} + \frac{X_{m_{PR}>0}c_{p,PC}}{c_{p,PR}}T_{PC} + \frac{-1}{c_{p,PR}M_{PR}}W_{loss,PR} + \frac{1}{c_{p,PR}M_{PR}}W_{heaty,PR}
$$

$$\tag{20}$$

$$
\mu = e^{\int\left(X_{m_{PR}>0}-1\right)dt}
$$

$$\tag{21}$$

$$
\mu = e^{\left(X_{m_{PR}>0}-1\right)t}
$$

$$\tag{22}$$

$$
\left(e^{\left(X_{m_{PR}>0}-1\right)t}\right)\left(T_{PR}+(X_{m_{PR}>0}-1)T_{PR}\right) = \frac{X_{m_{PR}>0}c_{p,PC}}{c_{p,PR}}T_{PC} + \frac{-1}{c_{p,PR}M_{PR}}W_{loss,PR} + \frac{1}{c_{p,PR}M_{PR}}W_{heaty,PR}
$$

$$\tag{23}$$

$$
\int T_{PR}\left(T_{PR}e^{\left(X_{m_{PR}>0}-1\right)t}\right) = \int \left(e^{\left(X_{m_{PR}>0}-1\right)t}\right)\left(\frac{X_{m_{PR}>0}c_{p,PC}}{c_{p,PR}}T_{PC} + \frac{-1}{c_{p,PR}M_{PR}}W_{loss,PR} + \frac{1}{c_{p,PR}M_{PR}}W_{heaty,PR}\right)dt
$$

$$\tag{24}$$

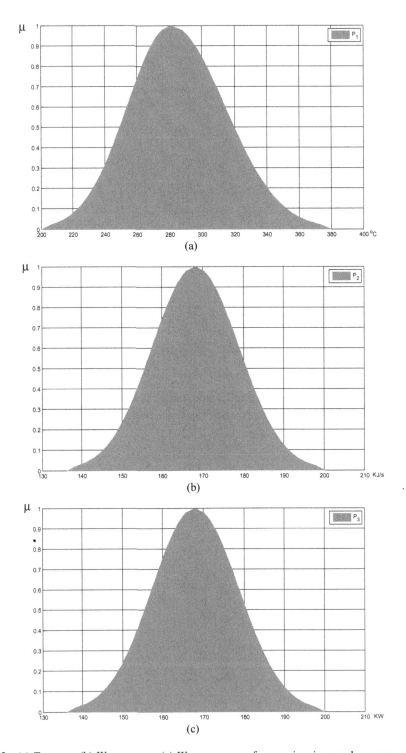

Fig. 3. (a) $T_{PC} = p_1$ (b) $W_{loss,PR} = p_2$ (c) $W_{heaty,PR} = p_3$ of pressurizer in a nuclear power plant.

$$T_{PR}e^{\left(X_{m_{PR}>0}-1\right)t} = \frac{\left(e^{\left(X_{m_{PR}>0}-1\right)t}\right)\left(\frac{X_{m_{PR}>0}c_{p,PC}}{c_{p,PR}}T_{PC} + \frac{-1}{c_{p,PR}M_{PR}}W_{loss,PR} + \frac{1}{c_{p,PR}M_{PR}}W_{heaty,PR}\right) + C}{\left(X_{m_{PR}>0} - 1\right)}$$

(25)

$$T_{PR} = \frac{\left(\frac{X_{m_{PR}>0}c_{p,PC}}{c_{p,PR}}T_{PC} + \frac{-1}{c_{p,PR}M_{PR}}W_{loss,PR} + \frac{1}{c_{p,PR}M_{PR}}W_{heaty,PR}\right)}{\left(X_{m_{PR}>0} - 1\right)} + Ce^{-\left(X_{m_{PR}>0}-1\right)t} \quad (26)$$

Substitute Eq. (26) into (19) yields

$$p_{PR} = p_*^T \left(\frac{\left(\frac{X_{m_{PR}>0}c_{p,PC}}{c_{p,PR}}T_{PC} + \frac{-1}{c_{p,PR}M_{PR}}W_{loss,PR} + \frac{1}{c_{p,PR}M_{PR}}W_{heaty,PR}\right)}{\left(X_{m_{PR}>0} - 1\right)} + Ce^{-\left(X_{m_{PR}>0}-1\right)t}\right)$$

(27)

3.1   Simulation Results of a Pressurizer System

The parameters of input pressurizer are expressed by $p_1 = T_{PC}$ (temperature), $p_2 = W_{loss,PR}$ (heat loss), and $p_3 = W_{heaty,PR}$ (heating power) whereby their membership functions are derived from Gaussian distribution as illustrated in Fig. 3.

The output variables $\tilde{q} = p_{PR}$ (pressure) is obtained and presented in Fig. 4. The fuzzy output p_{PR} does not show significant variations in distribution shape when compared to the original symmetric quasi-Gaussian distribution of the uncertain input parameters p_1, p_2, and p_3.

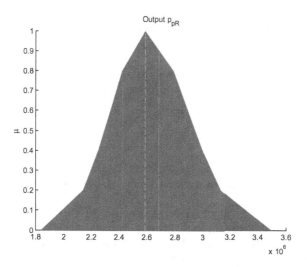

Fig. 4. The output variable $\tilde{q} = p_{PR}$ (pressure) of the pressurizer in a nuclear power plant.

3.2   Analysis Results of a Pressurizer System

The analysis of the pressurizer is given in Fig. 5. The output ρ_1 (T_{PC}) has the degree of influence is almost 100%. The measure of influence of the uncertainties for $\rho_2 = W_{loss,PR}$, and $\rho_3 = W_{heaty,PR}$ are less than 1% and can be negligible.

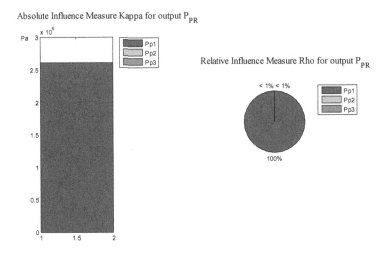

Fig. 5. The absolute and relative influence measure for the output p_{PR} of pressurizer in a nuclear power plant.

4   Conclusion

The reduced transformation method is applied on a pressurizer in a nuclear power plant. The influences of the $\rho_1 = T_{PC}$ (temperature), $\rho_2 = W_{loss,PR}$ (heat loss), and $\rho_3 = W_{heaty,PR}$ (heating power) are determined.

Acknowledgments. The authors are thankful to Universiti Teknologi Malaysia for providing necessary environment and thankful to the FRGS grant (4F756) for financial support.

References

1. Ashaari, A., Ahmad, T., Mohammad, W.M.W.: Transformation pressurized water reactor (AP1000) to fuzzy graph. In: International Seminar on Mathematics in Industry (ISMI), 1–2 August 2017 (2017)
2. Ashaari, A., Ahmad, T., Shamsuddin, M., Zenian, S.: Fuzzy state space model for a pressurizer in a nuclear power plant. Malays. J. Fundam. Appl. Sci. **11**, 57–61 (2015)
3. Ashaari, A., Ahmad, T., Shamsuddin, M., Mohammad, W.M.W., Omar, N.: Graph representation for secondary system of pressurized water reactor with autocatalytic set approach. J. Math. Stat. **11**(4), 107–112 (2015)

4. Ashaari, A., Ahmad, T., Shamsuddin, M., Mohammad, W.M.W.: An autocatalytic model of a pressurized water reactor in a nuclear power generation. In: Berry, M.W., Mohamed, A.H., Wah, Y.B. (eds.) SCDS 2015. CCIS, vol. 545, pp. 106–115. Springer, Singapore (2015). https://doi.org/10.1007/978-981-287-936-3_11
5. Gábor, A., Hangos, K., Szederkényi, G.: Modeling and identification of the pressurizer of a VVER nuclear reactor for controller design purposes. In: 11th International PhD Workshop on Systems and Control (2010)
6. Wan Mohamad, W.M., Ahmad, T., Ahmad, S., Ashaari, A.: Simulation of furnace system with uncertain parameter. Malays. J. Fundam. Appl. Sci. **11**, 5–9 (2015)
7. Wan Mohamad, W.M., Ahmad, T., Ashaari, A., Abdullah, A.: Modeling fuzzy state space of reheater system for simulation and analysis. In: Proceedings of the AIP Conference, vol. 1605, pp. 488–493 (2014)
8. Hanss, M.: The transformation method for the simulation and analysis of systems with uncertain parameters. Fuzzy Sets Syst. **130**, 277–289 (2002)
9. Hanss, M.: Applied Fuzzy Arithmetic. Springer, Heidelberg (2005)
10. Hanss, M., Oliver, N.: Simulation of the human glucose metabolism using fuzzy arithmetic. In: 19th International Conference of the North American Fuzzy Information Processing Society, NAFIPS, pp. 201–205. IEEE (2000)
11. Hanss, M., Oliver, N.: Enhanced parameter identification for complex biomedical models on the basis of fuzzy arithmetic. In: IFSA World Congress and 20th NAFIPS International Conference, Joint 9th, vol. 3, pp. 1631–1636. IEEE (2001)
12. Wan Mohamad, W.M., Ahmad, T., Ashaari, A.: Modeling steam generator system of pressurized water reactor using fuzzy arithmetic. In: Berry, M.W., Hj. Mohamed, A., Yap, B. W. (eds.) SCDS 2016. CCIS, vol. 652, pp. 237–246. Springer, Singapore (2016). https://doi. org/10.1007/978-981-10-2777-2_21
13. Wan Mohamad, W.M., Ahmad, T., Ahmad, S., Ashaari, A.: Identification of the uncertain model parameter of a steam turbine system. In: International Conference on Statistics in Science, Business and Engineering, Kuala Lumpur, pp. 150–155 (2015)
14. Hanss, M., Turrin, S.: A fuzzy-based approach to comprehensive modeling and analysis of systems with epistemic uncertainties. Struct. Saf. **32**, 433–441 (2010)
15. Fazekas, C., Szederkényi, G., Hangos, K.M.: A simple dynamic model of the primary circuit in VVER plants for controller design purposes. Nucl. Eng. Des. **237**, 1071–1087 (2007)

Nitrogen Fertilizer Recommender
for Paddy Fields

Mohamad Soffi Abd Razak, Shuzlina Abdul-Rahman,
Sofianita Mutalib$^{(\boxtimes)}$, and Zalilah Abd Aziz

Faculty of Computer and Mathematical Sciences, Universiti Teknologi MARA,
40450 Shah Alam, Selangor, Malaysia
{shuzlina, sofi, zalilah}@tmsk.uitm.edu.my

Abstract. There are many factors and ways to increase the quality and quantity
of paddy yields. One of the factors that can affect the quality of paddy is the
amount of fertilizer used. The optimum amount of fertilizer for any field in any
year cannot be determined with absolute certainty, thus in this project, we aim to
find the optimum amount of nitrogen fertilizer required in paddy fields. Problems that are characterized by uncertainty can be solved by using fuzzy expert
system. We develop fuzzy expert system prototype that utilizes Mamdani-style
inference where the combination of nitrogen fertilizer data contain factors and
rules, would produce results based on user's input. The data which were in form
of paddy fields images were captured by an Unmanned Aerial Vehicle (UAV) or
commonly known as drone and variables applied in fuzzy rules are obtained
from a thorough analysis made with team of agriculture experts from Malaysian
Agricultural Research and Development Institute (MARDI).

Keywords: Agriculture · Fuzzy expert systems · Fertilizer
Aerial images and paddy

1 Introduction

Global rice industry has produced 672.0 million tonnes of paddy in 2010 [1]. In
Malaysia, rice is vital in terms of economic activity, food supplements and culture for
some ethnic groups. Ninety percent of the people in this country consume rice.
Although Malaysia produces rice, the quantity of paddy yields is not enough to support
the needs of people in the country. There are many factors and ways to increase the
quality and quantity of the paddy yields. One of the factors that can affect the quality of
the paddy is the amount of fertilizer used. There is a relatively higher correlation
between fertilizer usage and paddy price than the correlation between fertilizer usage
and fertilizer price as been mentioned in [2, 3].

A new tool known as the Leaf Color Chart (LCC) described in [4] is used to identify
the optimum amount of nitrogen fertilizer that is needed for rice crops. It is an easy tool
where the nitrogen is used as a key indicator for the leaf's color. LCC is a color panel
that has four green strips, as shown in Fig. 1. It has a range of colors from yellow green
to dark green. This panel will measure the greenness of the rice leaf, which indicates its
nitrogen content. The current practices of using the LCC to determine the leaf color is

© Springer Nature Singapore Pte Ltd. 2017
A. Mohamed et al. (Eds.): SCDS 2017, CCIS 788, pp. 230–240, 2017.
https://doi.org/10.1007/978-981-10-7242-0_20

time consuming. Only one person is allowed to check the leaf color in one block of paddy field in order to avoid multiple results that might confuse farmers. This process would require many farmers as they need to cover many blocks of paddy fields. Besides, given the farmer's limited knowledge and expertise, the technique is hard to be employed by most farmers. Hence, there are only few people in Malaysia willing to use the LCC tool. More than that, the color range of the paddy leaves is very hard to be detected as there are many factors that can mislead the eyes such as sunlight and shadow.

Fig. 1. Using LCC to access paddy leaf [5]. (Color figure online)

This is a twofold study, which proposes the use of aerial images to cover a wide expanse of paddy fields, and the use of fuzzy logic to counter the problem of ambiguous knowledge and further develop an expert system to store the farmer's knowledge about the LCC. Our final aim is to develop an engine that can recommend the optimum amount of nitrogen fertilizer based on the selected region. In Sect. 2, we describe the background studies. In Sect. 3, the method and experimental setup is reported and the results are presented in Sect. 4. The discussion of the result is elaborated in Sect. 5 and finally, we conclude and summarize this study in Sect. 6.

2   Background Studies

Paddy is the staple food for most of the population in the world, including Malaysia. In Malaysia, paddy plays an important role in our country's agrarian economy and diet intake. As Malaysia's staple food and cereal crops, efforts such as crop management is done to ensure its continuous supply [6]. The increase in the rate of crop defect is one of the real concerns faced by farmers. According to the International Rice Research Institute (IRRI), there are over 99 paddy defects consisting of paddy diseases, pests and many more. One of the most common causes for paddy defects is the poor fertilizer management in paddy fields. According to [7], LCC is widely used by these three most established institutions, International Rice Research Institute, Zhejiang Agricultural University, and University of California Cooperative Extension. LCC is usually a plastic, ruler-shaped strip containing four or more panels that range in color from yellowish green to dark green. LCC is a simple and non-destructive, farmer friendly tool which can be used to monitor leaf colour. The amount of nitrogen is measured by visual comparison looking at the panel value closest in color to a leaf indicates whether nitrogen is deficient, sufficient, or in excess. The approach of LCC is highly dependent on the user's relative color perception and the incorrect threshold leaf greenness may not lead to accurate amount of Nitrogen [5, 8–10]. LCC is highly dependent on the person's relative color perception and it limit to a time of eight to ten in the morning and three onwards in the afternoon for better reading and LCC helps guide farmers for real-time nitrogen management in rice farming [11].

2.1 Fuzzy Set and Fuzzy Rules

Fuzzy set theory was introduced by Zadeh to handle uncertainties in real world and it is a precise logic of imprecision and approximate reasoning [12]. A fuzzy set can be simply defined as a set with fuzzy boundaries [13]. Let's X be the universe of discourse and its elements be denoted as x. In classical set theory, crisp set A of X is defined as function $f_A(x)$ called the characteristic function of A:

$$f_A(x) : X \rightarrow 0, 1, \tag{1}$$

where

$$f_A(x) = \begin{cases} 1, & \textit{if } x \in A \\ 0, & \textit{if } x \notin A \end{cases}$$

In fuzzy theory, fuzzy set A of universe X is defined by function $\mu_A(x)$ called the membership function of set A:

$$\mu_A(x) : X \rightarrow [0, 1] \tag{2}$$

where

$\mu_A(x) = 1$ if x is totally in A;
$\mu_A(x) = 0$ if x is not in A;
$0 < \mu_A(x) < 1$ if x is partly in A.

A fuzzy rule can be defined as a conditional statement in the form:

IF x is A THEN y is B

where x and y are linguistic variables while A and B are linguistic values which are determined by fuzzy sets on the universe of discourse X and Y respectively. Reasoning in fuzzy rules involve two distinct parts in which the rule antecent (the IF part of the rule) is evaluated first followed by the application of the result to the consequent (the THEN part of the rule).

2.2   Fuzzy Inference

Fuzzy inference can be defined as a process of mapping a given input to an output based on the theory of fuzzy sets [13]. Mamdani method is the most commonly used fuzzy inference technique which was introduced in 1975 [12]. This method was performed in four steps: fuzzification of the input variables, rule evaluation, aggregation of the rule and defuzzification. The fuzzification process transforms the crisp input and determines the degree of membership for the linguistic terms of fuzzy sets which it belongs to. The crisp input is always a numerical value limited to the universe of discourse which usually determined by the expert. Once the crisp input is fuzzified, it will be applied to the antecedents of the fuzzy rules. The fuzzy operator, AND or OR is used to obtain a single number that represents the result of the antecedent evaluation if a given fuzzy rule has multiple antecedents. The resulted number is then applied to the consequent membership function by clipping to the level of the truth value of the rule antecedent. The aggregation of the rule is a process of unification of the outputs of all rules in which the membership functions of all rule consequents are combined into a single fuzzy set. The last process is the defuzzification in which the aggregated output fuzzy sets is transformed into a crisp number. The commonly use method for defuzzification is the centre of gravity (COG) as expressed below:

$$\text{COG} = \frac{\int_a^b \mu_A(x)x\,dx}{\int_a^b \mu_A(x)\,dx} \tag{3}$$

The COG method finds a point that represents the centre of gravity of the fuzzy set, A, on the interval, ab.

3   Method

This section discusses the main activities, consisting of data collection of aerial images of paddy fields, image pre-processing, feature extraction, the flow of processes and experimental setup. The variables used in this research are the results from a thorough analysis made with a team of agriculture experts from the Malaysian Agricultural Research and Development Institute (MARDI).

The paddy fields images were collected twice time involving paddy fields in Kepala Batas, Pulau Pinang and those in Sungai Besar, Kuala Selangor, Selangor. The images were captured by an Unmanned Aerial Vehicle (UAV) or commonly known as drone. The drone used in this study has a flying camera that is able to capture 12 megapixel photos with 1280×720 dimensions and 72 dpi resolution. It is also equipped with professional lens with f/2.8 lens that has a $94°$ field of view, virtually eliminating unwanted distortion that may occur and a stable footage for securing and stabilizing the camera. The paddy field images and the altitude of which each image was taken were recorded. All of these images were subjected to the image pre-processing techniques involving image cropping, resizing, mean shift color generalization and grey scaling [14]. The input variables later were set as explained below:

- Color – refers to the color of the paddy leaves extracted from the images taken from the drone.
- Height – refers to the altitude of the flying drone.
- Light intensity – refers to the intensity of the light measured using the Lux Meter.
- Size – refers to the size of the paddy field.

The maximum altitude for the UAV has been set at 30 m and the time of the image taken is between 8 am to 1 pm. The output variable refers to the amount of nitrogen fertilizer. Forty image samples of paddy fields were used in this experiment and the sample of images is shown in Fig. 2.

Fig. 2. Sample of images that is captured by UAV. (Color figure online)

The images were divided into four groups that represent four color ranges of the Leaf Color Chart (LCC), which are very dark green, dark green, green and light green, with 10 samples of images for each range. Due to the difficulty in locating the field that is lacking in nitrogen fertilizers, a few modified samples was generated using image editing tools by altering the field's color. This process involved two panel colors, very dark green color and light green, with reference to LCC panel. In order to test whether the output of the model is accurate, we calculate how many times the model give correct or similar nitrogen amount (with threshold in 5 kg) as compared to the actual amount needed with the knowledge of the farmers or experts.

3.1 Fuzzy Linguistic Variables

In gaining the knowledge on representing the fuzzy sets using linguistic variable, expert from Paddy Excellent Center, Malaysian Agricultural Research and Development Institute (MARDI) in Kepala Batas, Pulau Pinang were identified. Based on discussion with them, we identified the representation of light intensity, height of drone and size of paddy field as in Table 1. Eventually, LCC color representation is using the rules of thumb in image processing which is based on gradient scale of grayscale. Next, farmers from Sungai Besar were also interviewed in understanding the amount of fertilizer to given for different colour range in LCC.

Table 1. Representation of height of drone, light intensity, size of paddy field and LCC color.

Linguistic variable	Height (m)	Light intensity (lux)	Paddy field size (hectare)	LCC color
Low	30–50	100–160	1.0–1.2	NA
Medium	45–84	150–200	1.15–1.35	NA
High	80–100	190–250	1.3–1.5	NA
Very dark green	NA	NA	NA	1–60
Dark green	NA	NA	NA	40–120
Green	NA	NA	NA	100–210
Light green	NA	NA	NA	190–255

*NA is referred to no fuzzy sets with the linguistic variable for that column.

3.2 Fuzzy Rules

In this model, 108 rules have been constructed based on the four linguistic variables with their fuzzy linguistic values using Mathlab Fuzzy Toolbox. Fuzzy rules work in the precedent and consequential approaches that refer to the prepositions that contain linguistic variables. The Mamdani fuzzy rules were applied in the form of IF *<fuzzy preposition for input_variable1>* AND *<fuzzy preposition for input_variable2>* THEN *<fuzzy preposition for output_variable>*. The following Table 2 shows two examples of inference rules. The fuzzy output is defuzzified using the Center of Gravity (COG) method in the Mathlab Fuzzy Toolbox.

Table 2. Fuzzy set rules

Sample	Representation
Rule 1	IF color is green, AND light_intensity is high, AND height is medium, AND size is low, THEN nitrogen is low
Rule 2	IF color is very dark green, AND light_intensity is high, AND height is medium, AND size is high, THEN nitrogen is very high

3.3 Process Flow in Using the Recommender

Figure 3 shows the process flow with three stages of automated nitrogen recommendation process, which are image pre-processing and user input value, fuzzy logic processing and output display. The first stage involves defining the inputs required for the engine. It involves image selection and pre-processing to obtain the color values which are the color of the leaf in terms of the average grayscale value, while the other three inputs involving height, light intensity and paddy field size are obtained from the user. The second stage is the construction of the inference engine that begins with the fuzzification of the input, followed by rule evaluation, then aggregation of the rules and finally, defuzzification process. The last stage is obtaining the output in terms of the crisp value of the proposed suitable amount of nitrogen fertilizer to be used. A prototype based on Fig. 3 has been developed for the purpose of this research.

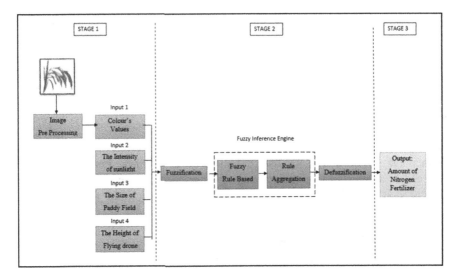

Fig. 3. Flow of the process fuzzification with four variables in determining the nitrogen intake in fertilizer

4   Fuzzification and COG Calculation

This section presents the implementation of the proposed model which has been developed based on the method described in Sect. 3. Firstly we discussed the fuzzy evaluation for four main variables. As can be seen in Table 3, the crisp output for color corresponds to the membership functions (μ) of v_2 and v_3 (darkgreen and green respectively), where v describes the color. The crisp output for light intensity (x) corresponds to the membership function of x_2 (medium) and the crisp output for height (y) corresponds to the membership functions of y_2 (medium) and meanwhile, the crisp output for size, (z) corresponds to the membership functions of z_1 (low). The input and its membership values are shown in Table 3 and visualized in Fig. 4. The next step is to take the fuzzified inputs, $\mu(v_2) = 0.13$, $\mu(v_3) = 0.38$, $\mu(x_2) = 0.85$, $\mu(y_3) = 1$, $\mu(z_1) = 1$ and apply them to the antecedents of the fuzzy rules. Once those values are fuzzified, the aggregation process will take place in which the membership functions of all rule consequents are combined into a single set. Since input color has multiple antecedents, the fuzzy operator (AND) is used to obtain a single number that represents the results of the antecedent evaluation. This number is then applied to the consequent membership function.

Table 3. Fuzzy variables and the correspondings crisp input.

Fuzzy variables	Input values	Fuzzy set	Ranges	Fuzzy set membership value, μ
Color	110	Very dark green	1–60	$\mu(v_1) = 0$
		Dark green	**40–120**	$\mu(v_2) = \mathbf{0.13}$
		Green	**100–210**	$\mu(v_3) = \mathbf{0.38}$
		Light green	190–255	$\mu(v_4) = 0$
Light intensity/lux	178 lux	Low	100–160	$\mu(x_1) = 0$
		Medium	**150–200**	$\mu(x_2) = \mathbf{0.85}$
		High	190–250	$\mu(x_3) = 0$
Height/m	100 m	Low	30–50	$\mu(y_1) = 0$
		Medium	45–85	$\mu(y_2) = 1$
		High	**80–100**	$\mu(y_3) = \mathbf{1}$
Size/	1.1 hectare	**Low**	**1.0–1.2**	$\mu(z_1) = \mathbf{1}$
		Medium	1.15–1.35	$\mu(z_2) = 0$
		High	1.3–1.5	$\mu(z_3) = 0$

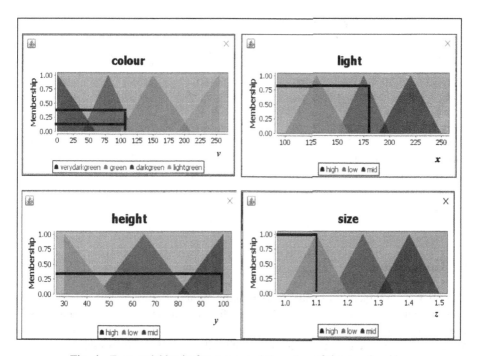

Fig. 4. Four variables in fuzzy sets and the value of the membership.

The last step in the fuzzy inference process is the defuzzification in which the aggregated output fuzzy values are transformed into a single number. The Eq. 3 is applied in getting the COG based on the membership value, μ, shown in Table 3 and the output of nitrogen in kg is 65.13.

5 Result and Discussion

Forty images were used for testing in this study and the results of the experiment are shown in Table 4. In determining the nitrogen amount as fertilizer, we had made suggestion based on the farmers and expert feedback, and the range of nitrogen amount is between 50 to 100 kg. If the color of the leaves is too dark, or nearly black, the plant might be unhealthy and it needs to be vanished. In this study, the expected nitrogen amount is shown in Table 4, while the average calculated amount of nitrogen is the amount produced by the model. The expected result for nitrogen amount (in kilogram) is given in the range of 10. The accuracy is the correctness of nitrogen amount for each given image samples over the total samples. The result shows that, most of samples with very dark green and dark green color can be correctly recommended with the right amount of nitrogen in fertilizer, with accuracy at 90% for both groups of images. However, for samples with green and light green would be lesser accurate at 70–75%. So, the average accuracy for this model is 81.25%.

Table 4. Accuracy of the model for each sample of images as in LCC color panel.

LCC chart panel	Grayscale value	Light intensity	Height of drone	Size of paddy field	Expected result for Nitrogen amount (kg)	Average calculated result for Nitrogen amount (kg)	Accuracy (%)
Very dark green	0–70	178	100	1.5	50–60	58.35	90
Dark green	71–150	178	100	1.5	70–80	76.5	90
Green	151–200	178	100	1.5	80–90	81.33	70
Light green	201–255	178	100	1.5	90–100	93.15	75
Average							**81.25**

6 Conclusion

This study is designed to measure an optimum amount of nitrogen fertilizer that needs to be used in any given paddy field. It mainly focuses on the color of the paddy leaves. It uses the same concept like the Leaf Color Chart (LCC) where the color of the paddy leaves will determine the optimum amount of fertilizer needed. The results are based on the four inputs, which are the color of paddy leaves, the intensity of sunlight, the height of the flying drone and the size of the paddy field. This model will help paddy farmers to increase the quality and quantity of the paddy produced by giving the right amount of fertilizer. Other limitations, the images were collected within a limited time and without any weather disruption. It could be studied and resolved in future research, thus opening opportunities for future works to be undertaken in order to increase the performance of the model.

Acknowledgement. The authors would like to thank Research Management Centre of Universiti Teknologi MARA and REI 19/2015 in supporting the research.

References

1. Food and Agriculture Organization of the United Nations: FAOSTAT statistics database. FAO, Rome (1998)
2. Roslan, N.A., Abdullah, A.M., Ismail, M.M., Radam, A.: Determining risk attitudes of paddy farmers in KETARA Granary, Malaysia. Int. J. Soc. Sci. Hum. **2**, 225–231 (2012)
3. Ziadi, N., Bélanger, G., Claessens, A., Lefebvre, L., Tremblay, N., Cambouris, A.N., Nolin, M.C., Parent, L.-É.: Plant-based diagnostic tools for evaluating wheat nitrogen status. Crop Sci. **50**, 2580–2590 (2010). All rights reserved. No part of this periodical may be reproduced or transmitted in any form or by any means, electronic or mechanical, including photocopying, recording, or any information storage and retrieval system, without permission in writing from the publisher. Permission for printing and for reprinting the material contained herein has been obtained by the publisher
4. Alagurani, K., Suresh, S., Sheela, L.: Maintaining ecological integrity: real time crop field monitoring using leaf colour chart. Int. J. Adv. Res. Electr. Electron. Instrum. Eng. **3**, 593–601 (2014)
5. Witt, C., Buresh, R.J., Peng, S., Balasubramanian, V., Dobermann, A.: Nutrient management. In: Fairhurst, T.H., Witt, C., Buresh, R., Dobermann, A. (eds.) Rice: A Practical Guide to Nutrient Management, 2nd edn, pp. 1–45. International Rice Research Institute (IRRI), Philippines, International Plant Nutrition Institute (IPNI) and International Potash Institute (IPI). Los Baños (Philippines) and Singapore (2007)
6. Kahar, M.A.A., Mutalib, S., Abdul-Rahman, S.: Early detection and classification of paddy diseases with neural networks and fuzzy logic. In: Proceedings of the 17th International Conference on Mathematical and Computational Methods in Science and Engineering, MACMESE 2015. Recent Advances in Mathematical and Computational Methods, vol. 44, pp. 248–257 (2015)
7. Yang, W.-H., Peng, S., Huang, J., Sanico, A.L., Buresh, R.J., Witt, C.: Using leaf color charts to estimate leaf nitrogen status of rice. Agron. J. **95**, 212–217 (2003)
8. Bijay-Singh, Varinderpal-Singh, Yadvinder-Singh, Kumar, A., Jagmohan-Singh, Vashistha, M., Thind, H.S., Gupta, R.K.: Fertilizer nitrogen management in irrigated transplanted rice using dynamic threshold greenness of leaves. Agric. Res. **5**, 174–181 (2016)
9. Zhou, Q., Wang, J.: Comparison of upper leaf and lower leaf of rice plants in response to supplemental nitrogen levels. J. Plant Nutr. **26**, 607–617 (2003)
10. Chen, L., Lin, L., Cai, G., Sun, Y., Huang, T., Wang, K., Deng, J.: Identification of nitrogen, phosphorus, and potassium deficiencies in rice based on static scanning technology and hierarchical identification method. PLoS One **9**, e113200 (2014)

11. Abergos, V.J., Boreta, P., Comprado, R., Soltes, S., Tatel, A.: Android-based image processing application for rice nitrogen management. Bachelor of Science in Electronics Engineering, Ateneo de Naga University (2012)
12. Zadeh, L.A.: Fuzzy sets. Inf. Control **8**, 338–353 (1965)
13. Negnevitsky, M.: Artificial Intelligence: A Guide to Intelligent Systems. Addison Wesley Longman, Boston (2011)
14. Wang, Y., Wang, D., Shi, P., Omasa, K.: Estimating rice chlorophyll content and leaf nitrogen concentration with a digital still color camera under natural light. Plant Methods **10**, 36 (2014)

Ranking Performance of Modified Fuzzy TOPSIS Variants Based on Different Similarity Measures

Izyana Athirah Selemin, Rosma Mohd Dom[✉],
and Ainon Mardhiyah Shahidin

Centre of Mathematical Studies, Faculty of Computer and Mathematical
Sciences, Universiti Teknologi MARA (UiTM), 40450 Shah Alam, Malaysia
izyanaathirah93@gmail.com,
{rosma,ainon}@tmsk.uitm.edu.my

Abstract. Fuzzy TOPSIS has been widely used in solving multi criteria decision making problems. The performance of Fuzzy TOPSIS is dependent on its ability to rank alternatives thus enables decision makers to reach more accurate decisions. Three main steps in the algorithm of Fuzzy TOPSIS include criteria weights determination, calculation of similarity between alternatives and also the closeness coefficient measurements for alternatives in terms of negative and positive ideals. Enhancement at any of these three major steps would lead to a more accurate ranking of alternatives. This paper highlights the ranking ability of five (5) variants of Fuzzy TOPSIS. This is done by applying five different techniques in calculating similarities and/or differences between alternatives. In all the five Fuzzy TOPSIS variants constructed, the data and criteria weights used are similar to the data and criteria weights used in the original Fuzzy TOPSIS proposed by Chen (2000). Thus, the difference in ranking is solely due to the differences when different similarity measures were utilized. The final ranking by the five variants of Fuzzy TOPSIS are observed and analyzed. The finding shows that ranking performance of four Fuzzy TOPSIS variants are consistent to the findings reported in other research. The difference in ranking could be due to the differences of similarity formulation since each similarity measure technique is developed based on different concept thus its ability to find the similarity between any two fuzzy numbers are determined by different criteria involved and this is reflected in the ranking of alternatives. Hence, it can be said that similarity measure calculations have significant impact in the ranking process via Fuzzy TOPSIS.

Keywords: Criteria weights · Fuzzy TOPSIS · Similarity measure
Multi criteria decision making

1 Introduction

Multi criteria decision making (MCDM) would normally depend on comparison of decisions made by several decision makers which are usually vague. Thus, application of fuzzy concept is often used in ranking of criteria and attributes to achieve a more accurate and reliable decision. When the criteria or attributes are represented by fuzzy

© Springer Nature Singapore Pte Ltd. 2017
A. Mohamed et al. (Eds.): SCDS 2017, CCIS 788, pp. 241–252, 2017.
https://doi.org/10.1007/978-981-10-7242-0_21

numbers, the ranking, classifying or clustering of the criteria or attributes may not be easy. This is because determining the similarity between any two fuzzy numbers is not a very straight forward process since fuzzy numbers cannot be orderly arranged like real numbers. Instead, fuzzy numbers are compared in terms of their similarities. Over the years several similarity measures have been proposed to help find similarities between any two generalized fuzzy numbers [6].

Fuzzy TOPSIS (Technique for Order of Preference by Similarity to Ideal Solution) is a multi-criteria decision making technique proposed by [1]. It is popularly used to rank several alternatives based on multiple criteria or attributes. Many recent successful applications of Fuzzy TOPSIS have been reported by [7, 8, 9, 10]. Many more related researches are being carried out as to identify possible ways of enhancing the ranking ability of Fuzzy TOPSIS [6].

The performance of Fuzzy TOPSIS is dependent on its ability to rank alternatives thus enables decision makers to reach more accurate decisions. Three main steps in the algorithm of Fuzzy TOPSIS include criteria weights determination, calculation of similarity between criteria evaluation and also closeness coefficient measurement for alternatives in terms of negative and positive ideals. Research done by [6] to rank R&D projects as investment stated that the ranking of projects is dependent on the choice of ideal solution and on the choice of the similarity measure used [6]. In other word, a better similarity measure would lead to more accurate ranking of alternatives [6]. A fuzzy similarity measure is noted to be better when compared to the other fuzzy similarity measure based on the similarity results tested on a set of fifteen different types of generalized fuzzy numbers as analyzed by [3].

Each of the similarity measure developed has its own strength and limitation since each is developed based on different properties of the fuzzy numbers [3]. Four new similarity measures have been developed using combinations of geometric distance, centre of gravity (COG) points, area, perimeter and/or height [6]. However, their applications in Fuzzy TOPSIS have not been fully explored before. In this paper, the ranking ability of five (5) Fuzzy TOPSIS variants are investigated whereby four different similarity measures are incorporated inside Fuzzy TOPSIS to replace the currently used vertex method in existing Fuzzy TOPSIS. Hence this study will investigate the impact of using five different similarity measures in Fuzzy TOPSIS. The five Fuzzy TOPSIS variants are used to rank several alternatives based on a set of multiple criteria and their ranking results were analysed. In all the five Fuzzy TOPSIS variants constructed, the data and criteria weights used are similar to the data and criteria weights used in the original Fuzzy TOPSIS proposed by [1]. Thus, the difference in ranking is solely due to the differences in similarity measurements only. This is to ascertain that any changes to the ranking by the various Fuzzy TOPSIS experimented are due to the changes in the similarity measurements alone.

2 Fuzzy Similarity Measures

[11] had defined fuzzy similarity measure as a distance between fuzzy numbers in which the similarity between two triangular fuzzy numbers in a vector space is used to evaluate the similarities between them. [1] had used a distance measure to calculate

similarity between objects in fuzzy environment. Calculation of fuzzy similarity measurements are very important step in Fuzzy TOPSIS. To date five new similarity measures have been proposed. However, only four similarity measures will be highlighted in this paper. The analysis on the fifth newly introduced similarity measure is still in progress and will be reported in future work. In this paper five variants of Fuzzy TOPSIS were set up using four different similarity measures discussed in [6] in addition to the original Fuzzy TOPSIS by [1]. In the last step of Fuzzy TOPSIS, closeness coefficient measurement proposed by [1] was used as to complete the ranking process via Fuzzy TOPSIS method. The formulation for distance measure by [1] and the four similarity measurements used in this study are shown below.

2.1   Fuzzy Similarity Measure by [4]

The formulation of similarity measure by [4] is as follows:

$$S_1(A, B) = \frac{1}{1 + d(A, B)} \tag{1}$$

Where

$$d(A, B) = |P_1(A) - P_1(B)| \tag{2}$$

And

$P_1(A)$ and $P_1(B)$ represent the mean graded integration representation of A and B.

For trapezoidal fuzzy numbers, they are defined as below:

$$P_1(A) = \frac{a_1 + 2a_2 + 2a_3 + a_4}{6} \tag{3}$$

$$P_1(B) = \frac{b_1 + 2b_2 + 2b_3 + b_4}{6} \tag{4}$$

2.2   Fuzzy Similarity Measure by [2]

[2] proposed another formulation for similarity measure for fuzzy numbers by taking into consideration the centre of gravity points known as COG points in finding the distance for trapezoidal or triangular fuzzy numbers.

The formulation of similarity measure by [2] is as follows:

$$S_2(A, B) = \left(1 - \frac{\sum_{i=1}^{4} |a_i - b_i|}{4}\right) \left[(1 - |X_a - X_b|)^{\frac{S_a - S_b}{2}}\right] \times \left[\frac{\min(Y_a, Y_b)}{\max(Y_a, Y_b)}\right] \tag{5}$$

Where

Y_a, Y_b, X_a and X_b represent the centre of gravity (COG) points.

Meanwhile, w_a and w_b represent the heights of fuzzy numbers A and B respectively.

Thus,

$$Y_a = \begin{cases} w_a\left(\frac{a_3-a_2}{a_4-a_1}\right), & \text{if } a_1 \neq a_4 \text{ and } 0 < w_a \leq 1 \\ \frac{w_a}{2}, & \text{if } a_1 = a_4 \text{ and } 0 < w_a \leq 1 \end{cases} \tag{6}$$

$$Y_b = \begin{cases} w_b\left(\frac{b_3-b_2}{b_4-b_1}\right), & \text{if } b_1 \neq b_4 \text{ and } 0 < w_{ab} \leq 1 \\ \frac{w_b}{2}, & \text{if } b_1 \neq b_4 \text{ and } 0 < w_b \leq 1 \end{cases} \tag{7}$$

$$X_a = \frac{Y_a(a_3+a_2)+(a_4+a_1)(w_a-Y_a)}{2w_a} \tag{8}$$

$$X_b = \frac{Y_b(b_3+b_2)+(b_4+b_1)(w_b-Y_b)}{2w_b} \tag{9}$$

2.3 Fuzzy Similarity Measure by [10]

[10] proposed a similarity measure which takes into account the geometric distance, the perimeter and the height of generalized fuzzy numbers.

The formulation of similarity measure by [10] is as follows:

$$S_2(A,B) = \left(1 - \frac{\sum_{i=1}^{4}|a_i-b_i|}{4}\right) \times \left(\frac{\min(P_2(A),P_2(B))+\min(w_a,w_b)}{\max(P_2(A),P_2(B))+\max(w_a,w_b)}\right) \tag{10}$$

Where

$P_2(A)$ and $P_2(B)$ represent the perimeters of generalized fuzzy numbers A and B respectively.

Meanwhile, w_a and w_b represent the heights of fuzzy numbers A and B respectively.

Thus,

$$P(A) = \sqrt{(a_1-a_2)^2+w_A^2} + \sqrt{(a_3-a_4)^2+w_a^2} + (a_3-a_2)+(a_4-a_1) \tag{11}$$

$$P(B) = \sqrt{(b_1-b_2)^2+w_B^2} + \sqrt{(b_3-b_4)^2+w_b^2} + (b_3-b_2)+(b_4-b_1) \tag{12}$$

2.4 Fuzzy Similarity Measure by [3]

[3] developed a new similarity measure by adding the concept of area of generalized fuzzy numbers to the similarity measure proposed by [10]. Thus similarity measure proposed by [3] includes distance of geometric with area, perimeter and height of generalized fuzzy numbers.

The following is the formulation for similarity measure by [3]:

$$S_2(A, B) = \left(1 - \frac{\sum_{i=1}^{4} |a_i - b_i|}{4}\right) \left[\frac{\min(P_2(A), P_2(B))}{\max(P_2(A), P_2(B))}\right]$$
$$\times \left[\frac{\min(\text{Area}(A), \text{Area}(B)) + \min(w_a, w_b)}{\max(\text{Area}(A), \text{Area}(B)) + \max(w_a, w_b)}\right] \tag{15}$$

Where

$P_2(A)$ and $P_2(B)$ represent the perimeters of generalized fuzzy numbers A and B respectively.

Meanwhile, w_a and w_b represent the heights of fuzzy numbers A and B respectively. Thus,

$$\text{Area}(A) = \frac{1}{2} w_a (a_3 - a_2 + a_4 - a_1) \tag{16}$$

$$\text{Area}(B) = \frac{1}{2} w_b (b_3 - b_2 + b_4 - b_1) \tag{17}$$

3 Numerical Illustration

To illustrate the impact on fuzzy similarity measures in ranking by Fuzzy TOPSIS, the ranking performance of five variants of Fuzzy TOPSIS were explored and investigated. A numerical experimentation was carried out whereby five variants of Fuzzy TOPSIS were set up and applied to the data used by [1]. Their ranking abilities are measured, tabulated and analyzed. There are eight steps in Fuzzy TOPSIS. Changes made to Step 6 and Step 7 leads to the setting up of five variants of Fuzzy TOPSIS as discussed below:

Step 1. Assessment of the importance of the criteria.

To assess the importance of the criteria, linguistic values for the importance weights of each criterion used by [1] is being used in this study as shown in Table 1. The importance weight of the criteria is shown in Table 2.

Table 1. Linguistic values for the importance weights of each criterion

Linguistic terms for important weight	Fuzzy numbers
Very Low (VL)	(0,0,0.1)
Low (L)	(0,0.1,0.3)
Medium Low (ML)	(0.1,0.3,0.5)
Medium (M)	(0.3,0.5,0.7)
Medium High (MH)	(0.5,0.7,0.9)
High (H)	(0.7,0.9,1.0)
Very High (VH)	(0.9,1.0,1.0)

Table 2. The importance weights of the criteria

	D_1	D_2	D_3
C_1	H	VH	MH
C_2	VH	VH	VH
C_3	VH	H	H
C_4	VH	VH	VH
C_5	M	MH	MH

Step 2. Evaluate the rating of alternatives with respect to each criterion.

To evaluate the rating of alternatives with respect to each criterion, we used the linguistic values for the ratings given in Table 3 which is similar to the linguistic values used by [1].

Table 3. Linguistic values for the ratings

Linguistic terms for rating	Fuzzy numbers
Very Poor (VP)	(0,0,1)
Poor (P)	(0,1,3)
Medium Poor (MP)	(1,3,5)
Fair (F)	(3,5,7)
Medium Good (MG)	(5,7,9)
Good (G)	(7,9,10)
Very Good (VG)	(9,10,10)

The rating of alternatives with respect to each criterion shown in Table 4 was used by [1] and once again being used in this study.

Step 3. In this study weights of criteria used are taken from [1] for all the five variants of Fuzzy TOPSIS.

This is to ensure that any changes in the ranking results of the five Fuzzy TOPSIS variants are due to similarity measure alone. In other words everything is the same in all the five Fuzzy TOPSIS variants studied here except for the steps involving fuzzy similarity measure which are Step 6 and Step 7.

Step 4. Construct the normalized fuzzy decision matrix.

The normalized fuzzy decision matrix is found using:

$$R = \left(r_{ij}\right)_{m \times n}$$

Where

R represents the criteria or alternatives of fuzzy decision matrix.
B and C are the set of benefit criteria and cost criteria respectively.

Table 4. Rating of alternatives with respect to each criterion

Criteria	Candidates	Decision-makers		
		D_1	D_2	D_3
C_1	A_1	MG	G	MG
	A_2	G	G	MG
	A_3	VG	G	F
C_2	A_1	G	MG	F
	A_2	VG	VG	VG
	A_3	MG	G	VG
C_3	A_1	F	G	G
	A_2	VG	VG	G
	A_3	G	MG	VG
C_4	A_1	VG	G	VG
	A_2	VG	VG	VG
	A_3	G	VG	MG
C_5	A_1	F	F	F
	A_2	VG	MG	G
	A_3	G	G	MG

Thus,

$$r_{ij} = \left(\frac{a_{ij}}{c_j^*}, \frac{b_{ij}}{c_j^*}, \frac{c_{ij}}{c_j^*} \right), \quad j \in B; \tag{18}$$

$$r_{ij} = \left(\frac{a_j^-}{c_{ij}}, \frac{a_j^-}{b_{ij}}, \frac{a_j^-}{a_{ij}} \right), \quad j \in B \tag{19}$$

$$c_j^* = \max_i c_{ij} \quad \text{if} \quad j \in B; \tag{20}$$

$$a_j^- = \min_i a_{ij} \quad \text{if} \quad j \in C; \tag{21}$$

Definitions for R and B are similar to the ones used in [1].

Step 5. Construct the weighted normalized fuzzy decision matrix.

By considering the different importance of each criterion, the weighted normalized fuzzy decision matrix can be constructed using the formula given below:

$$V = (v_{ij})_{m \times n} \tag{22}$$

where $v_{ij} = r_{ij} \times w_j$

The next two steps are where the modification in existing Fuzzy TOPSIS is being introduced. Change done in Step 6 and Step 7 leads to the setup of four variants of Fuzzy TOPSIS.

Step 6. To select the point of reference to be used in the calculation of FPIS and FNIS.

Two different ways of selecting the point of reference to be used in the calculation of FPIS and FNIS are utilized as proposed by [1, 5].
    The point of reference used by [1, 5] are shown below:

(i) Selection of point of reference proposed by [5] to find the FPIS and FNIS is as follows:

$$A^{\oplus} = \left[v_1^+, v_2^+, \ldots, v_n^+ \right] \tag{23}$$

$$A^{\ominus} = \left[v_1^-, v_2^-, \ldots, v_n^- \right] \tag{24}$$

Where

Element of $A^{\oplus}$ is the maximum for all i weighted normalized values
Element of $A^{\ominus}$ is the minimum for all i weighted normalized values

Thus

$$v_j^+ = \left(\max_i v_{ij1}, \ \max_i v_{ij2}, \ \max_i v_{ij3}, \ \max_i v_{ij4} \right) \tag{25}$$

$$v_j^- = \left(\min_i v_{ij1}, \ \min_i v_{ij2}, \ \min_i v_{ij3}, \ \min_i v_{ij4} \right) \tag{26}$$

Table 5 shows the reference points calculated using method proposed by [5].

Table 5. Reference point for FPIS and FNIS found using [5]

	C_1	C_2	C_3	C_4	C_5
v_j^+	(0.46,0.77,1)	(0.81,1,1)	(0.64,0.9,1)	(0.81,1,1)	(0.3,0.57,0.83)
v_j^-	(0.41,0.71,0.93)	(0.45,0.7,0.9)	(0.44,0.72,0.9)	(0.63,0.9,1)	(0.13,0.32,0.58)

(ii) Selection of point of reference proposed by [1] to find the FPIS and FNIS is as follows:

$$A^{\oplus} = [(1,1,1),[(1,1,1),[(1,1,1),[(1,1,1),(1,1,1)]$$
$$A^{\ominus} = [(0,0,0),(0,0,0),(0,0,0),(0,0,0),(0,0,0)]$$

Table 6 shows the values of FPIS and FNIS using method proposed by [1, 5].

Table 6. Values of FPIS and FNIS by [1, 5]

Method used	$A^{\oplus} = \left[v_1^+, v_2^+, \ldots, v_n^+\right]$	$A^{\ominus} = \left[v_1^-, v_2^-, \ldots, v_n^-\right]$
[2] [3] [4] [10]	[(0.46,0.77,1), (0.81,1,1), (0.64,0.9,1), (0.81,1,1), (0.3,0.57,0.83)]	[(0.41,0.71,0.93), (0.45,0.7,0.9), (0.44,0.72,0.9), (0.63,0.9,1), (0.13,0.32,0.58)]
[1]	[(1,1,1), [(1,1,1), [(1,1,1), [(1,1,1), (1,1,1)]	[(0,0,0), (0,0,0), (0,0,0), (0,0,0), (0,0,0)]

Step 7. Calculate distance of measurement by using four different fuzzy similarity measures proposed by [2–4, 10]. In this step, the vertex method used in the original Fuzzy TOPSIS by [1] is replaced with the four fuzzy similarity measures proposed by [2–4, 10] respectively.

The values of distance measurements found for five variants of Fuzzy TOPSIS are tabulated in Table 7 below.

Table 7. Distance measurements for five fuzzy TOPSIS variants

	[4]: $S_1(A, B)$		[2]: $S_2(A, B)$		[10]: $S_3(A, B)$		[3]: $S_4(A, B)$		[1]			
	$A^{\oplus} = \left[v_1^+, v_2^+, \ldots, v_n^+\right]$ $A^{\ominus} = \left[v_1^-, v_2^-, \ldots, v_n^-\right]$										Distance found using vertex method	
	$S_1^{\oplus}$	$S_1^{\ominus}$	$S_2^{\oplus}$	$S_2^{\ominus}$	$S_3^{\oplus}$	$S_3^{\ominus}$	$S_4^{\oplus}$	$S_4^{\ominus}$	$S^{\oplus}$	$S^{\ominus}$	$S^{\oplus}$	$S^{\ominus}$
A_1	6.0	4.9	3.9	4.8	4.2	4.9	3.9	4.8	0.1	0.8	3.4	1.9
A_2	5.0	4.3	4.9	4.0	5.0	4.0	5.0	3.8	0.8	0.0	4.1	1.3
A_3	5.4	4.6	4.5	4.3	4.5	4.4	4.3	4.4	0.5	0.4	3.8	1.6

Step 8. Calculate the closeness coefficient for each alternative using [1].

The following is the formulation to calculate the closeness coefficient which was introduced by [1].

$$CCS_i = \frac{S_i^{\ominus}}{S_i^{\oplus} + S_i^{\ominus}}, \quad i = 1, 2, 3, \ldots, m \qquad (27)$$

4 Analysis of Findings

The ranking of alternatives when Fuzzy TOPSIS is used will be done in reference to the CCS_i values. Alternatives with the bigger CCS_i values would be at the higher ranking and vice-versa.

The closeness coefficients values calculated, CCS_i are shown in Table 8 below.

Table 8. Values of CCS_i for five variants of fuzzy TOPSIS

Method to find CCS_i	CCS_i	[4]	[2]	[10]	[3]	[1]	Selection point by [1]
		Selection point of FPIS and FNIS by [5]					
[1] $CCS_i = \dfrac{S_i^{\ominus}}{S_i^{\oplus} + S_i^{\ominus}}$	A_1	0.452	0.549	0.540	0.549	0.118	0.640
	A_2	0.463	0.447	0.447	0.430	0.996	0.766
	A_3	0.460	0.490	0.496	0.505	0.564	0.708
	Top Ranked	A_2	A_1	A_1	A_1	A_2	A_2

Table 8 shows that the CCS_i values for Fuzzy TOPSIS proposed by [1] using selection point of reference proposed by [5] gives similar ranking of alternatives. Hence the selection point of reference proposed by [5] provide a good substitute method for finding the positive ideal (FPIS) and the negative ideal (FNIS) distances.

In the final step of Fuzzy TOPSIS, the ranking of alternatives by five variants of Fuzzy TOPSIS are determined by the values of CCS_i as shown in Table 9.

Table 9. Ranking by five fuzzy TOPSIS variants.

Fuzzy TOPSIS variants	Fuzzy TOPSIS utilizing similarity measure proposed by [4]	Fuzzy TOPSIS utilizing similarity measure proposed by [2]	Fuzzy TOPSIS utilizing similarity measure proposed by [10]	Fuzzy TOPSIS utilizing similarity measure proposed by [3]	Fuzzy TOPSIS utilizing distance measure proposed by [1]
Alternatives	A2	A1	A1	A1	A2
	A3	A3	A3	A3	A3
	A1	A2	A2	A2	A1

Results of ranking by Fuzzy TOPSIS variants tabulated in Table 9 reveal the following findings:

(i) Ranking of Fuzzy TOPSIS utilizing similarity measure by [4] equals to the ranking of Fuzzy TOPSIS utilizing distance measure used in [1] while the other three Fuzzy TOPSIS variants utilizing similarity measures by [2, 3, 10] produced similar ranking. It seems that it exists two different sets of ranking. In order to decide on which set of ranking is a more reliable ranking we need to refer to a research done by [3]. In their report analysis, [3] had stated that similarity measure proposed by [3] showed the highest consistency when used on fifteen sets of fuzzy number as compared to the other three similarity measures proposed by [2, 4, 10]. In other words similarity measure [3] has better ability in finding similarities of generalized fuzzy numbers compared to the other three similarity measures [2, 4, 10]. On top of that [3] also reported that [2] had made the most mistakes (five mistake) in deciding on similarities on the fifteen types of fuzzy numbers while similarity measure by [3] made the least mistakes (only one mistake). Similarity measures proposed by [2, 10] made two mistakes each. With this argument in mind we conclude that the ranking of Fuzzy TOPSIS variant utilizing similarity measures proposed by [3] is the most reliable and is very much consistent to the ranking given by Fuzzy TOPSIS variants when similarity measures proposed by [2, 10] were utilized. This finding is in agreement by [3] since number of errors made by similarity measures [2, 10] is very much closer to the number of error made by similarity measure [3]. It is not a surprise that ranking by Fuzzy TOPSIS variant utilizing similarity measure by [4] is different by the ranking given by Fuzzy TOPSIS variants when similarity measures [2, 3, 10] were utilized since similarity measure proposed by [4] had the most number of errors thus its ranking ability is expected to be different.

(ii)   It is interesting to note that ranking results of original Fuzzy TOPSIS by [1] is the same as ranking given by Fuzzy TOPSIS variant utilizing similarity measure [4]. Thus, a thorough investigation would deem necessary in order to really comprehend this result. It could suggest that distance measure needs to be enhanced, in order to produce a more accurate ranking.

(iii) All the five Fuzzy TOPSIS variants had picked similar alternative (A3) to be in the middle rank. This finding would probably open a new perspective on discussion of similarity measure where the various similarity measurements techniques may share some similar findings while differ in some other area since each similarity measure was developed based on different properties.

5   Conclusion

In summary, similarity measures have an impact on the ranking ability of Fuzzy TOPSIS variants. Thus, it is essential that the most reliable similarity measure is used in MCDM. Failing to do so will lead to a not so accurate decision being made. From the numerical illustration given in this study, ranking performance enhancement of MCDM method such as Fuzzy TOPSIS is feasible via modification of important steps of the algorithm in this particular study, the modification is done by replacing the existing distance measure with a similarity measure which have been found to produce less errors when tested on many types of fuzzy numbers.

In this study weights of criteria used are taken from [1] for all the five variants of Fuzzy TOPSIS. This is to ensure that any changes in the ranking ability of the Fuzzy TOPSIS variants are only due to the utilization of similarity measures alone. For future research, it is recommended that criteria weights could be determined using other methods to be used in the five variants of Fuzzy TOPSIS proposed here.

References

1. Chen, C.T.: Extensions of the TOPSIS for group decision-making under fuzzy environment. Fuzzy Sets Syst. **114**(1), 1–9 (2000)
2. Chen, J., Chen, S.: A new method to measure the similarity between fuzzy numbers, pp. 1123–1126 (2001)
3. Hejazi, S.R., Doostparast, A., Hosseini, S.M.: An improved fuzzy risk analysis based on a new similarity measures of generalized fuzzy numbers. Expert Syst. Appl. **38**(8), 9179–9185 (2011). https://doi.org/10.1016/j.eswa.2011.01.101
4. Hsieh, C.H., Chen, S.H.: Similarity of generalized fuzzy numbers with graded mean integration representation. In: Proceedings of 8th International Fuzzy Systems Association World Conference, vol. 2, pp. 551–555 (1999)
5. Luukka, P.: Fuzzy similarity in multicriteria decision-making problem applied to supplier evaluation and selection in supply chain management. Adv. Artif. Intell. **2011**, 1–9 (2011). https://doi.org/10.1155/2011/353509. Article ID 353509
6. Collan, M., Lukka, P.: Evaluating R&D projects as investments by using an overall ranking from four fuzzy similarity measure-based TOPSIS variants. IEEE Trans. Fuzzy Syst. **22**(3), 505–515 (2014)
7. Onu, P.U., Quan, X., Xu, L., Orji, J., Onu, E.: Evaluation of sustainable acid rain control options utilizing a fuzzy TOPSIS multi-criteria decision analysis model frame work. J. Clean. Prod. **141**, 612–625 (2017). https://doi.org/10.1016/j.jclepro.2016.09.065
8. Peng, C., Du, H., Liao, T.W.: A research on the cutting database system based on machining features and TOPSIS. Robot. Comput. Integr. Manuf. **43**, 96–104 (2017). https://doi.org/10.1016/j.rcim.2015.10.011
9. Walczak, D., Rutkowska, A.: Project rankings for participatory budget based on the fuzzy TOPSIS, 1–9 (2017). http://doi.org/10.1016/j.ejor.2016.12.044
10. Wei, S.-H., Chen, S.-M.: A new approach for fuzzy risk analysis based on similarity measures of generalized fuzzy numbers. Expert Syst. Appl. **36**(1), 589–598 (2009). https://doi.org/10.1016/j.eswa.2007.09.033
11. Zhang, L., Xu, X., Tao, L.: Some similarity measures for triangular fuzzy number and their applications in multiple criteria group decision-making. J. Appl. Math. **2013**, 1–7 (2013)

Prediction Models and E-Learning

Prediction of New Bioactive Molecules of Chemical Compound Using Boosting Ensemble Methods

Haslinda Hashim[1,2] and Faisal Saeed[2,3(✉)]

[1] Kolej Yayasan Pelajaran Johor, KM16, Jalan Kulai-Kota Tinggi,
81900 Kota Tinggi, Johor, Malaysia
haslinda@kypj.edu.my
[2] Information Systems Department, Faculty of Computing,
Universiti Teknologi Malaysia, 81310 Skudai, Johor, Malaysia
faisalsaeed@utm.my
[3] College of Computer Science and Engineering,
University of Taibah, Medina, Saudi Arabia

Abstract. Virtual screening (VS) methods can be categorized into structure-based virtual screening (SBVS) that involves knowledge about the target's 3D structure and ligand-based virtual screening (LBVS) approaches that utilize information from at least one identified ligand. However, the activity prediction of new bioactive molecules in highly diverse data set is still less accurate and the result is not comprehensive enough since only one approach is applied at one time. This paper aims to recommend the boosting ensemble method, MultiBoost, into LBVS using the well-known chemoinformatics database, the MDL Drug Data Report (MDDR). The experimental results were compared with Support Vector Machines (SVM). The final outcomes showed that MultiBoost ensemble classifiers had improved the effectiveness of the prediction of new bioactive molecules in high diverse data.

Keywords: AdaBoost · Ensemble methods · MultiBoost · Virtual screening

1 Introduction

Data mining is defined as the procedure of discovering recent interesting and valuable patterns and knowledge from big data sources which include databases, data warehouses, the Web, or any other information repositories and applying algorithms to mine the hidden information [1, 2]. The two primary goals of data mining are prediction and description. Briefly the first goal, prediction, entails employing some variables in the dataset to forecast other unrevealed variables of concern. In contrast, the second goal, description, aims to obtain patterns, defining the data in a way that can be understood by people. In other words, the aim of predictive data mining is to build a model which can be applied to achieve classification, prediction, or to reproduce a further identical process, while the target of descriptive data mining is to acquire comprehension of the examined system by revealing patterns and associations in huge collections of data [3]. Data mining techniques play an important role in chemoinformatics [4]. For instance,

© Springer Nature Singapore Pte Ltd. 2017
A. Mohamed et al. (Eds.): SCDS 2017, CCIS 788, pp. 255–262, 2017.
https://doi.org/10.1007/978-981-10-7242-0_22

Brown [5] defines chemoinformatics as the combination of information sources to transform data into information, and later information into knowledge, in order to make better decisions rapidly in the field known as drug lead identification and optimization. A main focus in the chemoinformatics is virtual screening (VS), which employs data mining approaches that looks in massive chemical databases for the latest drug leads.

VS methods are generally categorized into two types: structure-based (SBVS) and ligand-based (LBVS), where SBVS methods involve knowledge about the target molecule's 3-D structure [6], while, LBVS approaches the information utilized from the aspect of at least one identified ligand. Ligand is any molecule that binds to a specific site on a protein or other molecule (from Latin ligare, to bind) [7]. The method of choice frequently relies on the amount of information that is accessible [8]. The goal of LBVS methods is to explore chemical databases to discover the compounds that are most relevant to a certain inquiry [9]. These LBVS methods are also computationally economical and effortless to employ. Machine-learning approaches [10, 11] that are becoming increasingly well-known in LBVS include support vector machines (SVM), naive Bayesian methods, k-nearest neighbors (k-NN), artificial neural networks (ANN) and also decision trees (DT).

LBVS for target or activity prediction of new bioactive molecules in highly diverse data sets still has low accuracy rates and the results are not comprehensive enough, since only one approach is applied at a time [12]. In addition, there are several issues regarding the target prediction approaches. In several studies [13–15], distinct approaches were found to forecast a dissimilar subset of targets for a similar compound. Furthermore, although a certain method may work better on particular targets or databases, it is difficult to foresee which method will be successful on a certain combination of database and query [12]. Therefore, it is important to highlight that employing numerous techniques at one time for target prediction can provide a wider comprehension of the data than utilizing merely one approach solely [12, 16]. Therefore, by combining multiple approaches or classifiers on highly diverse data set we can obtain better decisions and good results. This paper assists in improving the accuracy rates in the prediction of biological activity in highly diverse datasets, as required in the discovery of new bioactive molecules, towards novel drug discovery based on chemical and biological data. Secondly, using the boosting ensemble methods gave better decisions and provided good results. This approach was comprehensive, using multiple methods on highly diverse datasets instead of only employing a single method.

2 Materials and Methods

In this study, a subset from the MDL Drug Data Report (MDDR) dataset [17] was used, which contains a total of 8569 heterogeneous active molecules from 10 classes, as shown in Table 1. AdaBoostM1 [18] and MultiBoostAB [19] are the boosting ensemble methods that were used for this study and combined with bagging [20], JRip [21], PART [22], J48 [23], RandomForest (RF) [24] and REPTree [25].

The AdaBoost [18] algorithm stands for ADAptive BOOSTing. AdaBoost was considered as one of the best-performing combined classifier and AdaBoost.M1 was

Table 1. Activity class for dataset MDDR

Activity index	Activity class	Activity molecules	Pairwise similarity (mean)
09249	Muscarinic (M1) agonists	900	0.257
12455	NMDA receptor antagonists	1400	0.311
12464	Nitric oxide synthase inhibitors	505	0.237
31281	Dopamine β-hydroxylase inhibitors	106	0.324
43210	Aldose reductase inhibitors	957	0.37
71522	Reverse transcriptase inhibitors	700	0.311
75721	Aromatase inhibitors	636	0.318
78331	Cyclooxygenase inhibitors	636	0.382
78348	Phospholipase A2 inhibitors	617	0.291
78351	Lipoxygenase inhibitors	2111	0.365

the enhanced edition of AdaBoost. This is because of its extremely precise predictions, sound theoretical basis, and effortlessness, along with its wide and successful previous applications [26]. According to Wu et al. [27], AdaBoost – was recognized - as one of top the ten data mining algorithms by the IEEE International Conference on Data Mining (ICDM) in 2006. The top ten data mining algorithms were considered to be k-Means, SVM, C4.5, Apriori, AdaBoost, EM, kNN, Naive Bayes, CART as well as PageRank.

The MultiBoost algorithm is an expansion to the extremely successful AdaBoost method for shaping decision committees, which can be envisioned as joining AdaBoost with wagging and it is beneficial to acquire more precision than AdaBoost and wagging independently [19]. Wagging (Weight Aggregation) is a variant of bagging that it seeks to repeatedly perturb the training set as in Bagging, but instead of sampling from it, Wagging adds Gaussian noise to each weight with mean zero and a given standard deviation (e.g., 2), which for each trial, starting with uniformly weighted instances, noise is added to the weights, and a classifier is induced. The method has the nice property that one can trade off bias and variance: by increasing the standard deviation of the noise introduced, more instances can have their weight decrease to zero and disappear, thus increasing bias and reducing variance [26]. Table 2 shows the details of MultiBoost algorithm.

In addition, a quantitative approach was used using the Kendall's W test of concordance [28]. According to Abdo et al. [12], this test was developed to quantify the level of agreement between multiple sets ranking the same set of objects. This approach was used to rank the performance of the tested algorithms. The outputs of the Kendall's W test were the Kendall's coefficient of concordance (W), chi-square ($\chi2$) and the significance level (p_value). The value of the Kendall's coefficient of concordance (W), ranging from 0 (no agreement between set of ranks) to 1 (complete agreement), and the associated significance level (p_value), indicated whether this value of the coefficient could have occurred by chance. If the value is significant (use cutoff values of 0.01 or 0.05), then it is possible to give an over-all ranking of the objects that have been ranked [12].

Table 2. The MultiBoost algorithm [18]

Input:	
S,	a sequence of m labeled examples $((x_1; y_1), ..., (x_m, y_m))$ with labels $y_i \in Y$.
BaseLearn	Base learning algorithm.
integer T	Specifying the number of iterations.
vector of integers I_i	Specifying the iteration at which each subcommittee $i \geq 1$ should terminate.

1. $S' = S$ with instance weights assigned to be 1.
2.   Set $k = 1$.
3.   For $t = 1$ to T {
4.   If $I_k = t$ then
5.   Reset S' to random weights drawn from the continuous Poisson distribution.
6.   Standardize S' to sum to n.
7.   Increment k.
8. $C_t = BaseLearn(S')$.
9. $\epsilon_t = \dfrac{\sum_{x_j \in S : C_t(x_j) = y_i} weight(x_j)}{m}$ *[the weighted error on the training set]*
10.   If $\epsilon_t > 0.5$ then
11.   Reset S' to random weights drawn from continuous Poisson distribution.
12.   Standardize S' to sum to n.
13.   Increment k.
14.   Go to Step 8.
15.   Otherwise if $\epsilon_t = 0$ then
16.   Set β_t to 10^{-10}
17.   Reset S' to random weights drawn from continuous Poisson distribution.
18.   Standardize S' to sum to n.
19.   Increment k.
20.   Otherwise,
21. $\beta_t = \dfrac{\epsilon_t}{(1 - \epsilon_t)}$
22.   For each $x_j \in S'$,
23.   Divide *weight*(x_j) by $2\epsilon_t$ if $C_t(x_j) \neq y_j$ and $2(1 - \epsilon_t)$ otherwise.
24.   If *weight*$(x_j) < 10^{-8}$, set weight(x_j) to 10^{-8}.
25. }

Output the final classifier:
$$C^*(x) = \underset{y \in Y}{avg\ max} \sum_{t : C_t(x) = y} \log \frac{1}{\beta_t}$$

The experiments used WEKA-Workbench version 3.6.14 to run LSVM, AdaBoost and MultiBoost with optimal parameters. SVC, as one of LSVM learning tasks, supports two-class and multiclass classifiers. Therefore, here, the LSVM type selected in WEKA-Workbench, was C-SVC. The simple linear kernels have been used successfully [5] and thus the kernel type applied in these experiments was a linear kernel. Discovering the optimal parameters for a classifier was a time consuming task. Nevertheless, WEKA-Workbench offers the possibility of automatically finding the best possible setup for the LSVM classifier. The values of 1.0, 0.1, and 0.001 were given to the Cost, Gamma and Epsilon parameters, correspondingly, while the default values available in WEKA-Workbench for the other parameters.

In this study, six AdaBoost ensemble classifiers were applied: AdaBoostM1 +Bagging (Ada_Bag), AdaBoostM1+Jrip (Ada_Jrip), AdaBoostM1+J48 (Ada_J48), AdaBoostM1+PART (Ada_PART), AdaBoostM1+RandomForest (Ada_RF) and AdaBoostAB+REPTree (Ada_RT) and four MultiBoost ensemble classifiers were employed, which were MultiBoostAB+J48 (Multi_J48), MultiBoostAB+RandomForest (Multi_RF), MultiBoostAB+REPTree (Multi_RT) and MultiBoostAB +Bagging (Multi_Bag). After that, a ten-fold cross-validation was carried out and the results were evaluated using the Area Under the receiver operating characteristic curve (AUC) and the Accuracy. These criteria were formulated as follows:

$$\text{Sensitivity (Sens)} = \frac{TP}{TP + FN}$$

$$\text{Specificity (Spec)} = \frac{TN}{TN + FP}$$

$$\text{AUC} = \frac{Sens + Spec}{2}$$

$$\text{Accuracy} = \frac{(TP + TN)}{(TP + TN + FP + FN)}$$

where, True Positive (TP), False Negative (FN), True Negative (TN), and False Positive (FP).

3 Results and Discussion

The results of LSVM, six AdaBoost classifiers, and four MultiBoost classifiers were shown in Table 3 for the AUC measure. In this experiment, Multi_RT, as one of the MultiBoost classifiers used in these experiments, was the best performing classifier.

In addition, from Fig. 1, the highest accuracy was obtained by Multi_RT, with **97.7%**. Thus, it can be concluded that MultiBoost ensemble classifiers outperformed all other classifiers. The Multi_RT, as the MultiBoost ensemble classifier, had improved the effectiveness of the prediction of new bioactive molecules in highly diverse data compared to LSVM with the accuracy rates of **97.7%** and **93.7%** respectively.

Table 3. AUC measure for the prediction of new bioactive molecules using MDDR heterogeneous data set

Class	Activity index	LSVM	Ada_Bag	Ada_Jrip	Ada_J48	Ada_PART	Ada_RF	Ada_RT	Multi_J48	Multi_RF	Multi_RT	Multi_Bag
1	09249	0.989	0.984	0.988	0.983	0.982	0.989	0.984	0.985	**0.992**	0.960	0.986
2	12455	0.973	**0.978**	0.967	0.965	0.967	**0.978**	0.967	0.960	0.982	0.953	0.974
3	12464	0.953	0.949	0.954	0.951	0.953	0.954	0.945	0.953	0.959	**0.996**	0.945
4	31281	**0.986**	0.977	0.967	0.934	0.943	0.958	0.948	0.938	0.948	**0.986**	0.957
5	43210	0.973	0.977	0.965	0.971	0.969	0.976	0.966	0.971	0.979	**0.987**	0.976
6	71522	0.954	0.958	0.957	0.954	0.946	0.954	0.938	0.945	**0.961**	0.853	0.944
7	75721	0.989	0.987	0.979	0.971	0.974	0.984	0.977	0.972	0.985	**0.997**	0.981
8	78331	0.914	0.925	0.894	0.899	0.911	0.919	0.914	0.904	0.922	**0.980**	0.920
9	78348	0.945	0.954	0.937	0.948	0.941	0.944	0.930	0.946	0.948	**0.998**	0.944
10	78351	0.960	0.968	0.957	0.957	0.960	0.968	0.960	0.961	0.972	**0.995**	0.967
Means		0.963	0.965	0.956	0.953	0.954	0.962	0.953	0.953	0.965	**0.970**	0.959

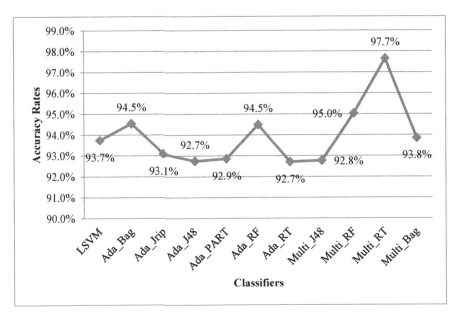

Fig. 1. Accuracy rates using boosting ensemble methods for the prediction of new bioactive molecules.

In addition, Table 4 shows that the Kendall's coefficients (using the AUC measure) were significant (p < 0.05, χ^2 > 18.31), and the performance of Multi_RF considerably outperformed all of the other methods. The overall ranking for MDDR dataset is: Mul-ti_RF > Ada_Bag > Multi_RT > Ada_RF > LSVM > Multi_Bag > Ada_Jrip > Multi_J48 > Ada_PART > Ada_J48 > Ada_RT.

Table 4. Kendall's W test results using the MDDR dataset by AUC measure

W	χ^2	p		Ranks										
			Technique	LSVM	Ada_Bag	Ada_Jrip	Ada_J48	Ada_PART	Ada_RF	Ada_RT	Multi_J48	Multi_RF	Multi_RT	Multi_Bag
0.42	41.97	0.00	Mean Ranks	7.15	8.25	4.90	3.55	3.80	7.60	3.40	4.25	9.30	7.95	5.85

4 Conclusion

To improve the effectiveness of the prediction of biological activity for highly diverse data set, MultiBoost ensemble method was used and the results showed that it gave better decisions. The combination, MultiBoostAB+REPTree, was found to be the best approach to improve the accuracy rates in highly diverse data set. Future work could investigate the performance of boosting methods on different chemical data sets for the prediction of new bioactive molecules in chemical compounds. In addition, different ensemble methods could be used to improve the prediction accuracy.

Acknowledgments. This work is supported by the Ministry of Higher Education (MOHE) and Research Management Centre (RMC) at the Universiti Teknologi Malaysia (UTM) under the Research University Grant Category (VOT Q.J130000.2528.16H74).

References

1. Han, J., Kamber, M., Pei, J.: Data Mining: Concepts and Techniques, 3rd edn. Elsevier, Cambridge (2012)
2. Chen, F., Deng, P., Wan, J., Zhang, D., Vasilakos, A.V., Rong, X.: Data mining for the internet of things: literature review and challenges. Int. J. Distrib. Sens. Netw. **2015**(2015), 14 p. (2015)
3. Kantardzic, M.: Data Mining: Concepts, Models, Methods, and Algorithms, 2nd edn. Wiley, Hoboken (2011)
4. Geppert, H., Vogt, M., Bajorath, J.: Current trends in ligand-based virtual screening: molecular representations, data mining methods, new application areas, and performance evaluation. J. Chem. Inf. Model. **50**(2), 205–216 (2010)
5. Brown, F.K.: Chemoinformatics: what is it and how does it impact drug discovery. Annu. Rep. Med. Chem. **33**, 375–384 (1998)
6. Lavecchia, A.: Machine-learning approaches in drug discovery: methods and applications. Drug Discov. Today **20**(3), 318–331 (2015)

7. Alberts, B., Johnson, A., Lewis, J., Raff, M., Roberts, K., Walter, P.: Molecular Biology of the Cell, 4th edn. Garland Science, New York (2002)
8. Svensson, F., Karlen, A., Sköld, C.: Virtual screening data fusion using both structure-and ligand-based methods. J. Chem. Inf. Model. **52**(1), 225–232 (2011)
9. Venkatraman, V., Perez-Nueno, V.I., Mavridis, L., Ritchie, D.W.: Comprehensive comparison of ligand-based virtual screening tools against the DUD data set reveals limitations of current 3D methods. J. Chem. Inf. Model. **50**(12), 2079–2093 (2010)
10. Mitchell, J.B.: Machine learning methods in chemoinformatics. Wiley Interdiscipl. Rev.: Comput. Mol. Sci. **4**(5), 468–481 (2014)
11. Melville, J.L., Burke, E.K., Hirst, J.D.: Machine learning in virtual screening. Comb. Chem. High Throughput Screen. **12**(4), 332–343 (2009)
12. Abdo, A., Leclère, V., Jacques, P., Salim, N., Pupin, M.: Prediction of new bioactive molecules using a Bayesian belief network. J. Chem. Inf. Model. **54**(1), 30–36 (2014)
13. Sheridan, R.P., Kearsley, S.K.: Why do we need so many chemical similarity search methods? Drug Discov. Today **7**(17), 903–911 (2002)
14. Ding, H., Takigawa, I., Mamitsuka, H., Zhu, S.: Similarity-based machine learning methods for predicting drug–target interactions: a brief review. Brief. Bioinform. **15**(5), 734–747 (2014)
15. Jenkins, J.L., Bender, A., Davies, J.W.: In silico target fishing: predicting biological targets from chemical structure. Drug Discov. Today: Technol. **3**(4), 413–421 (2007)
16. Harper, G., Bradshaw, J., Gittins, J.C., Green, D.V., Leach, A.R.: Prediction of biological activity for high-throughput screening using binary kernel discrimination. J. Chem. Inf. Comput. Sci. **41**(5), 1295–1300 (2001)
17. Chang, C.C., Lin, C.J.: LIBSVM: a library for support vector machines. ACM **2**(3), 27 (2011)
18. Freund, Y., Schapire, R.E.: Experiments with a new boosting algorithm. In: Proceedings of 13th International Conference on Machine Learning, pp. 148–156. Morgan Kaufmann (1996
19. Webb, G.: Multiboosting: a technique for combining boosting and wagging. Mach. Learn. **40**(2), 159–196 (2000)
20. Breiman, L.: Bagging predictors. Mach. Learn. **24**(2), 123–140 (1996)
21. Cohen, W.W.: Fast effective rule induction. In: Proceedings of 12th International Conference on Machine Learning, pp. 115–123 (1995)
22. Frank, E., Witten, I.H.: Generating accurate rule sets without global optimization. In: Proceedings of 15th International Conference on Machine Learning. Department of Computer Science, University of Waikato (1998)
23. Smusz, S., Kurczab, R., Bojarski, A.J.: A multidimensional analysis of machine learning methods performance in the classification of bioactive compounds. Chemometr. Intell. Lab. Syst. **128**, 89–100 (2013)
24. Breiman, L.: Random forests. Mach. Learn. **45**(1), 5–32 (2001)
25. Witten, I.H., Frank, E., Hall, M.A.: Data Mining: Practical Machine Learning Tools and Techniques. Elsevier (2005)
26. Bauer, E., Kohavi, R.: An empirical comparison of voting classification algorithms: bagging, boosting, and variants. Mach. Learn. **36**(1), 105–139 (1999)
27. Wu, X., Kumar, V., Quinlan, J.R., Ghosh, J., Yang, Q., Motoda, H., McLachlan, G.J., Ng, A., Liu, B., Philip, S.Y., et al.: Top 10 algorithms in data mining. Knowl. Inf. Syst. **14**(1), 1–37 (2008)
28. Siegel, S., Castellan Jr., N.J.: Nonparametric Statistics for the Behavioral Sciences, 2nd edn. Mcgraw-Hill Book Company, New York (1988)

Pragmatic Miner to Risk Analysis for Intrusion Detection (PMRA-ID)

Samaher Al-Janabi[(⊠)]

Department of Computer Science, Faculty of Science for Women (SCIW),
University of Babylon, Babylon, Iraq
samaher@uobabylon.edu.iq

Abstract. Security of information systems and their connecting networks has become a primary focus given that pervasive cyber-attacks against information systems are geometrically increasing. Intrusion Detection and Prevention Systems (IDPS) effectively secure the data, storage devices and the systems holding them. We will build system consist of five steps: (a) description the orders that required to archives the event by five fuzzy concepts as input and three fuzzy concepts as output, then save it in temporal bank of orders, (b) Pre-processing that order by convert from the description to numerical values and compute the Membership function for that values. (c) applied the association data mining techniques on these database after compute the correlation among their features, this lead to generation thirty two rules but not all this rules is salsify the confidence measures (i.e., we take only the rules that satisfy the purity 100%) (d) Building the Confusion matrix for all the samples using in training processing (e) Testing the Pragmatic Miner to Risk Analysis (PMRA) model and verification from the accuracy of their results by press new samples to model not used in training stage then compute the values of error and accuracy measures, in addition of correct. The existing systems employing firewalls and encryptions for data protection are getting outdated. IDPS provides a much improved detection system that can prevent the intrusions to attack the system. However, as effective as it is in preventing intrusions, which can disrupt the retrieval of desired information as the system sometimes perceives it as an attack. The base aim of this work is to determine a way to risk analysis of IDPS to an acceptable level while detecting the intrusions and maintaining effective security of a system. Experimental results clearly show the superficiality of the proposed model against the conventional IDPS system.

Keywords: Information security · Intrusion Detection and Prevention
Fuzzy descriptive · Data mining · Error and accuracy measures · Risk analysis

1 Introduction

One of the obligations of government, corporations, private business, financial and healthcare institutions, the military and several other organizations is to gather a detailed information of their customers, products, researchers, employees and their financial status. Informatics takes the key responsibility for providing the security, privacy and confidentiality of the primary digital data. Collection, processing and

© Springer Nature Singapore Pte Ltd. 2017
A. Mohamed et al. (Eds.): SCDS 2017, CCIS 788, pp. 263–277, 2017.
https://doi.org/10.1007/978-981-10-7242-0_23

storage of most of the information on computers, necessarily takes place before transmission of the information across networks. Likewise, protection of a patient's medical history is very taxing.

The growth in the use of information management systems has become more powerful and widely distributed [1, 15] allowing the number of threats facing computers networks and the files stored in them grow and diversify, as a result. Researchers are encouraged to intensify and focus on intrusion detection mechanisms so that novel attacks can be detected through anomaly detection since ways of attacks are increasing and misuse detection functions often fail to detect no signature of novel attacks [3].

Intrusion Detection and Prevention system (IDPS) can perform an early detection of malicious activities and therefore, an organization can easily monitor events occurring within its computer network or system. Then, it can shield itself from misuse of its information, analyze the network for signs of intrusion and can prevent serious damage to already protected systems [10].

IDPS can disclose abnormal patterns of behaviors through the establishment of baselines of normal usage patterns and anything that differs from the norm is considered as a possible intrusion [10]. Unfortunately, the false alarms challenge this system, which causes inadvertent system behaviour and unnecessary data processing [7], This paper is similar with [7] in stage of using Fuzzy Logic as pre-processing stage for raw data but it different with it in terms of its use of data mining techniques in a constructive rules and then it'll take the rules that satisfy t the percentage of confidence 100% and ignore the rest. Also, the current research evaluated the results reached by a precision and error scales.

IDPS triggers a positive alarm when it mistakenly interprets a benign activity as a malicious one. On the other hand, sometimes IDPS does not detect an attack or intrusion; still it flags a false positive alarm. Once the basic criteria is used to distinguish IPSs from IDSs, the intrusion prevention system of the former tries to prevent the success of detected threats, unlike the latter [2, 12]. The IPS can change the content of the attack or change the security environment into a countermeasure. It can also alter the configuration of other security controls in order to stop an attack by denying access to the attacker or disabling the target so that the attack cannot proceed. The host-based firewall settings can be reconfigured so that the incoming attack is completely hindered. Some IPSs can take away the malicious parts of the attack and render those as impotent to attack [5].

IDPSs fail to provide a complete and an accurate detection of attacks because they do not employ proper risk analysis and assessment techniques [5]. This paper presents a new way of detection and prevention of intrusions based on risk analysis and risk assessment in a way that the false alarm rate will be minimized in an IDPS. This approach employs a Pragmatic Miner-risk analysis technique to analyze the generated alarms. The Pragmatic Miner technique, also known as Pragmatic Miner to Risk Analysis (PMRA) computes the significance and impact severity of the detected activities. By doing so, the system will adapt itself while evaluating an activity should be regarded as an attack attempt or a possible normal behavior.

The organization of the paper is as follows: Sect. 2 contains the latest trends in the PMRA researches. Section 3 discloses the most significant limitations in the preexisting intrusion detection and prevention methods. Section 4 reviews the proposed architectural model of the PMRA system. Section 5 presents the mechanics and experimental results obtained from the implementation of PMRA. The result proves that the proposed architecture is helpful in order to predict the vulnerabilities and array countermeasures that can be used against reasonable and manageable alarm rates. Finally, Sect. 6 contains a brief discussion and the conclusion of this paper, with suggested possible future plans that can yield better results.

2 Related Works

IDPS has been a major topic of interest in terms of research work to develop sophisticated systems. In this section, some of the latest research that relate to soft computing techniques regarding false alarm rates in IDPS are presented. [9] used the Self-Organizing Map (SOM) neutral networks to develop a two-stage classification system. By this way, it is possible to reduce the false alerts to a level, which can be over 50% of the entire false alarm occurrences. Conversely, [6] used a technique for mining the data based on a Growing Hierarchal Self-Organized Map (GHSOM). The GHSOM can modify its design if necessary, based upon the aspects of the input alarm data. This map reorganizes the alarms according to the incoming data and classifies them as true or false, providing aid to the network administrator.

[8] proposed a post-processing filter that reduces the false positives via a "network-based intrusion detection system." This system employs specific feature of detections that align itself with true attacks to filter alarms and limit false positives.

In another approach, [11] proposed a New Intrusion Detection Method Based on Antibody Concentration (NIDMBAC) to reduce the false alarm rate. Definitions of self, non-self, antigen and detector in the intrusion domain are utilized. An antigen recognition scheme is employed as a part of the clone proliferation. The intrusion detectors are processed to alter the number of antibodies in relation to the overall intensity of the antigen intrusion. In this case, the probabilistic calculation method was used for the intrusion alarm production based on the correlation between the antigen intrusion intensity and the antibody concentration factors. The analysis and results in this proposed method provided better results compared to the traditional methods.

The strategy proposed by [1] is based on statistical analysis, detection of both attacks and normal traffic patterns with respect to a hybrid statistical approach, which utilizes a data mining and decision classification techniques. However, the outcomes of the statistical analysis can be adjusted in order to minimize the miscalculation of false positive alarms and to distinguish between real attacks and false positive alarms of the traffic data.

Each of the above-mentioned schemes provides some solution against intrusion and reduces the rate of false alarm. Yet none of them alone was successful in taking a complete security risk of intrusions into account as a serious issue and is able to reduce the false positive alarms. They all failed to propose a uniform and robust architecture which can secure the whole system and be an aid for the administrator.

In this proposed work, it is assumed that unwanted false positive alarms can be eliminated by using a combination of soft computing techniques ranging from fuzzy logic (FL) for objective risk analysis, knowledge-based systems for determining intrusion patterns, artificial intelligence for machine and reinforced learning. The application of FL is considered appropriate for performing for good risk analysis and risk assessment. FL ensures that a set of complex related variables are grouped when making a final decision risk assessment and reducing unwanted false positive alarms and protecting the system not just effectively, but also efficiently, which is what PMRA is. In short, by employing PMRA, information and systems can be protected effectively without undue false positives.

3   Limitations of the Current Systems

The occasions when IPS accidentally or sincerely confirms a good activity as a malicious one or vice-versa hence flagging, a false alarm is not palatable to the users. Anomaly detection may not be able to detect an attack sometimes, but can still trigger a high false alarm. For instance, a legitimate system behavior may sometimes can be seen as an abnormal operation; hence, an alarm is triggered with respect to that. Furthermore, anomaly-based IDS systems are prone to false positive alarms, where the prevailing attack is based on changes to the evaluation of "normal" and false attack. Therefore, the application of risk analysis on the detected attacks and the measurement of their exposure factor on impact can help to confirm the validity of the alerts, as well as help to reduce the false alarms to a minimal acceptable level.

Information and computer technology is complicated and dynamic in nature. This nature consequently does not only pose a big problem to limit false alarms, but also for the usage of IDPS, that is also complex. Formulation of a Collaborative-IDPS (CIDPS) with soft computing elements is the suggestion in 2013 to overcome these complexities. This however does not, overcome the other challenges, such as newly injected attacks. Figure 1 shows the CIDPS architecture. The management function flows from the CIDPS in the intermediate section/layer of various components like the fuzzy risk manager, which controls the triggering of false alarms. The knowledge and multi-agent manager manages the intrusions from host computers and network elements that in turns provides the right feeds for the countermeasure operations. It also allows the monitoring and enabling the IDPS to detect hardware and software seamlessly and automatically.

The automatic manager includes four types of agents to cover the four segments of computing: Self-configuration, self-healing, self-optimization and self-protection. Self-configuration is important as the automatic manager makes the rules at runtime. Self-healing allows the system to operate in a cycle to detect the anomalous behavior while a solution is developed. The self-optimization allows the search without compromising other resources. Self-protection allows for the detection of bad functions and updates the knowledge base (KB) to limit its future recurrences. The checker monitors related resources through the consultation of sub ontology. It also detects the abnormal behavior. The ontology is normally updated so that it will be able to identify any non-expected change. At the same time, the checker reports the status to the analyzer agent that determines the system's current state by modeling, often complex behavior of the data to predict future anomalies. An estimated risk tool is used to consult the KB to find the best recourse and take the most appropriate actions. In addition, the KB is also updated to aid in future analysis. The planner structures actions should make sure the goals are achieved and should produce a series of changes that will help the element under protection. The executor performs the healing action, such as updating the policies by following the instructions given to it.

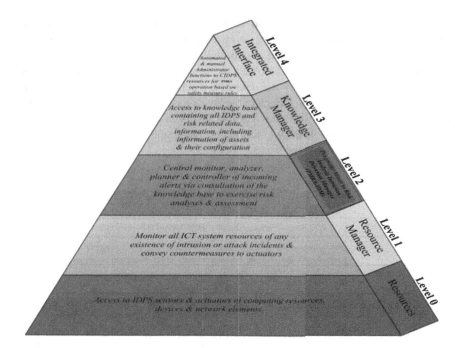

Fig. 1. Information and processes management layers integrating from the lowest to the highest layers in the proposed PMRA model

Furthermore, as soon as a threat is spotted, the affected systems are inspected deeper by the vulnerability scanner. The vulnerability is later assessed, while the scanner makes a real-time picture of all the ongoing attacks in order to determine the possible impact of the attack on the target system. The domain ontology, including high-level concepts (attacks, vulnerabilities and incidents, etc.) is acquired and the risk calculator then assigns a critical rating to the assets. After this step, any intrusion prevention solutions can be taken to evade intrusions, ensure appropriate system operation and limit operational overheads. For example, in the instance that intrusion prevention rules cannot be used on certain systems, a range can be disabled on a specific IP address, reducing false positive alarms. Furthermore, this dynamic protection/prevention allows the system to maintain a constant state of monitoring, assessment and optimization. A proper analysis of the false alarm reduction strategy requires the actual risk exposure to the attacked assets to be quantified.

4   Proposed Pragmatic Miner to Risk Analysis (PMRA) Model

To minimize the exposure of vulnerable resources on information systems and network services is the main purpose of using IDPS [10, 13, 14]. This paper proposed the framework as a way of reducing false alarm rates in intrusion detection systems by the implementation of Pragmatic Miner to Risk Analysis (PMRA) model. PMRA can compute the significant and the impact severity of each suspected activity in terms of malicious objects. In the process, the system will be capable of making more accurate decisions whether an activity or an object is an attack attempt or a normal system behavior that is misinterpreted by the detection mechanism. Figure 1 describes a five layered organization of the model. Each of the layers has a distinct function and services.

The hierarchy in Fig. 1 illustrates the five layers of the proposed PMRA model. These layers are explained in the following paragraph, while Table 1 shows a description to those management layers.

Table 1. Description of the five management layers of the proposed PMRA model

LAYER	MODULE	FUNCTIONALITY	
0	Resources	All the computing resources, transmission network, network elements, monitoring devices and IDPS sensors and actuators accessed by the resource manager that come under its purview and responsibilities.	
1	Resource Manager	The resource manager manages the total number of sensors under its purview that monitors the ICT system resources of any existence of intrusion or attack incidents.	
2	Pragmatic Miner to Risk Analysis Intrusion Detection Manager (PMRA-IDM)	Monitor	This sub-layer is responsible for monitoring, gathering incoming security of related information and analyzing same for indications of the potential incidents, malicious activities or policy violations.
		Analyzer	This sub-layer is responsible for evaluating and computing the risk of the detected attacks by using Pragmatic Miner technique as well as the predefined policies, rules and standards. On the other side, the guidelines and information from the knowledge-base are administered and controlled by the knowledge manager layer. In conventional systems, managing the knowledge-base would be provided by the system administrator in an ad hoc manner from time of initiation and subsequent updates.
		Planner	This sub-layer provides a way to observe and analyze the problems to better determine if any modifications need to be made, considering the risk assessment obtained from the analyzer module.
		Controller	This sub-layer provides the mechanism to schedule and carry out the necessary changes to protect the elements, which are under attack.
3	Knowledge Manager	The knowledge manager is not only the source of all pertinent knowledge via the knowledge-base that harvest and host details of data, information and existing rules and defines new rules as the operational case of the overall system warrants but also gives the general information of all assets under its control. This knowledge and information are formed by facts, beliefs, rules, norms and contracts. All the essential experiences, learning and knowledge are stored in the knowledge-base as a central repository, which is populated with and include the: 1) rules, policies, guidelines, results from the previous IDPS actions. 2) The security management functions, associated formulas and algorithms that leverage through the integrated interface layer by the system administrator functions (where the knowledge, information and data are the key inputs used in the risk analysis and risk assessment processes).	
4	Integrated Interface	The integrated interface is a unique bridging point between the system administrator functions (both in an automated and manual mode of operation) and the CIDPS. The learned experiences and system operation knowledge are important in defining and updating the system policies, rules and guidelines in the knowledge-base for subsequent consumption by various CIDPS/PMRA components in order to be continuously computed and operated with the latest information.	

5 Deployment and Analysis of the Pragmatic Miner to Risk Analysis (PMRA) Model

The fundamental parts of PMRA are (Preprocessing stage using FL, Mining Stage using Classify association rules, Evaluation stage using accuracy and error measures). Fuzzy linguistic variables or fuzzy expressions also termed as input and output parameters. Low, medium, high, very high, and critical respectively are the membership functions that were used for each input variable. The output functions have three variables (countermeasures) and they are 'Avoidance', 'Transference' and 'Acceptance'. Table 2 shows the characteristics of the input and output variables.

Table 2. Shows the input and output variables.

INPUT	
Residual risk	Low, Medium, High, Very High, Critical
Explore risk	
OUTPUT	
Countermeasures	Avoidance, Transference, Acceptance

The first input is the RR that can be defined as an "indicator of risk conveyed by an asset" and the second input is the ER that can be defined as an "indicator to risk generated by the attack." The ranges of inputs are divided into five classes (fuzzy sets) for each of the residual and exposed risks. The ranges of risk extend from 'Critical' to 'Low' with 'Very High', 'High' and 'Medium' falls in between of them.

5.1 Membership Functions for Input and Output Pragmatic Miner Model

As far as "fuzzifications" are concerned, relevant events determine the type of membership functions that should be used in the experiment [4]. The trapezoidal-shaped membership function is used to describe the fuzzy sets for input and output variables.

Furthermore, three basic membership functions for the countermeasure output are defined in the fuzzy sets First, the membership function denotes 'Avoidance' that is a high-risk exposure requiring some action to eliminate the threat. Second, denotes the 'Acceptance' that is a low- risk exposure that not necessarily requires action against the threat. Finally, the 'Transference' as the final step requires an expert judgment that needs to be taken by the system administrator.

X is a function of the trapezoidal curve x. The curve also depends on a, b, c and d as shown below.

$$f(x; a, b, c, d) = \begin{cases} 0, & x \leq a \\ \frac{x-a}{b-a}, & a \leq x \leq b \\ 1, & b \leq x \leq c \\ \frac{d-x}{d-c}, & c \leq x \leq d \\ 0, & d \leq x \end{cases} \tag{1}$$

If the expression is compressed, it becomes;

$$f(x; a, b, c, d) = \max\left(\min\left(\frac{x-a}{b-a}, \ 1, \ \frac{d-x}{d-c}\right), \ 0\right) \tag{2}$$

The Parameters a and d trace the "feet" the trapezoid while the parameters b and c trace the "shoulders".

5.2   Structure of the Pragmatic Miner Rules

Practically, residual and exposed risks determine the type of response the IDPS will trigger. For example, if the residual and exposed risks are very high, then the correct response is to apply safeguards in order to reduce the effects of the attack. However, if they are very low, the correct response to the trigger is first to understand the effects and acknowledge the risk without any attempts to control or mitigate them as illustrated in Table 3.

Table 3. Description risk analysis matrix

Types of risks		Exposed risk (ER)				
		Critical	Very high	High	Medium	Low
Residual risk (RR)	Critical	Avoidance	Avoidance	Avoidance	Transference	Transference
	Very high	Avoidance	Avoidance	Transference	Transference	Transference
	High	Avoidance	Transference	Transference	Transference	Transference
	Medium	Transference	Transference	Transference	Transference	Acceptance
	Low	Transference	Transference	Transference	Acceptance	Acceptance

Obviously, the matrix in Table 2 makes the fact that the upper and lower parts of the triangle are equivalent (meaning that they lead to the same result). Therefore, the cross between RR-VERY HIGH and ER-HIGH leads to the same transference as between ER-VERY.HIGH and RR-HIGH. There is a possibility of twenty-five combinations of inference rules in the fuzzy sets as shown in Table 2. Table 4 shows that a set of fifteen rules constructed based on the actual experimental qualitative analysis and the characteristics of the input and output variables (i.e., this result refute the hypothesis that appear in (Qassim and Mohd-Zin [7]).

5.3   Defuzzification

The term "defuzzification" implies the conversion of a fuzzy quantity into a precise value. It is the exact opposite process of fuzzification, which is the conversion of a precise value to a fuzzy quantity [4, 16]. The union of the output of each rule is used to develop the resultant membership functions.

Table 4. The description rules

1.	IF (RR is Critical) and (ER is Critical) Then (countermeasure is Avoidance)
2.	IF (RR- Critical) and (ER_ is Very High) OR (ER- is Critical) and (RR _ is Very High) Then (countermeasure is voidance)
3.	IF (RR is Critical) and (ER is High) OR(ER is Critical) and (RR is High) Then (countermeasure is Avoidance)
4.	IF (RR is Critical) and (ER is Medium) OR(ER is Critical) and (RR is Medium) Then (countermeasure is Transference)
5.	IF (RR is Critical) and (ER is Low) OR (ER is Critical) and (RR is Low) Then (countermeasure is Transference)
6.	IF (RR is Very High) and (ER is Very High)OR (ER is Very High) and (RR is Very High) Then (countermeasure is Avoidance)
7.	IF (RR is Very High) and (ER is High) OR (ER is Very High) and (RR is High) Then (countermeasure is Transference)
8.	IF (RR is Very High) and (ER is Medium) OR (ER is Very High) and (RR is Medium) Then (countermeasure is Transference)
9.	IF (RR is Very High) and (ER is Low) OR (RR is Very High) and (ER is Low) Then (countermeasure is Transference)
10.	IF (RR is High) and (ER is High) Then (countermeasure is Transference)
11.	IF (RR is High) and (ER is Medium) OR (ER is High) and (RR is Medium) Then (countermeasure is Transference)
12.	IF (RR is High) and (ER is Low) OR (ER is High) and (RR is Low) Then (countermeasure is Transference)
13.	IF (RR is Medium) and (ER is Medium) Then (countermeasure is Transference)
14.	IF (RR is Medium) and (ER is Low) OR (ER is Medium) and (RR is Low) Then (countermeasure is Acceptance)
15.	IF (RR is Low) and (E R is Low) OR (ER is Low) and (RR is Low) Then (countermeasure is Acceptance)

However, a close ascent of the countermeasure values clearly reveals and confirms that the pragmatic miner model can be used to predict countermeasure values under consideration.

Table 5 shows some of samples while Table 6 shows the correlation matrix that is calculated after the defuzzification carried out on the original database. Statistical analysis can be used to determine the active attributes of each class, while some attributes in linguistic terms can be expressed through the use of FL.

Each class of the acquired pattern classified can be inspected after classification has been made through PMRA. Statistical analysis is not only the best method of inspection on each class, but also can help one to generate the rules that control each class, attribute. This can be done through the use of the measures of central tendency and dispersion. Although there are many ways to measure the variability of data, in this work, we will use measures of dispersion, Standard deviation gives the average distance with which each element deviated from the mean. However, besides standard deviation, there are other important techniques prevails which are discussed below.

Range refers to the difference between the highest outcome in the data and the lowest outcome; it can be calculated by using the formula Xmax−Xmin. However, this range only uses two values from the entire set of data, making it unreliable. It cannot take into consideration the fact that extreme values can be very large and at the same time, many elements may be very close to each other. Looking at the range of this set of data 1, 1, 2, 4, 7, the range is 7−1 = 6, since 1 is the lowest while 7 is the highest outcome.

There are a few books indicate that the statistical range is the same as mathematical range. Therefore, the interval over which a data occurred is more important than a single number. From the above example, the range is from 1 to 7 or [1, 7]. Most of these measures will be used in this work.

Table 5. Sample of generation database

Residual risk					Exposed risk					Countermeasures
RR_Low	RR_Medium	RR_High	RR_Very High	RR_Critical	ER_Low	ER_Medium	ER_High	ER_Very High	ER_Critical	
3.518423248	28.97838246	44.70769534	82.51340374	99.51084659	6.491937775	11.49547943	44.86631873	87.0750865	96.75724227	1
5.99512786	25.09239558	58.91119317	76.77269645	98.93912277	2.940579222	24.30951448	51.02016885	73.14588889	97.7655921	2
7.847275971	36.99791336	58.27590313	82.44395231	95.28481309	3.423143681	14.22235216	56.42649048	88.63707334	97.60051147	3
6.542029219	39.83819901	47.78704478	80.22383222	93.82694231	7.774794347	28.97202866	51.60354076	75.4583978	92.04814068	1
5.861745268	19.68527951	69.54887876	84.23679779	96.53790496	0.638851319	35.24259225	67.1518145	73.75239924	93.4298592	2
9.867828263	20.02176915	58.01326439	74.17035904	94.39298796	7.544338868	33.46747283	54.07889779	88.95311385	92.96186557	3
8.547584561	21.43947004	54.77429516	81.71850608	95.30333806	6.37563549	27.17654586	53.17090207	74.36135939	97.42440043	1
2.752772132	31.05953781	46.05527981	78.49724646	98.56625556	9.969384472	12.98368693	67.63718875	76.14097827	97.70969963	2
8.592575547	23.38343425	55.25247732	78.0142526	97.34718736	3.271051575	21.30597164	62.80206931	87.14585915	98.92727356	3
0.605905878	12.93375694	57.94187465	85.94892174	93.8641164	4.9916625	32.75716133	55.61123628	72.14424367	96.15211615	1
7.7861302	32.35252307	62.0362408	83.88373015	96.88168811	0.218471278	19.52531705	69.96555856	85.30930202	98.62832459	2
5.036512958	13.05588823	64.94809088	87.64271193	99.18226391	6.539912209	33.92214584	63.46221583	80.82448607	99.47560229	3

Table 6. Correlation matrix after the Defuzzification stage

Variables	RR_Low	RR_Medium	RR_High	RR_Very High	RR_Critical	ER_Low	ER_Medium	ER_High	ER_Very High	ER_Critical
RR_Low	1.000	0.040	0.067	0.205	0.006	0.208	−0.025	−0.038	−0.049	−0.003
RR_Medium	0.040	**1.000**	−0.068	0.094	0.025	−0.088	−0.102	−0.094	−0.023	0.084
RR_High	0.067	−0.068	**1.000**	−0.082	0.114	0.172	0.027	0.048	0.016	0.073
RR_Very High	0.205	0.094	−0.082	**1.000**	0.054	0.092	0.006	−0.064	−0.153	0.094
RR_Critical	0.006	0.025	0.114	0.054	**1.000**	−0.055	0.054	0.194	0.067	0.098
ER_Low	0.208	−0.088	0.172	0.092	−0.055	**1.000**	−0.028	−0.019	0.075	−0.201
ER_Medium	−0.025	−0.102	0.027	0.006	0.054	−0.028	**1.000**	0.152	−0.124	−0.117
ER_High	−0.038	−0.094	0.048	−0.064	0.194	−0.019	0.152	**1.000**	0.007	−0.029
ER_Very High	−0.049	−0.023	0.016	−0.153	0.067	0.075	−0.124	0.007	**1.000**	−0.070
ER_Critical	−0.003	0.084	0.073	0.094	0.098	−0.201	−0.117	−0.029	−0.070	**1.000**

The real rules of a database, which contains 100 samples and ten inputs (i.e., five features related to RR and five features related to ER), are selected. Because the accuracy is one of the main goals of this work as explained in Table 5, therefore the only rules that satisfy 100% of accuracy are selected as shown in Table 7. The confusion matrix of the estimating samples is shown in Table 7 and the final decision surface of the PMRA System is shown in Fig. 2.

5.4   Pragmatic Miner Accuracy and Error

After the formation of classify association rules, five experimental tests were done from separate experiments and the proposed mining model is used to recognize the abnormality of the system at the same conditions as shown in Table 2, so that the investigation of mining accuracy and error can be accomplished. The error is computed in order to measure the gap between the predicted and the measured values. The individual error percentage can be determined by dividing the absolute difference between the predicted and the measured values is given by:

$$e_i = \left(\frac{|A_m - A_p|}{A_m} \right) * 100\% \tag{3}$$

where, e_i is the individual error, Am is the measured value and Ap is the predicted value.

The accuracy, however, measures the closeness of the predicted value to the measured value. The average of the individual accuracies is the model accuracy as shown in Eq. 3.

$$a = \frac{1}{N} \sum\nolimits_{i=1}^{N} \left(1 - \frac{|A_m - A_p|}{A_m} \right) * 100\% \tag{4}$$

where a is the model accuracy and N are the total number of the tested data sets.

Table 7. The actual rules for PMRA

Purity	Rules
34.00%	
37.08%	If ER_High in [41.3, 67.624[then Countermeasures = 1 in 37.1% of cases
81.82%	If ER_High in [67.624, 69.901[then Countermeasures = 2 in 81.8% of cases
37.18%	If ER_Very High in [71.414, 87.212[and ER_High in [41.3, 67.624[then Countermeasures = 3 in 37.2% of cases
72.73%	If ER_Very High in [87.212, 89.855[and ER_High in [41.3, 67.624[then Countermeasures = 1 in 72.7% of cases
35.94%	If RR_Critical in [91.123, 98.704[and ER_Very High in [71.414, 87.212[and ER_High in [41.3, 67.624[then Countermeasures = 2 in 35.9% of cases
64.29%	If RR_Critical in [98.704, 99.915[and ER_Very High in [71.414, 87.212[and ER_High in [41.3, 67.624[then Countermeasures = 3 in 64.3% of cases
35.00%	If ER_Very High in [71.414, 86.173[and RR_Critical in [91.123, 98.704[and ER_High in [41.3, 67.624[then Countermeasures = 1 in 35% of cases
100.00%	If ER_Very High in [86.173, 87.212[and RR_Critical in [91.123, 98.704[and ER_High in [41.3, 67.624[then Countermeasures = 2 in 100% of cases
40.00%	If ER_Low in [0.227, 7.714[and ER_Very High in [71.414, 86.173[and RR_Critical in [91.123, 98.704[and ER_High in [41.3, 67.624[then Countermeasures = 2 in 40% of cases
60.00%	If ER_Low in [7.714, 9.975[and ER_Very High in [71.414, 86.173[and RR_Critical in [91.123, 98.704[and ER_High in [41.3, 67.624[then Countermeasures = 3 in 60% of cases
66.67%	If RR_Low in [0.452, 8.611[and RR_Critical in [98.704, 99.915[and ER_Very High in [71.414, 87.212[and ER_High in [41.3, 67.624[then Countermeasures = 3 in 66.7% of cases
50.00%	If RR_Low in [8.611, 9.186[and RR_Critical in [98.704, 99.915[and ER_Very High in [71.414, 87.212[and ER_High in [41.3, 67.624[then Countermeasures = 2 in 50% of cases
72.73%	If RR_Low in [0.452, 7.955[and RR_Critical in [98.704, 99.915[and ER_Very High in [71.414, 87.212[and ER_High in [41.3, 67.624[then Countermeasures = 3 in 72.7% of cases
100.00%	If RR_Low in [7.955, 8.611[and RR_Critical in [98.704, 99.915[and ER_Very High in [71.414, 87.212[and ER_High in [41.3, 67.624[then Countermeasures = 1 in 100% of cases
100.00%	If RR_High in [47.424, 58.355[and RR_Low in [8.611, 9.186[and RR_Critical in [98.704, 99.915[and ER_Very High in [71.414, 87.212[and ER_High in [41.3, 67.624[then Countermeasures = 3 in 100% of cases
100.00%	If RR_High in [58.355, 69.285[and RR_Low in [8.611, 9.186[and RR_Critical in [98.704, 99.915[and ER_Very High in [71.414, 87.212[and ER_High in [41.3, 67.624[then Countermeasures = 2 in 100% of cases
88.89%	If RR_Low in [0.033, 6.778[and ER_Very High in [87.212, 89.855[and ER_High in [41.3, 67.624[then Countermeasures = 1 in 88.9% of cases
100.00%	If RR_Low in [6.778, 9.762[and ER_Very High in [87.212, 89.855[and ER_High in [41.3, 67.624[then Countermeasures = 3 in 100% of cases
100.00%	If ER_Low in [1.448, 7.651[and RR_Low in [0.033, 6.778[and ER_Very High in [87.212, 89.855[and ER_High in [41.3, 67.624[then Countermeasures = 1 in 100% of cases
66.67%	If ER_Low in [7.651, 9.179[and RR_Low in [0.033, 6.778[and ER_Very High in [87.212, 89.855[and ER_High in [41.3, 67.624[then Countermeasures = 1 in 66.7% of cases

100.00%	If RR_High in [52.354, 58.22[and ER_Low in [7.651, 9.179[and RR_Low in [0.033, 6.778[and ER_Very High in [87.212, 89.855[and ER_High in [41.3, 67.624[then Countermeasures = 1 in 100% of cases
100.00%	If RR_High in [58.22, 60.819[and ER_Low in [7.651, 9.179[and RR_Low in [0.033, 6.778[and ER_Very High in [87.212, 89.855[and ER_High in [41.3, 67.624[then Countermeasures = 3 in 100% of cases
88.89%	If ER_Critical in [91.718, 98.217[and ER_High in [67.624, 69.901[then Countermeasures = 2 in 88.9% of cases
50.00%	If ER_Critical in [98.217, 98.653[and ER_High in [67.624, 69.901[then Countermeasures = 1 in 50% of cases
100.00%	If RR_Low in [0.861, 9.198[and ER_Critical in [91.718, 98.217[and ER_High in [67.624, 69.901[then Countermeasures = 2 in 100% of cases
50.00%	If RR_Low in [9.198, 9.581[and ER_Critical in [91.718, 98.217[and ER_High in [67.624, 69.901[then Countermeasures = 2 in 50% of cases
100.00%	If RR_Very High in [76.721, 79.65[and RR_Low in [9.198, 9.581[and ER_Critical in [91.718, 98.217[and ER_High in [67.624, 69.901[then Countermeasures = 2 in 100% of cases
100.00%	If RR_Very High in [79.65, 82.58[and RR_Low in [9.198, 9.581[and ER_Critical in [91.718, 98.217[and ER_High in [67.624, 69.901[then Countermeasures = 3 in 100% of cases
100.00%	If RR_Low in [3.089, 6.221[and ER_Critical in [98.217, 98.653[and ER_High in [67.624, 69.901[then Countermeasures = 1 in 100% of cases
100.00%	If RR_Low in [6.221, 9.353[and ER_Critical in [98.217, 98.653[and ER_High in [67.624, 69.901[then Countermeasures = 2 in 100% of cases

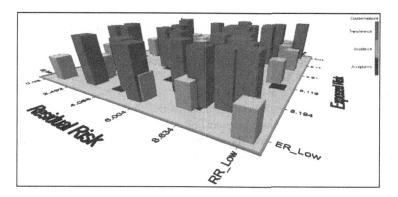

Fig. 2. The final decision surface of PMRA system

The model accuracy for PMRA_ID was determined after calculating the error of the data set. Table 8 shows the experimental condition, countermeasure results, and the miner model predicted values.

As we can observe from Table 9 that the highest percentage of error in the PMRA model prediction is 0.32%. This indicates and confirms that the PMRA prediction countermeasure results are low and very close to the real experimental where is this number countermeasures values. It also shows that the average accuracy of proposed PMRA model is 90.11%. The value of the accuracy shows that the proposed model can predict the vulnerability of a system as it can be observed from the graph trend lines.

Table 8. Confusion matrix for all the samples

From\To	Avoidance	Transference	Acceptance	Total	% Correct
Avoidance	10	16	8	34	29.41%
Transference	0	32	1	33	96.97%
Acceptance	0	11	22	33	66.67%
Total	10	59	31	100	

Table 9. The accuracy and error of the pragmatic miner model

Risk parameters (INPUTS)		Countermeasure parameter (OUTPUT)			STATISTICS				
Residual risk	Expose risk	1st epoch	2nd epoch	3rd epoch	Average	Standard deviation (σ)	Measured	Error %	Proposed pragmatic miner model
90.00	70.00	90.00	100.0	70.0	86.67	15.28	76.00	0.18	76.80
10.00	40.00	80.00	90.00	70.0	80.00	10.00	68.00	0.18	78.30
70.00	50.00	67.00	70.00	100.0	79.00	18.25	98.00	0.32	99.07
98.00	98.00	90.00	100.0	69.00	86.33	15.82	95.40	0.06	96.80
67.00	43.00	80.00	79.00	98.00	85.67	10.69	99.00	0.19	99.60
							Average accuracy of miner model = 90.11%		

6 Discussion and Conclusion

The combination of risk analysis mechanism with a developed PMRA for online IDS through the modification of an FL Controller with mining algorithm detects Distributed Denial of Service (DDoS) attack with 90.11% accuracy and that is superior to FL Controller IDS and D-SCIDS by themselves. The main parameters used to compute MF is a = 0.02, b = 0.05, c = 0.08, d = 1.2.

The calculation of Discretization, feature selection and accuracy are simultaneously handled in this work. This reduced the cost of computation and built the detection in a detailed manner. Observation has shown that the detection of continuous attack attribute by FLC when the same parameters are applied to all the attributes causes the classified association rules accuracy to vary widely. Conversely, the best result of classification accuracy is obtained when FLC is combined with the object Risk Analysis for different attributes in a different class.

Because of the increased level of computer information attacks, the necessity to provide an effective intrusion detection and prevention methods had increased. IDPS suffered from a several weaknesses including post event detection, overwhelming false alarms and a centralized analysis of intrusion. This paper introduced a centralized and automatic system called as PMRA, as a reliable substitute for conventional IDPS. The experimental result showed PMRA was more effective and consistent than the other IDPSs.

References

1. Anuar, N.B., Papadaki, M., Furnell, S., Clarke, N.: Incident prioritisation using analytic hierarchy process (AHP): Risk Index Model (RIM). Secur. Commun. Netw. **6**, 1087–1116 (2013). https://doi.org/10.1002/sec.673
2. Bajpai, S., Sachdeva, A., Gupta, J.P.: Security risk assessment: applying the concepts of fuzzy logic. J. Hazard. Mater. **173**, 258–264 (2010). https://doi.org/10.1016/j.jhazmat.2009.08.078
3. Catania, C.A., Garino, C.G.: Automatic network intrusion detection: current techniques and open issues. Comput. Electr. Eng. **38**(5), 1062–1072 (2012). https://doi.org/10.1016/j.compeleceng.2012.05.013
4. Chen, P.Y., Kataria, G., Krishnan, R.: Correlated failures, diversification and information security risk management. MIS Q. **35**, 397–422 (2011)
5. Liao, H.J., Lin, C.H.R., Lin, Y.C., Tung, K.Y.: Intrusion detection system: a comprehensive review. J. Netw. Comput. Appl. **36**(1), 16–24 (2013). https://doi.org/10.1016/j.jnca.2012.09.004
6. Mansour, N., Chehab, M., Faour, A.: Filtering intrusion detection alarms. Clust. Comput. **13**, 19–29 (2010)
7. Qassim, Q., Mohd-Zin, A.: Strategy to reduce false alarms in intrusion detection and prevention systems. Int. Arab J. Inf. Technol. (IAJIT) **11**(5) (2014)
8. Spathoulas, G.P., Katsikas, S.K.: Reducing false positives in intrusion detection systems. Comput. Secur. **29**, 35–44 (2010). https://doi.org/10.1016/j.cose.2009.07.008
9. Tjhai, G.C., Furnell, S.M., Papadaki, M., Clarke, N.L.: A preliminary two-stage alarm correlation and filtering system using SOME neural network and K-means algorithm. Comput. Secur. **29**, 712–723 (2010). https://doi.org/10.1016/j.cose.2010.02.001
10. Whitman, M.E., Mattord, H.J.: Principles of Information Security. Cengage Learning, Boston (2011)
11. Zeng, J., Li, T., Li, G., Li, H.: A new intrusion detection method based on antibody concentration. In: Huang, D.-S., Jo, K.-H., Lee, H.-H., Kang, H.-J., Bevilacqua, V. (eds.) ICIC 2009. LNCS, vol. 5755, pp. 500–509. Springer, Heidelberg (2009). https://doi.org/10.1007/978-3-642-04020-7_53
12. Zhou, Y.P., Fang, J.A.: Intrusion detection model based on hierarchical fuzzy inference system. In: The 2th International Conference on Information and Computing Science, ICIC 2009, vol. 2, pp. 144–147. IEEE (2009). http://dx.doi.org/10.1109/ICIC.2009.145
13. Al-Janabi, S., Al-Shourbaji, I., Shojafar, M., Shamshirband, S.: Survey of main challenges (security and privacy) in wireless body area networks for healthcare applications. Egypt. Inform. J. **18**, 113–122 (2017)
14. Al-Janabi, S., Al-Shourbaji, I.: A study of cyber security awareness in educational environment in the Middle East. J. Inf. Knowl. Manag. **15**, 1650007 (2016)
15. Ahamad, S.S., Al-Shourbaji, I., Al-Janabi, S.: A secure NFC mobile payment protocol based on biometrics with formal verification. Int. J. Internet Technol. Secur. Trans. **6**, 103–132 (2016)
16. Folorunso, O., Ayo, F.E., Babalola, Y.E.: Ca-NIDS: a network intrusion detection system using combinatorial algorithm approach. J. Inf. Priv. Secur. **12**, 181–196 (2016)

Intelligent System E-Learning Modeling According to Learning Styles and Level of Ability of Students

Utomo Budiyanto[1(✉)], Sri Hartati[2], S. N. Azhari[2], and Djemari Mardapi[3]

[1] Budi Luhur University, Jakarta 12260, Indonesia
utomo.budiyanto@budiluhur.ac.id
[2] Gadjah Mada University, Sekip Utara, Yogyakarta 55521, Indonesia
[3] Yogyakarta State University, Yogyakarta 55281, Indonesia

Abstract. Presenting learning materials that suited to learners is very important in e-learning. This is due to e-learning now provides the same learning materials for all learners. This study built intelligent system models that can detect learning styles and ability levels of learners. Categorizing and identifying styles are important to map abilities in accordance to provide learning material that matches preferences of the learners. There are many methods used to detect characteristics of learners this research used Felder Silverman Learning Style Model (FSLSM) which proven to be suitable for application in an e-learning environment. In addition to learning style, other elements that are required is to determine the level of abilities. The level of abilities is categorized into three categories known as Beginner, Intermediate, and Advanced which will be determined through an evaluation with Rasch Model (Model Rasch). To determine the suitability of learning materials with learning styles and ability level used artificial intelligence techniques that are rule-based. This research produces adaptive e-learning model that can present learning materials in accordance with learning style and ability level, to be able to improve the ability of learners.

Keywords: Intelligent E-learning system
Felder Silverman Learning Style Model · Rasch Model
Rule-based recommendation system

1 Introduction

The existing of E-learning system generally will present the material the same for each learner regardless of the characteristics of their learning styles. In fact, learners have different characteristics in learning styles, the level of understanding, background and more. Therefore learners not necessarily received learning materials appropriately.

The traditional approach "one-size-fits-all" learning methods are no longer suitable to meet the needs of learners [10]. Learners have different ways in their studying styles when learning styles don't match the way learning in environmental education; students will have a problem in the study [4]. Statically the results inferential show that different learning styles have a significant effect on the flow of experience,

© Springer Nature Singapore Pte Ltd. 2017
A. Mohamed et al. (Eds.): SCDS 2017, CCIS 788, pp. 278–290, 2017.
https://doi.org/10.1007/978-981-10-7242-0_24

different learning styles have a significant effect on the effectiveness of the learning overall, and the flow of experience students correlated with the effectiveness of learning [17]. Some of the pattern found in which students with preferences different learning styles show different behavior significantly in an online course. The results of this pattern of learning style seem to be important to provide the course which includes the features by different learning styles [6]. This determined how students know their styles of learning and therefore presents the material that combines learning style each student has the potential to make learning easier for students and can increase the progress of learning [7].

In the absence of the ability to present learning strategies that match the learning style of the learners as well as present the learning materials by the learners' abilities, it can lead to the absorption of the material to be less effective for learners [10]. The learning process that adapts to the needs of learners can be done by face-to-face one on one teachers with learners. This is hard to do in the real world because it requires teachers in large numbers and has competencies that match the needs of learners. Therefore it is necessary tool-shaped tool that can meet the needs of learning. To be effective, then this learning system must be accessible not limited space and time, wherever whenever learners can use it as needed.

This research proposes a model of intelligent e-learning system in accordance with the needs of learners. In this case, the needs of learners in question are the learning style as well as the level of ability. Learning style used is FSLSM because it is best suited to computer-based learning [3] and using Item Response Theory (IRT) to determine the level of ability of learners. From both parameters, the learners will get the material and learning strategies that match the learning style and level of ability. This parameter is used as input data for recommendation system using artificial intelligence that is rule-based. Also, this study will be able to improve the ability of learners; this has not been done in previous research.

2 Intelligent System E-Learning Model

Intelligent e-learning systems are a computerized learning system that supported by intelligent methods. Research in e-learning system that can deliver learning materials suit to learners developed using some methods. Using adaptive personalized learning path [2], psychological type and learning styles [8], a fuzzy tree matching recommender [23], case-based reasoning [22], rule-based adaptive test generation and assessment [11], data mining decision tree J48 [16], ontology [14], literature-based [5], rule-based [20], neuro-fuzzy [12, 18], three modules of student, domain and tutor [1], user profiling and learning resources adaptation [21].

The aim of intelligent e-learning system is to meet the needs of different students get different learning materials based on their needs. This research use rule-based to present learning materials for each student based on their learning style and level of ability. The model of intelligent E-learning system according to the needs of learners consists of several processes as shown in Fig. 1.

By the needs of learners rules regarding recommendations learning materials used for each learner based learning styles and the level of ability obtained from teachers and

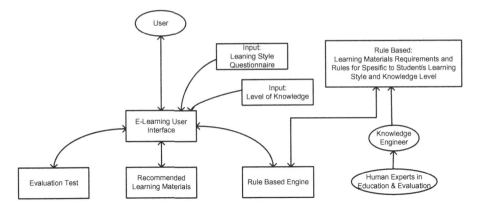

Fig. 1. The structure of intelligent e-learning model

teaching team. Processes of the model intelligent system E-learning divided into 6 phases of process (1) determine learning style (2) determine the ability levels (3) determine the suitability of the material (4) evaluation of learning process (5) improve and (6) evaluation and testing model.

2.1   Determining Learning Styles

Everyone has style in learning. Learning styles will determine the learning process, so different recommendations are needed for each learning style.

Some theories of learning styles such as Kolb [13], Honey and Mumford [9] and Felder and Silverman (FSLSM) [4] seem most appropriate for use in computer-based education systems [3]. Most learning style models classify learners into groups, while the FSLSM describes them in more detail by differentiating between learners' preferences into four dimensions: (1) the active-reflective dimension (2) the perception dimension (sensing-intuitive) (3) the dimensions Input (visual-verbal) and (4) dimensions of understanding (sequential-global).

Active-Reflective learning style. Learners with active learning styles tend to nurture and understand the best information by activating by discussing, applying it or explaining it to others. On the other hand, reflective learners choose to think it over calmly. Active learners prefer group learning instead reflective learners prefer self-study. Listening to lectures without physical activity unless writing difficult notes is done both, but especially for active learners.

Learning style Sensing-Intuitive (Dimensional Perception). Learners of type sensing tend to like learning facts, while intuitive learners often prefer to find possibilities and relationships. Learners of sensing type are easier to understand if given examples of concepts and procedures and find out how the concept is applied to practice, whereas intuitive learners try to understand an interpretation or theory linking facts. Therefore learners are sensing suitable given teaching with real applications, while intuitive learner suitable to be provided with concept maps.

Visual-Verbal learning style (Input Dimension). Visual learners remember what they see best, such as drawings, diagrams, timelines, movies or demonstrations, whereas verbal learners get more information from words and explanations spoken, so they like discussions and writing projects.

Sequential-Global learning style (Dimension of Understanding). Sequential learners gain an understanding through linear steps. Global students seek to understand the big picture and solve complex problems once they get the big picture.

In this process, the user will fill out the learning style questionnaire as contained on https://www.engr.ncsu.edu/learningstyles/ilsweb.html consisting of forty-four questions. The question is divided into four dimensions, so each dimension has eleven questions. With values ranging from +11 to −11 with addition/subtraction of ±2. Each question has a +1 value for answer an or −1 for answer b. Answer a refers to the first part of each dimension: active, sensing, visual or sequential whereas answer b will refer to the next part of each dimension: reflective, intuitive, verbal or global.

2.2   Determining Level of Ability of Learners

This process is used to determine the level of ability, known as Beginner, Intermediate and Advanced. What the learner does is choose the difficulty level of learning. After that learner will be tested whether the ability is enough to be able to do the learning process. For example when learners choose an Advanced difficulty level, they must pass the test given. If not passed then learners should choose the level of difficulty underneath and so on.

The test uses the CBT (Computerized Based Test) technique that presents the test by adjusting the ability level of participants to utilize the item characteristics in IRT (Item Response Theory). The IRT parameter used is 1-PL (Parameter Logistic) known as Rasch Model.

Item Response Theory (IRT) is a psychometric theory that provides a basis for measuring the scale of test participants and questions based on the responses given to the problem to provide similarities between the statistics of the problem and the learner's estimation of ability.

Classical measurement theory is also known as classical test theory. In the classical test, the level of difficulty items dependent (dependent) to the ability of respondents. For the high-ability respondents, the item is not difficult (easy). For low-ability respondents, items are difficult. Of the item is not difficult (easy), it appears the ability of respondents to be high. On the hard point, it appears the ability of the respondents to be low. The difficulty level of item depends on the ability of the respondent. The same item will feel heavy for those who are low-ability and light to those who are capable.

The ability of respondents to depend on the item difficulty, those who work on difficult items will appear to be inferior while those who work on the items will appear highly capable. Classical measurement theory (classical test theory) cannot be used for matching respondents' ability with item difficulty (because they are dependent). In classical theory, there is interdependence between the respondents' ability and the item difficulty. We recommend that the method of mentioning the measurement results is matched by the name of the measuring instrument. For example, 450 TOEFL, 630

SPMB [19]. Measuring results can be understood through relation to the measuring instrument used (TOEFL or SPMB). We recommend that the name of the measuring instrument be widely known by many people. On the test, modern measurement theory is known as modern test theory. In the modern measurement, the item difficulty level is not directly related to the ability of the respondent.

In the modern measurements with modern approaches, the item difficulty level is directly related to item characteristics. The item-difficult point of the modern measurement lies in: $P(\theta) = Pmin + 0.5$ (Pmaks − Pmin) = Pmin + 0.5 (1 − Pmin). The difficulty level of the items is given a notation b. In modern measurements, the direct item difficulty level is directly related to item characteristics. The high and low (θ) abilities have the same difficulty point b. The ability of the respondent and the item difficulty level to be independent. Modern measurements can be used to match the ability of respondents to item difficulty.

Item characteristics are determined by respondents' responses (both high and low ability) so that it is known as response theory item (Item Response Theory). This theory is also known by various names: Latent Trait Theory (LTT), Item Characteristic Curve (ICC), Item Characteristic Function (ICF). The item characteristics are the regression between the ability and the probability of answering correctly.

The Item response theory needs to determine the model of item characteristics used; it can take the form of one parameter (1P), two parameters (2P), three parameters (3P), or other models.

1. $P : P(\theta) = f(b, \theta)$
2. $P : P(\theta) = f(a, b, \theta)$
3. $P : P(\theta) = f(a, b, c, \theta)$

One, two and three are the number of item parameters. The parameter θ is a parameter of respondent ability. Parameter b is the parameter of the hardness point of the item. In 1P and 2P, $b = \theta$ when $(P(\theta) = 0.5)$. At 3P $b = \theta$ when $(P(\theta) = 0.5(1 + c))$. Parameter a is a different item power parameter. Parameter c is the correct guess parameter of the answer of the item. Rasch Model is the response theory of item that uses one parameter is the difficulty item level.

In order to prepare the questions to be presented in the evaluation, there are nine steps to be taken in preparing the standardized test result or achievement [15], namely: (a) Preparing the test specifications (b) Writing the test (c) Reviewing the test d) Testing the test (e) Analyzing the test items (f) Fixing the test (g) Assembling the test (h) Implementing the test (i) Interpreting the test results.

2.3   Rule-Based Engine

Rule-based is used as a way to store and manipulate knowledge to be manifested in information that can assist in solving problems, with rules based on determining conformity. Table 1. shows the rule-based that is formed from the combination of learning style, the level of ability and the learning materials to be presented.

The Combination of Learning Styles using FSLSM as in Table 2.

Presentation of learning materials based on FSLSM is divided into four dimensions, as in Table 3.

Table 1. Rule-based learning styles, ability levels, and learning materials

LS	Level			Subject	Ct	Learning presentation
	B	I	A			
LS-1	√			M-1	C-1	Activities, Example, Explanations, Theory, Additional Materials, Many Visual/Image, Step by Step
LS-2	√			M-1	C-1	Activities, Examples, Explanations, Theory, Additional Materials, Many Visual/Image, Global
LS-3	√			M-1	C-1	Activities, Examples, Explanations, Theory, Additional Materials, Many Text, Step by Step
LS-4	√			M-1	C-1	Activities, Examples, Explanations, Theory, Additional Materials, Many Text, Global
LS-5	√			M-1	C-1	Activity, Example, Explanation, Theory, (Abstract, Formula, Concept), Many Visual/Image, Step by Step
LS-6	√			M-1	C-1	Activities, Examples, Explanations, Theory, (Abstract, Formula, Concepts), Many Visual/Image, Global
LS-7	√			M-1	C-1	Activities, Examples, Explanations, Theory, (Abstract, Formula, Concepts), Many Text, Step by Step
LS-8	√			M-1	C-1	Activities, Examples, Explanations, Theory, (Abstract, Formula, Concepts), Many Text, Global
LS-9	√			M-1	C-1	Example, Explanation, Theory, Activity, Additional Materials, Many Visual/Image, Step by Step
LS-10	√			M-1	C-1	Example, Explanation, Theory, Activity, Additional Materials, Many Visual/Image, Global
LS-11	√			M-1	C-1	Example, Explanation, Theory, Activity, Additional Materials, Many Text, Step by Step
LS-12	√			M-1	C-1	Example, Explanation, Theory, Activity, Additional Materials, Many Text, Global
LS-13	√			M-1	C-1	Example, Explanation, Theory, Activity, (Abstract, Formulation, Concept), Many Visual/Image, Step by Step
LS-14	√			M-1	C-1	Example, Explanation, Theory, Activity, (Abstract, Formulation, Concept), Many Visual/Image, Global
LS-15	√			M-1	C-1	Example, Explanation, Theory, Activity, (Abstract, Formulation, Concept), Many Visual/Image, Global
LS-16	√			M-1	C-1	Example, Explanation, Theory, Activity, (Abstract, Formulation, Concept), Many Text, Global

The level of learning material is divided into three parts, as in Table 4.
Therefore formulated as an if-then rule below:

IF (
   Learning Style IS Active – Sensing – Visual – Sequential AND Level IS Beginner
)
THEN (
   Material IS M-1 AND Content IS C-1 AND Presentation IS Activities, Examples, Explanation, Theory, Additional learning Materials, Many Visual/Figures, Step by Step
)

Table 2. Combination of learning styles

ID	Learning style
LS-1	Active – Sensing – Visual – Sequential
LS-2	Active – Sensing – Visual – Global
LS-3	Active – Sensing – Verbal – Sequential
LS-4	Active – Sensing – Verbal – Global
LS-5	Active – Intuitive – Visual – Sequential
LS-6	Active – Intuitive – Visual – Global
LS-7	Active – Intuitive – Verbal – Sequential
LS-8	Active – Intuitive – Verbal – Global
LS-9	Reflective - Sensing – Visual – Sequential
LS-10	Reflective – Sensing – Visual – Global
LS-11	Reflective – Sensing – Verbal – Sequential
LS-12	Reflective – Sensing – Verbal – Global
LS-13	Reflective – Intuitive – Visual – Sequential
LS-14	Reflective – Intuitive – Visual – Global
LS-15	Reflective – Intuitive – Verbal – Sequential
LS-16	Reflective – Intuitive – Verbal – Global

Table 3. Presentation of material by dimension

ID	Styles	Presentation/display materials
Process	Active	Presentations begin with activities, examples, explanations and then theory
	Reflective	Presentation begins with an example, later explanations of theory and activity
Perception	Sensing	Additional materials
	Intuitive	It has abstract explanations, formulas and concepts
Reception/input	Visual	It has many visual/image views
	Verbal	It has many verbals text views
Understanding	Sequential	Step by step
	Global	Displays verbal, visual, explanations, formula and concepts

Table 4. Ability levels and learning materials

Level	Learning materials	Chapter
Beginner (B)	Learning Materials 1 (M-1) to Learning Materials 5 (M-5)	Chapter 1 (C-1) to Chapter 5 (C-5)
Intermediate (I)	Learning Materials 6 (M-6) to Learning Materials 10 (M-10)	Chapter 6 (C-6) to Chapter 10 (C-10)
Advanced (A)	Learning Materials 11 (M-11) to Learning Materials 15 (M-15)	Chapter 11 (C-11) to Chapter 10 (C-15)

2.4   Recommended Learning Materials

The rule-based engine produces learning materials that match the learning style and level of ability of the learners. The next process is to present the learning that is stored in the database to learners. The e-learning learning process begins.

2.5   Learning Evaluation

After the learning process, the next is to evaluate whether the learners have understood the material provided. Evaluation using CBT technique as in point 2.2. If the value is met, then the learner can proceed to the next material. Instead, learners must repeat the learning materials.

By using Rasch technique, it can be known what material is less controlled by learners. So the process of repeating does not have to be from the beginning of learning, just starting from the material that has not been mastered.

3   Model Design

In this smart E-learning system model, after entering into the system with each identifier as shown in Fig. 2. Learners will be given a questionnaire to be determined learning style as shown in Fig. 3. The results of the questionnaire are one of sixteen possible learning styles, as shown in Fig. 4.

After obtained learning styles of learners next is to determine the level of the ability of the students using CBT Rasch Model. Learners can choose level capabilities by the ability as in Fig. 5. Then appear questions that have been adapted to the level select as in Fig. 6. if learners can pass the selected tests it will appear as Fig. 7 and vice versa if not be able to pass through will be presented test level below.

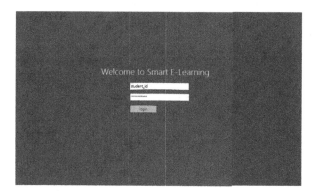

Fig. 2. Entering e-learning system

Learning styles and level of ability to be input for the recommendations of learning materials that will be displayed to learners. This recommendation uses artificial intelligence techniques with rule-based methods that produce the rules as discussed earlier.

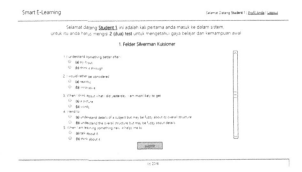

Fig. 3. Felder silverman questionnaires

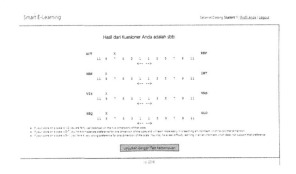

Fig. 4. Learning styles results

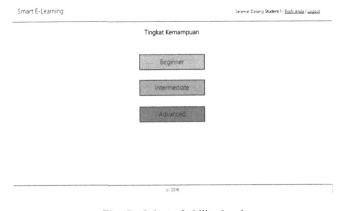

Fig. 5. Select of ability level

With this recommendation then each learner will get the learning materials according to his needs, as in Fig. 8.

Testing this model is done from the side of learners is to ask whether the material presented by the learning style and compare the output of the system with the expert opinion of education and learning evaluation in the field.

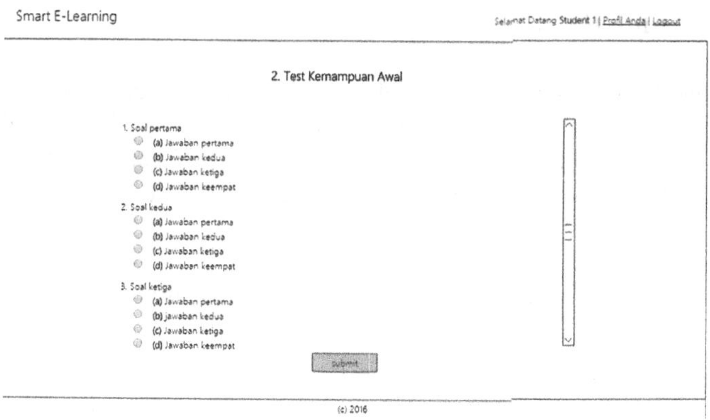

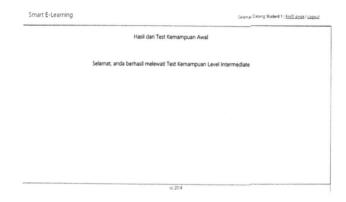

Fig. 6. Computerized based test according to student ability level

Fig. 7. Test ability level results

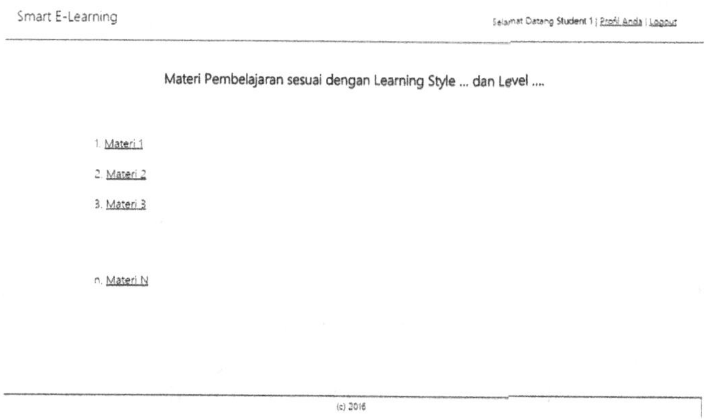

Fig. 8. Learning materials recommendation according to student's ability and styles

4 Analysis and Result

The system had been implemented in web programming subject and thirty students as learners. The system outputs learning style and level of ability of student then present the learning materials based on it. The output result of learning style showed in Table 5, the result of the ability of student showed in Table 6.

Table 5. Summary of Student's learning styles

ID	Styles	Number of student(s)	%
Process	Active	23	76.67
	Reflective	7	23.33
Perception	Sensing	25	83.33
	Intuitive	5	16.67
Reception/input	Visual	24	80.00
	Verbal	6	20.00
Understanding	Sequential	19	63.33
	Global	11	36.67

Table 6. Summary of students ability

Level	Number of student(s)	%
Beginner (B)	26	86.67
Intermediate (I)	4	13.33
Advanced (A)	0	0

The system presents the learning materials recommendation for each learner. Twenty-eight students confirmed that the learning materials suited for student's learning style and level of ability. The suitability of learning materials that recommended by the system is 93.33% based on student confirmation.

5 Discussion

E-learning models that can adapt to the needs of learners can be used to improve the ability of learners. This is because each learner will get different treatment tailored to their needs. E-learning that exists today can not do that.

This study produces models that can detect the needs of learners based on learning styles and their level of ability. The learning material is presented based on the input using the rule-based engine. The results of the learning will be known based on the evaluation using CBT with IRT 1-PL method (Rasch Model). If students passes the evaluation they will get the next learning material. Otherwise it will repeat the learning process. The process of repeating it only on material that has not been mastered.

References

1. Adrianna, K.: A method for scenario recommendation in intelligent e-learning systems. Cybern. Syst.: Int. J. **42**(2), 82–99 (2011)
2. Alshammari, M., Anane, R., Hendle, R.J.: An e-learning investigation into learning style adaptivity. In: 2015 48th Hawaii International Conference on System Sciences, Kauai, HI 2015, pp. 11–20 (2015)
3. Carver, C.A., Howard, R.A., Lane, W.D.: Enhancing student learning through hypermedia courseware and incorporation of student learning styles. IEEE Trans. Educ. **42**(1), 33–38 (1999)
4. Felder, R.M., Silverman, L.K.: Learning and teaching styles in engineering. Education **78** (June), 674–681 (2002)
5. Gaikwad, T., Potey, M.A.: Personalized course retrieval using literature based method in e-learning system. In: 2013 IEEE 5th International Conference on Technology for Education (t4e 2013), Kharagpur, pp. 147–150 (2013)
6. Graf, S., Kinshuk, K.: Considering learning styles in learning management systems : investigating the behavior of students in an online course. In: 1st International Workshop on Semantic Media Adaptation and Personalization, SMAP 2006 (2006)
7. Graf, S., Kinshuk, K., Liu, T.-C.: Identifying learning styles in learning management systems by using indications from students' behaviour. In: Kinshuk, K., Liu, T.-C. (eds.) 8th IEEE International Conference on Advanced Learning Technologies, pp. 482–486 (2008)
8. Halawa, M.S., Hamed, E.M.R., Shehab, M.E.: Personalized E-learning recommendation model based on psychological type and learning style models. In: 2015 IEEE 7th International Conference on Intelligent Computing and Information Systems (ICICIS), Cairo, pp. 578–584 (2015)
9. Honey, P., Mumford, A.: The Learning Styles Helper's Guide. Peter Honey (1986)
10. Huang, C., Ji, Y., Duan, R.: A semantic web-based personalized learning service supported by on-line course resources. In: 2010 6th International Conference on Networked Computing (INC), pp. 1–7 (2010)
11. Jadhav, M., Rizwan, S., Nehete, A.: User profiling based adaptive test generation and assessment in E-learning system. In: 2013 3rd IEEE International Advance Computing Conference (IACC), pp. 1425–1430 (2013)
12. Khodke, P.A., Tingane, M.G., Bhagat, A.P., Chaudhari, S.P., Ali, M.S.: Neuro fuzzy intelligent e-Learning systems. In: 2016 Online International Conference on Green Engineering and Technologies (IC-GET), Coimbatore, pp. 1–7 (2016)
13. Kolb, D.A.: Experiential Learning : Experience as the Source of Learning and Development. Prentice-Hall, Upper Saddle River (1984)
14. Kusumawardani, S.S., Prakoso, R.S., Santosa, P.I.: Using ontology for providing content recommendation based on learning styles inside E-learning. In: 2014 2nd International Conference on Artificial Intelligence, Modelling and Simulation, Madrid, pp. 276–281 (2014)
15. Mardapi, D.: Pengukuran Penilaian & Evaluasi Pendidikan, Nuha Medika (2015)
16. Nongkhai, L.N., Kaewkiriya, T.: Framework for e-Learning recommendation based on index of learning styles model. In: 2015 7th International Conference on Information Technology and Electrical Engineering (ICITEE), Chiang Mai, pp. 587–592 (2015)
17. Rong, W.J., Min, Y.S.: The effects of learning style and flow experience on the effectiveness of E-learning. In: 5th IEEE International Conference on Advanced Learning Technologies, ICALT 2005, pp. 802–805 (2005)

18. Sevarac, Z., Devedzic, V., Jovanovic, J.: Adaptive Neuro-Fuzzy Pedagogical Recommender, Expert Systems with Applications. Elsevier, Amsterdam (2012)
19. Sudaryono: Teori Responsi Butir, Graha Ilmu (2013)
20. Takács, O.: Rule based system of adaptive e-learning. In: 2014 IEEE 12th IEEE International Conference on Emerging eLearning Technologies and Applications (ICETA), Stary Smokovec, pp. 465–470 (2014)
21. Tzouveli, P., Mylonas, P., Kollias, S.: An Intelligent E-Learning System based on Learner Profiling and Learning Resources Adaptation, Computers and Education. Elsevier, Amsterdam (2008)
22. Wang, W., Wang, Y., Gong, W.: Case-based reasoning application in e-learning. In: 2012 9th International Conference on Fuzzy Systems and Knowledge Discovery, Sichuan, pp. 930–933 (2012)
23. Wu, D., Lu, J., Zhang, G.: A fuzzy tree matching-based personalized E-learning recommender system. IEEE Trans. Fuzzy Syst. **23**(6), 2412–2426 (2015)

Text and Sentiment Analytics

Mining Textual Terms for Stock Market Prediction Analysis Using Financial News

Asyraf Safwan Ab. Rahman, Shuzlina Abdul-Rahman[(⊠)],
and Sofianita Mutalib

Faculty of Computer and Mathematical Sciences, Universiti Teknologi MARA,
40450 Shah Alam, Selangor, Malaysia
asysafwan@gmail.com, {shuzlina, sofi}@tmsk.uitm.edu.my

Abstract. This study focuses on the use of machine learning algorithms to construct a model that can predict the movements of Bursa Malaysia stock prices. In this research, we concentrate on linguistics terms from financial news that can contribute movements of the prices. Our aim is to develop a prototype that can classify sentiments towards financial news for investment decision. We experimented with five blue-chip companies from different industries of the top market constituents in Bursa Malaysia KLCI. A total of 14,992 finance articles were crawled and used as the dataset. Support Vector Machine algorithm was employed and the accuracy recorded was at 56%. The findings of this research can be used to assist investors in investment decision making.

Keywords: Bursa Malaysia · Sentiment analysis · Stock market prediction
Support Vector Machine · Text mining

1 Introduction

Stock markets are where public companies issue or trade their shares on either exchanges or over-the-counter markets. Stock prices are determined by the supply and demand of a particular stock. Understandably, it can be argued that stock markets cannot be entirely predicted. Changes in these markets are often associated with the sentiments of an investor. In predicting stock markets, there are two analysis used, which are fundamental analysis and technical analysis. Fundamental analysis focuses on studying the business, competitors, and markets. Technical analysis, on the other hand, focuses on analyzing historical prices to determine the direction of stock prices [11]. Fundamental analysis is concerned with the availability of quantitative or qualitative information to investors in driving investment decisions.

News articles may carry useful information to market participants and by using fundamental analysis, an investment decision can be made. Basically, these articles usually come verbosely and lengthy in textual representation, thus extracting patterns manually can be time-consuming. There is a correlation between fluctuating stock prices and the publication of financial news [5]. Thus, changes in any stock market can be attributed to the publication of news. However, it is difficult to predict stock prices from textual data [12].

© Springer Nature Singapore Pte Ltd. 2017
A. Mohamed et al. (Eds.): SCDS 2017, CCIS 788, pp. 293–305, 2017.
https://doi.org/10.1007/978-981-10-7242-0_25

Nevertheless, financial news are still widely used by researchers to predict stock prices despite its complexity. These news are sourced mainly from Bloomberg.com and Quamnet.com [2], Twitter [8] and the US financial news [14]. In fact, consolidated articles from several newspapers might be essential to be used as research data, for instance, the LexisNexis Database as used by [15]. Besides, Bursa Malaysia sees minimal studies carried out to stress the use of algorithms to forecast the movements of stock prices based on news. Therefore, it is necessary to study the use of algorithms to predict stock prices focussing on selected companies listed in Bursa Malaysia.

This study aims to understand the use of machine-learning algorithms in predicting stock prices on the FTSE Bursa Malaysia KLCI (FBMKLCI) using financial news. The remainder of this paper is organized as follows: The next section i.e. Sect. 2 discusses the related work of machine-learning algorithms in this area of study. Section 3 describes the research methodology for this study, following which Sect. 4 presents the analysis and results of the experiments. Finally, Sect. 5 concludes the research with the recommendations for future work.

2   Related Machine Learning Studies

Machine-learning algorithms are usually employed by researchers to discover patterns from datasets. Specifically, machine-learning can be defined as a means of letting computers 'learn' by themselves. In addition, with the advancement of extensive computing capabilities, learning through a tremendous amount of data are made possible. In other words, machine-learning is practically suitable for analysis as humans are prone to errors. Machine-learning can be classified into various types of learning such supervised learning and unsupervised learning. Supervised learning is learning through data that are labeled corresponding to its expected output. Unsupervised learning, on the other hand, is learning through data that are not labeled [6]. Classical supervised learning is explained as the construction of a classifier that can classify training objects [3]. Problems in discovering patterns can be solved by using these supervised and unsupervised learning.

In stock market prediction, machine learning is usually put to work in discovering patterns mainly in unstructured data such as text. Prediction in financial domain usually consists of machine-learning, text mining and the other computing science techniques combined [9]. Several studies listed machine-learning approaches that are employed in predicting stock prices such as Artificial Neural Network (ANN) [1], Support Vector Machines (SVM) [5] and Multiple Kernel Learning (MKL) [10]. These studies focused on predicting the direction of stock market prices driven by the publication of news articles. Previous studies can be found focusing on learning through textual data or the combination of textual and time series data. Similarly, different approaches of machine learning can be applied to construct a model that can be used to predict stock prices. Above all, financial data that exist in huge amounts require the discovery of its patterns through the applications of machine learning.

Through studying and reviewing numerous research papers and journals, there are few studies conducted by other researchers and scholars that have been identified as the related works for this project. Discovering patterns in textual data for predicting stock market prices sees numerous machine-learning approaches used for the purpose. In addition, comparisons on these approaches toward its accuracy in predicting stock market prices were studied in other work [5, 13–15].

Multiple Kernel Learning (MKL) was used by [15] with lists of kernels in addressing prediction of multiple news categories. This is in contrast with the use of SVM and KNN for a single news category. Polynomial and Gaussian kernels used in the MKL approach saw significant results compared to SVM and KNN by dividing the news into five categories.

In addition, Support Vector Regression (SVR) was employed by [14] in predicting intraday stock prices using U.S. Financial News, which resulted in a 59.0% accuracy from their measurement. In comparison to their previous study [13] on the use of Support Vector Machine, the results obtained were at 58.2%. Meanwhile, comparisons of approaches by [7] on the prediction of Chinese stock market were carried out based on announcements by employing SVM, KNN and Naïve Bayes for learning. Naïve Bayes was measured at 80% on its accuracy from 10 samples and it can be used as a significant model for this research. KNN was measured poorly from their studies and the results were not promising.

SVM with a linear kernel was tested by [5] based on financial news to predict daily stock prices but were mainly focused on measuring lists of context-capturing features. SVR was used for prediction on the returns differently to binary classes used in SVM. Comparisons on the accuracy were measured based on different features employed instead of the learning approach capability. The chosen approaches were inspired by the accuracy reported by [14] for SVR and [13] for SVM. In addition, a prediction model was constructed by [4] using SVM by training this model to German Ad-Hoc announcements with reported accuracy at 56.5%. Meanwhile, their studies in 2011 on the same datasets using SVM learning approach for intraday market prediction were not comparable.

3 Material and Method

Methods for this study are primarily outlined as several phases mentioned below. All the phases are sketched into a model architecture that serves as the basis for research deliverables. Figure 1 depicts the model architecture in this research. The first component highlights the data collection process. This system manipulates two types of data, which are the textual data that are taken from online news, and Bursa Malaysia historical prices taken from Yahoo! Finance. For the online news, crawlers are incorporated to find the relevant news and scrap the articles from the web.

The second component i.e. the textual data pre-processing involves several activities. First, data analysis emphasizes the use of simple statistical analysis to observe and understand the distribution of data. Secondly, data cleansing works by discarding unnecessary articles, for instance, redundant and mandarin articles. Thirdly, feature extraction is employed to extract features (words) by tokenizing the articles into bigram of features. In fact, during this feature extraction phase, stop words are filtered out of the text articles, all tokens are lowercased and tokens are stemmed to its root forms. Next, in feature selection, several features need to be selected by computing the

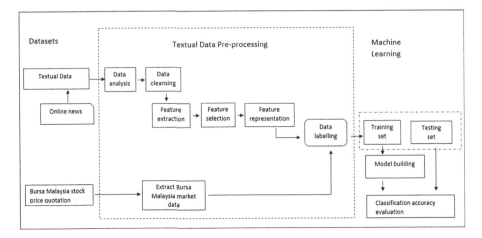

Fig. 1. Model architecture of this research.

Chi-square values for all features and select the highest values for 500 features. Then, feature representation transforms the selected features into vectors of statistical weights of Term Frequency-Inverse Document Frequency. Finally, data labelling works by calculating the rate of change based on the opening and closing prices from the historical prices according to the label of positive, for closing price is higher than opening or negative, for closing price is lower than opening. Finally, in the machine learning component, the process of modelling the prediction system with Support Vector Machine (SVM) algorithm was performed by splitting data into training and testing sets. The optimum parameters for the model were obtained with a series of experiments prior to testing the model. This system architecture is capable of providing an acceptable performance accuracy for stock prices prediction.

3.1 Taxonomy of Industries

In experimenting with predictive modelling, we proposed a taxonomy for our experiment. Companies in Bursa Malaysia FBMKLCI trade in different industries and sectors thus opens up opportunities in broadening our taxonomy. For this study, we opted to look at five blue-chip companies in four different commercial sectors for experimentation. These five companies are the Axiata Group, CIMB Group Holdings, Malayan Banking, Petronas Chemicals Group and Sime Darby.

3.2 Data Collection

In this study, news articles released in a three-year period dating from Jan 2014 to Feb 2017 were selected for experimentation. For practical reference, we chose The Edge Markets as they publish news on the Malaysia financial markets. Hence, news articles with these taxonomies mentioned were entirely collected from the publisher's website.

Crawlers were scripted to extract these news articles from the website. In fact, all articles gathered contain the necessary information such as the title, text article and time published. All articles are stored in JSON Lines format separately for each of the taxonomy. In total, the dataset amounted to 14,992 articles downloaded by using the two crawlers. In the meantime, historical prices for these taxonomies were downloaded. They are publicly accessible in Yahoo! Finance mainly to be addressed in data labelling that is to label data instances according to its classes. Meaningful attributes are extracted, in particular, the opening prices and closing prices throughout the market trading period to analyze the intraday returns for each of the taxonomy.

3.3   Data Analysis and Cleansing

In data analysis, simple analysis on data through simple observations and statistical tools is carried out to understand the distribution of the textual dataset obtained. In addition, it enables discovery of possible outliers or noises in the data. Observations of the data show that there exist redundant articles caused by corrections made by the authors, articles that are published in mandarin and unrelated articles to the experiment. In text mining, data that exist in textual format must be transformed into structured format before learning can be done. In attaining the structured format, the data was cleansed so that unnecessary information inside the data can be discarded, and the data will be functional for the research's project.

In data cleansing, we dropped redundant articles by looking at the similarity of these articles. Simple algorithm was coded to test the similarity of the current article with the next article in the file. In addition, the value of similarity was experimented to ensure that none of necessary articles are discarded out. Thus, we chose the value of 0.90 to be used in this similarity algorithm. The Mandarin articles were dismissed from the dataset as the research was aimed at constructing a prediction model for English financial news articles. Similarly, simple algorithm was coded to observe Mandarin words in the articles and the articles are dropped if the Mandarin words appear.

3.4   Data Labelling

Data labelling focuses on classifying the instances or data points in the dataset into its respective classes or labels. It is essential in supervised learning to have datasets that are labelled into its classes or to its expected output. Labelling names that were used for these data are positive and negative. Fundamentally, relevant financial information from the historical prices can be used to determine the labels. In particular the class can be labelled as 'y' by the following equation Eq. (1) and represents the simple return by extracting opening and closing prices which can be measured by Eq. (2).

$$y = \begin{cases} Positive, & r \geq 0 \\ Negative, & r < 0 \end{cases} \tag{1}$$

$$r = \frac{Close - Open}{Open} \tag{2}$$

3.5   Data Pre-processing

In the construction of predictive models, it is essential for the textual data to be pre-processed. In data pre-processing, the dataset goes through numerous text mining techniques in attaining suitable data for mining.

Feature Extraction: In feature extraction, we opted to choose bigrams by tokenizing sentences to two words per token. Table 1 elaborates the mechanisms of N-gram approaches in tokenizing textual contents.

Table 1. Mechanisms of N-Gram approaches

Feature extraction approaches	Tokenization
Unigram	{Financial, quarter}
Bigram	{Financial quarter}
Trigram	{Financial quarter reports}

Therefore, bigram is sufficient enough in permitting the actual message of article contents to be well represented. Accordingly, stop word filtering was employed to smooth out word vectors that convey less information because they are frequently used in English. For instance, familiar words like "the" and "we" are treated as stop words. In detail, NLTK Corpus is equipped with lists of stop words that can be filtered out thus ease out the process significantly.

In the same way, it is essential to lowercase all words to guarantee consistency across word vectors. Commonly, the capitalization of words attributable to the words exist as proper nouns or positioned at the beginning of any sentence. Similarly, it is important to restore word vectors to its root forms and this can be done through stemming. Word vectors exist differently in singulars and plurals yet they carry similar meanings that need stemming. In stemming, Porter's stemming algorithm is employed in pre-processing of these textual data.

Feature Selection: Specifically, feature selection stresses on the removal of attributes that carry irrelevant sentiments thus resulting in greater accuracy model. Chi-square is opted for this experiment and it is used to look at the explanatory power of the features. Thus, this would result in smaller datasets. Smaller datasets might be functional for the learning process as it would reduce the computational time throughout the development. As mentioned earlier, the feature selection technique that is chosen in this research is Chi-square. In other words, features that appear consistent in positive and negative documents can result in zero Chi-square values. Contrarily, the features that exist more in either positive or negative can carry higher values. In this study, we chose 500 features with the highest computed Chi-square values to be used in the training and testing datasets.

Feature Representation: Features are transformed and represented into vectors of Term Frequency-Inverse Document Frequency (TF-IDF). TF-IDF determines the statistical weights of features by measuring the importance of these features in the whole collection of corpus. The TF-IDF of features can be measured by Eq. (3),

$$TF - IDF = tf_{i,j} \times \log\left(\frac{N}{df_i}\right) \tag{3}$$

where, the TF-IDF weight of a feature is calculated by the number of occurrences of the feature 'i' in the document 'j' multiplied by the log of the total number of articles 'N' divided with the number of articles containing the feature 'i'.

Modelling with SVM: SVM model is built using Python Scikit-Learn. In brief, the dataset is separated into training, testing and validation accordingly. News articles published from the date January 2014 to December 2015 were used as training set. On the contrary, testing set uses news articles published from the date January 1 2016 to December 2016, and the remaining news articles were reserved for validation purposes.

Testing and Parameter Tuning of SVM: Several parameters exist in SVM algorithm and these parameters need to be wisely selected. Thus, the selection of the best parameters that gives the best predictive power requires parameter tuning. It is wise to test the model with test set that are not used in the training. Different parameters can be selected and tuned using Grid Search such as the kernels, gamma (γ) and C. Default parameters for the initial training is tabulated as in the Table 2.

Table 2. List of parameters used in parameter tuning

Kernel	C	Gamma
Linear	1, 5, 10, 100, 1000, 10000	N/A
Radial Basis Function (RBF)		$2^{-9}, 2^{-5}, ..., 2^{-1}$
Polynomial		

Classifier accuracy on a test set is measured by its ability to classify them accurately.

There are error measures in machine learning that can be used in validation activity and for this classification task, the best and commonly used error measure is classification error rate. Simply described, the accuracy score of the model can be measured as Eq. (4);

$$Accuracy = \frac{TP + TN}{TP + TN + FP + FN} \tag{4}$$

The best accuracy score from several tuned models is selected for the prototype. Further, the best trained model is encapsulated and bundled into the prototype for the prediction of recently published news.

Engine Development: In engine development, we focus on two principal stages. The first stage is the construction of a web-based prototype that can be fed and classify news recently published, and the second stage is data visualization on the textual articles. The dashboard displays the prediction probability of the experimented companies during the trading and non-trading hours. During the trading hours, the prototype is expected to predict the closing prices of the companies. In addition, the engine can perform text visualization for new articles.

4 Analysis and Testing

4.1 Data Analysis

The purpose of the data analysis is to understand the distribution of the dataset obtained from The Edge Markets. In other words, numerous measures are expected to be discovered during this data analysis such as the frequency of news publication based on days and years. Figure 2 explains the distribution of articles according to year.

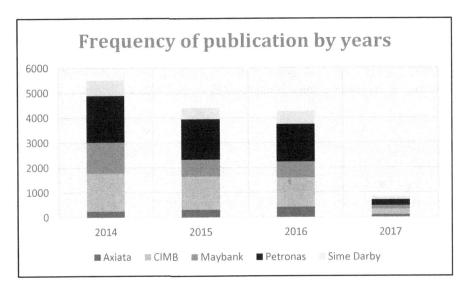

Fig. 2. Distribution of articles in years

From the chart, it can be observed that the publication of articles for these companies declined by several percent throughout the years. In addition, it is illustrated that the first two years constitute 65% of the total dataset and was chosen as the training period of this research. The frequency of distribution based on days for the whole years can be represented as in the Fig. 3.

In short, the chart above summarizes the distribution of publication based on days for these companies. By looking at the chart, observations can be focused towards the frequency of article publication throughout weekends. This small population of data can be attributed that the market closes during weekends. Thus, this research dismisses articles published on weekends as these articles might contain noises that can distort the accuracy of the prediction.

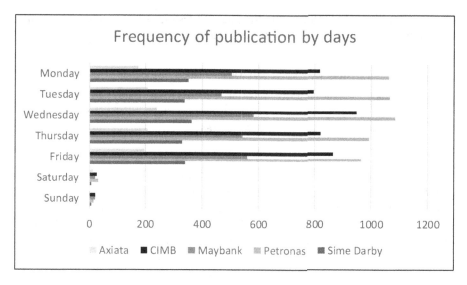

Fig. 3. Frequency of article publication by days

4.2 Data Pre-processing Analysis

The redundant articles, Mandarin articles and unnecessary articles were dismissed during data pre-processing. This enables the generation of reliable and useful data for the training of the prediction model. The results of the data cleansing is tabulated as shown in Table 3. The discarded data amounted to 3,871 articles from the categories of similar, mandarin and unnecessary articles for these five experimented companies. But, the remaining dataset comprised of 11,121 articles that can be functionally used in the research.

Table 3. Results on data cleansing

Companies	Similar articles	Mandarin articles	Unnecessary articles	Remaining articles	Total
Axiata	76	173	NA	793	1042
CIMB	481	278	116	3423	4298
Maybank	397	69	329	1906	2701
Petronas	714	801	NA	3702	5217
Sime Darby	215	222	NA	1297	1734
Total	1883	1543	445	11121	14992

4.3 Training and Testing Analysis

SVM which was employed to train and test the datasets and the evaluation was aimed at measuring the accuracy scores of the trained models. The dataset is divided into training, testing and validation in chronological order with percentages of 65%, 30%

and 5% respectively. In addition, the training and testing datasets consist of positive and negative distribution as in the following Table 4.

Table 4. Positive and negative labels distribution

Label	Axiata	CIMB	Maybank	Petronas	Sime Darby	Total
Positive	406	1911	1139	2041	701	6198
Negative	387	1512	767	1661	596	4923

By observation, the positive and negative labels are balanced in both the training and testing datasets. In fact, this enables the data to be learnt concisely/accurately. Scikit-learn SVM implementation was employed with different parameters. The costing and gamma parameters were selected and filtered from the previous research recommendations. However, the computational costs to train the data using the widest range of parameters as recommended caused us to select a subset of parameters. Grid search was carried out by searching the parameters through grid. Table 5 below reports the experimental results obtained during the SVM grid search on the parameters.

Table 5. Model scoring and parameter selection

Machine learning approaches	Model performance and parameters			
	Accuracy, %	Precision, %	Costing	Gamma
SVM, Linear	52	51	10,000	–
SVM, RBF	**56**	**52**	**10**	**0.03125**
SVM, Polynomial	55	51	1000	0.5

As can be seen in Table 5 above, the best accuracy was obtained with Costing and Gamma values of 10 and 0.03125 respectively. By comparison, the application of linear kernel resulted in a 51% accuracy out of the tuned parameters by utilizing the cost of misclassification, Costing parameter gave the value of 10,000. Similarly, the polynomial kernel recorded a considerable accuracy of 55% with Costing parameter value set at 1,000. The overall final accuracy was approximately 60% and this is consistent with the accuracies recorded by other researchers in similar research domains [13, 14].

4.4   Screenshot of Results

Figure 4 of the initial screen dashboard below displays the probability for either positive or negative of the predicted articles. Fundamentally, if the dashboard displays positive, the closing price for that company would be higher than the opening price of the company and vice versa. In addition, the positive probability of 58.8% reports the certainty of the model in predicting the article into its class.

In this text visualization, the technique chosen is wordle. Briefly described, wordle summarizes the text articles by counting the frequency of each word appearing in the

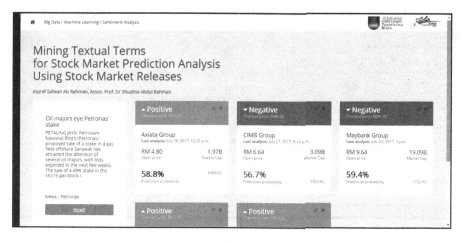

Fig. 4. Dashboard of the initial screen

article. Words that are larger the more they appear in the articles compared to the others. Thus, in Fig. 5 below, it shows that the words 'participate' and 'shareholder' appear more than the other words in that particular article. The following Fig. 6 represents the prediction gauge for the predicted article. Briefly, this section focuses on the certainty of the model in predicting the respective classes. For instance, the screen depicts that the positive model was predicted at 58% percent while the negative article gets the remaining percentage at 41%.

Fig. 5. Text-visualization screen

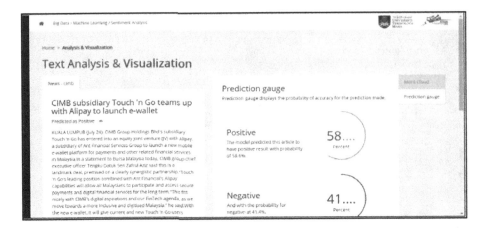

Fig. 6. Screen of the prediction gauge

5 Conclusions and Future Works

This research was designed to enable the prediction of stock prices using financial news for Bursa Malaysia listed companies. It experimented with top market constituents in FBMKLCI. This research predicts and classify the stock prices of these companies labeling them either positive or negative. In fact, Support Vector Machine (SVM) demonstrates promising prediction capability for recently published articles. Thus, this allows investors to expect positive return on investment from their investments and enable them to understand the probabilities of the movements in the stock prices. However, accuracy of the trained model was at 56% and this can be attributed to the uncertainty of the stock markets and the challenging finance domain. But, the accuracy is consistent with some of the accuracies obtained by other researches in the similar research interests.

In addition, comparisons of the several machine learning approaches should be measured to understand the best approaches that can be used for this research. In textual data pre-processing, different techniques for feature extraction, selection and representation can be experimented to observe and understand the best pre-processing solution for this research case. Further researches are needed to study other companies beyond this research. Consequently, this will allow more companies to be predicted using the prototype. In addition, in future works, this research can be further tuned by carrying out predictions based on different industries. This research can also be extended by integrating with portfolio management to monitor the real returns based on the prediction.

Acknowledgement. The authors would like to thank Faculty of Computer and Mathematical Sciences, and Research Management Centre of Universiti Teknologi MARA for supporting this research with LESTARI 111/2017 grant.

References

1. Bucur, L., Florea, A.: Techniques for prediction in chaos–a comparative study on financial data. UPB Sci. Bull. Ser. C **73**(3), 17–32 (2011)
2. Chan, S.W., Franklin, J.: A text-based decision support system for financial sequence prediction. Decis. Support Syst. **52**(1), 189–198 (2011)
3. Dietterich, T.: Machine learning for sequential data: a review. Struct. Syntact. Stat. Pattern Recogn. 227–246 (2002)
4. Heidari, M., Felden, C.: Financial footnote analysis: developing a text mining approach. In: Proceedings of International Conference on Data Mining (DMIN), p. 10. The Steering Committee of the World Congress in Computer Science, Computer Engineering and Applied Computing (WorldComp), January 2015
5. Hagenau, M., Liebmann, M., Neumann, D.: Automated news reading: stock price prediction based on financial news using context-capturing features. Decis. Support Syst. **55**(3), 685–697 (2013)
6. Jain, A.K., Murty, M.N., Flynn, P.J.: Data clustering: a review. ACM computing surveys (CSUR) **31**(3), 264–323 (1999)
7. Jiang, Z., Chen, P., Pan, X.: Announcement based stock prediction. In: 2016 International Symposium on Computer, Consumer and Control (IS3C), pp. 428–431. IEEE, July 2016
8. Koppel, M., Shtrimberg, I.: Good news or bad news? Let the market decide. In: Shanahan, J. G., Qu, Y., Wiebe, J. (eds.) Computing Attitude and Affect in Text: Theory and Applications, pp. 297–301. Springer, Dordrecht (2006). https://doi.org/10.1007/1-4020-4102-0_22
9. Kumar, B.S., Ravi, V.: A survey of the applications of text mining in financial domain. Knowl.-Based Syst. **114**, 128–147 (2016). https://doi.org/10.1016/j.knosys.2016.10.003
10. Luss, R., d'Aspremont, A.: Predicting abnormal returns from news using text classification. Quant. Financ. **15**(6), 999–1012 (2015)
11. Murphy, J.J.: Technical analysis of the financial markets: a comprehensive guide to trading methods and applications. Penguin (1999)
12. Schumaker, R., Chen, H.: Textual analysis of stock market prediction using financial news articles. In: AMCIS 2006 Proceedings, vol. 185 (2006)
13. Schumaker, R.P., Chen, H.: A quantitative stock prediction system based on financial news. Inf. Process. Manag. **45**(5), 571–583 (2009)
14. Schumaker, R.P., Zhang, Y., Huang, C.N., Chen, H.: Evaluating sentiment in financial news articles. Decis. Support Syst. **53**(3), 458–464 (2012)
15. Shynkevich, Y., McGinnity, T.M., Coleman, S.A., Belatreche, A.: Forecasting movements of health-care stock prices based on different categories of news articles using multiple kernel learning. Decis. Support Syst. **85**, 74–83 (2016)

Content Quality of Latent Dirichlet Allocation Summaries Constituted Using Unique Significant Words

Muthukkaruppan Annamalai[1,2(✉)] and Amrah Mohd Narawi[1]

[1] Faculty of Computer and Mathematical Sciences,
Universiti Teknologi MARA (UiTM), 40450 Shah Alam, Selangor, Malaysia
mk@tmsk.uitm.edu.my, amrahnarawi@gmail.com
[2] Knowledge and Software Engineering Research Group,
Advanced Computing and Communication Communities of Research,
Universiti Teknologi MARA (UiTM), 40450 Shah Alam, Selangor, Malaysia

Abstract. The accessibility to the Big Data platform today has raised hopes to analyse the large volume of online documents and to make sense of them quickly, which has provided further impetus for automated unsupervised text summarisation. In this regards, Latent Dirichlet Allocation (LDA) is a popular topic model based text summarisation method. However, the generated LDA topic word summaries contain many redundant words, i.e., duplicates and morphological variants. We hypothesise that duplicate words do not improve the content quality of summary, but for good reasons, the morphological variants do. The work sets out to investigate this hypothesis. Consequently, a unique LDA summary of significant topic words is constituted from the LDA summary by removing the duplicate words, but retaining the distinctive morphological variants. The divergence probability of the unique LDA summary is compared against the LDA baseline summary of the same size. Short summaries of 0.67% and 2.0% of the full text size of the input documents are evaluated. Our findings show that the content quality of unique LDA summary is no better than its corresponding LDA baseline summary. However, if the duplicate words are removed from the baseline summary, producing a compressed version of itself with unique words, i.e., a unique LDA baseline summary; and, if the compression ratio is taken into consideration, it will appear that the content quality of a LDA summary constituted using unique significant words have indeed improved.

Keywords: Latent Dirichlet Allocation · Text summarisation · Unique keywords · Content quality

1 Introduction

Internet has given rise to the explosive growth of online documents. The accessibility to the Big Data platform today has raised hopes to analyse these large volume of documents and to make sense of them quickly. This recent development has provided

© Springer Nature Singapore Pte Ltd. 2017
A. Mohamed et al. (Eds.): SCDS 2017, CCIS 788, pp. 306–316, 2017.
https://doi.org/10.1007/978-981-10-7242-0_26

further impetus for text summarisation. Automated text summarisation aims to produce a concise representative substitute of the analysed documents.

There are two summarisation techniques, extractive and abstractive to capture the essence of text content [9]. Extractive summarisation attempts to select representative words or sentences from the text content and arranges them in some order to form a summary, while abstractive summarisation attempts to 'comprehend' the text content and outlines them in fewer words through generalisation of what is important in them. This research focuses on the amenable extractive summarisation technique.

Latent Dirichlet Allocation (LDA) is a popular topic model based extractive summarisation method. This generative probabilistic model assumes that document(s) can be represented as a mixture of various topics where a topic is viewed as a distribution of words [1]. LDA extracts co-occurring words in documents to form topics covered by these documents according to the multinomial distribution of these words within topics and the Dirichlet distribution of the words over topics [2]. These generated LDA topical words constitute a LDA summary, which can be regarded as the bones a text summary.

A conspicuous drawback of LDA summary is redundancies. Even though LDA is said to select information that has less repetition, but because LDA assigns higher probability for words that are repeated throughout documents [2], the generated topic summaries contain many redundant words, i.e., duplicates and morphological variants [3]. The gravity of this matter is apparent in short summaries constituted from significant (weighty) topic words of the LDA summaries. For example, Table 1 shows the first four numbered topics and their corresponding set of words for a collection of documents on *Durian* fruit. LDA assigns weights to each topic (*tw*). The set of words that constitute each topic is shown on the right hand side column. LDA assigns weights to each topic word (*ww*). The topics and their words are sorted based on the LDA assigned weights in Table 1. It can be seen that there are redundant weighty topic words like *fruit, durian, ripe* and *tree*, within and across topics in the summary.

In general, duplicates do not add value to content, but more likely downgrades it. At first glance, morphological variants also appear as redundant words. However, a morphological variant may represent different sense or the same sense that differs in tense or plurality. In any case, it usefully conveys different effects in the text. Take for example the morphological variants: *plants, planting, planter* and *plantation*. First, it is unclear which of these words best describe a particular sense; it may be one of the variants or their root word (*plant*) that may not even be present in the summary. Second, if we ignore the variants, we might be losing the essence of the summary, i.e., the notion of the process of planting, the role of planter, and so on. Third, the word *plant* exhibits part-of-speech ambiguity, i.e., as to whether it refers to a noun or a verb. Forth, words like *plant* have also multiple senses, as in industrial plant or flora. Therefore, we argue the need to retain the morphologically related significant words in the summary, as long as they are unique, thus avoiding the loss of pertinent information. We hypothesise that it suffices to remove the duplicate words from the summary, but not the different morphological variants of the significant words; the resulting summary constituted using unique words will signify a higher quality LDA summary. The purpose of the research is to test the hypothesis.

Table 1. The first four topics in a LDA summary about *Durian* fruit

Topic	Topic Weight *(tw)*	Topic Word (Word Weight *(ww)*)
0	0.15765	fruits(51.17) fruit(49.17) durian(20.17) maturity(17.17) storage (15.17) days(15.17) mature(15.17) harvested(14.17) cultivars (13.17) harvesting(11.17) anthesis(9.17) life(9.17) abscission(8.17) market(7.17) ripe(7.17) stage(7.17) good(7.17) time(7.17) dehiscence(6.17)
1	0.13409	durian(167.17) fruit(55.17) tree(47.17) durians(34.17) fruits(31.17) trees(31.17) season(26.17) years(15.17) articles(15.17) good(12.17) smell(12.17) year(11.17) strong(10.17) part(10.17) news(10.17) consumers(10.17) production(10.17) photo(9.17) ripen(9.17)
2	0.10159	durian(28.17) mg(27.17) fruit(27.17) vitamin(15.17) flesh(15.17) nutrition(13.17) seeds(13.17) malaysia(10.17) fruits(9.17) cm(8.17) body(6.17) popular(6.17) durio(6.17) cooked(5.17) source(5.17) back(5.17) health(5.17) variety(5.17) hotels(5.17)
3	0.08468	chang(15.17) durians(13.17) history(11.17) penang(10.17) varieties (9.17) taste(9.17) red(7.17) bak(7.17) ang(7.17) pulau(7.17) rm (7.17) found(7.17) sheng(6.17) bao(6.17) experience(6.17) portal (6.17) creamy(6.17) wet(5.17) www(5.17)

2 Tools and Techniques

In this section, we discuss the tools and techniques behind creating unique LDA summaries and the means to measure the content quality of the resulting summaries.

2.1 Latent Dirichlet Allocation

LDA is a generative unsupervised probabilistic model is applied in topic modelling [1]. In LDA, the data is in the form of a collection of documents, where each document is considered as a collection of words. LDA attempts to provide a brief description of the words in the documents by pre-serving the essential statistical relationships between the words. For this, LDA assumes that each document is represented as a mixture of latent topics, and each topic is represented as a mixture over words. These mixture distributions are assumed to be Dirichlet-distributed random variables, which are inferred from the data itself. Blei et al. outlines the generative process as follows:

   I. For each topic, sample a distribution over words from a Dirichlet prior.
  II. For each document, sample a distribution over topics from a Dirichlet prior.
 III. For each word in the document
     – Sample a topic from the document's topic distribution.
     – Sample a word from the topic's word distribution.
     – Observe the word.

2.2   Conflation

Conflation is a process of normalising word in order to group the morphological variants in content [4]. Two types of conflation methods are used in this work: Stemming and Lemmatisation.

2.2.1   Stemming

Stemming is the process of reducing plural and singular forms and inflective variants of verbs or morphological variant words to their base or root form. Among the popular stemmers for English words are Lovins, Dawson, Porter and Paice/Husk. Of these, Paice/Husk is the strongest stemmer whereas Porter is the lightest [5].

2.2.2   Lemmatisation

Lemmatisation is closely related to stemming. While stemming operates on a single word without knowledge of the context, lemmatisation determines the root form by applying complex linguistic process such as context and part-of-speech of the word in a sentence. The Stanford Lemmatiser [14] is a popular online tool for determining the lemma of an English word. The lemmatiser is helpful when dealing with irregular nouns and verbs.

A set of 343 ambiguous English words and their morphological variants were gathered from the Online Merriam-Webster English Dictionary for preliminary analysis. We first applied Stanford Lemmatiser on the input, which outputs noun and/or verb root words, referred to as their lemmas. The lemmatiser helped to stem the irregular verbs. Then, the lemmas were independently run through each of the stemmers mentioned in the previous section. The stemmers helped to conflate the noun and verb lemmas.

The preliminary analysis indicates that the best conflation method is the combination of the Stanford Lemmatiser and the Paice/Husk Stemmer, which was consistent with our expectation.

2.3   Summary Evaluation

When evaluating summaries, there are two measures that are often considered: Retention and Compression ratios [10].

2.3.1   Retention Ratio

Retention ratio (RR) determines how much of the information from the full text of the documents is retained in the summary text as described by Eq. (1).

$$Retention\ ratio,\ RR = \frac{information\ in\ summary}{information\ in\ full\ text} \tag{1}$$

The retention ratio or omission ratio is hard to determine because measuring information that is relevant from a text is not a straight-forward task.

Steinberger and Jezek classify the summarisation evaluation methods into two classes: extrinsic and intrinsic [11]. In the extrinsic evaluation, the quality of summary is adjudged subjectively by human experts against a set of external criteria to determine whether the summary retains sufficient information to accomplish certain task that relies upon it. Document categorisation and Question game are examples of extrinsic evaluation. On the other hand, the intrinsic evaluation is done by analysing the textual structure of the summary, and can be viewed as a more objective indication of information retention [12]. The Divergence Probability measure is a good example of intrinsic evaluation [13], which has been adopted in this work.

Kullback Leibler Divergence Probability (KLDP)
KLDP characterises the differences between the summary text and the full text of the documents by gauging the information lost when summary is used to approximate the text [7]. In a way, it assesses the retention of relevant text information in the summary by measuring the theoretical divergence between the probability distributions of the words in the full text and the summary text. Equation (2) defines the KLDP measure K between two probability distributions R (full text) and S (summary text).

$$K(R \parallel S) = \sum_i ln\left(\frac{R_i}{S_i}\right) R_i \qquad (2)$$

However KLDP is a non-symmetric measure of the difference between two distributions, i.e., the KLDP from R to S is generally not the same as that from S to R.

Jensen-Shannon Divergence Probability (JSDP)
JSDP extends KLDP by incorporating the idea that the distance between two distributions shall not be very different from the average of distances from their mean distribution. Equation (3) defines the JSDP measure J, where P and Q are two probability distribution; K refers to KLD measure between two probability distributions and A is the mean distribution of P and Q (i.e., $A = [P + Q]/2$).

$$J(P \parallel Q) = \frac{1}{2}[K(P \parallel A) + K(Q \parallel A)] \qquad (3)$$

In this research, we applied JSDP to measure the content quality of the resulting summaries. We used the SIMetrix [13] to calculate the JSDP divergence values. A lower divergence value signifies a higher quality summary, and vice versa.

2.3.2 Compression Ratio

Compression is an important factor in deciding the size of the compressed text as a function of the full text size [6]. The compression ratio (CR) of summarisation refers to ratio of the length of the summary text over the length of the full text of the documents. It is represented by Eq. (4).

$$Compression\ Ratio,\ CR = \frac{length\ of\ summary}{length\ of\ full\ text} \qquad (4)$$

The shorter a summary, the higher is its CR. However, a high CR implies less information is retained, which suggests that some relevant text information may have been disregarded.

3   The Experiment

The purpose of the experiment is to test the hypothesis that content quality of Unique LDA summaries is better than their corresponding LDA Baseline summaries. A short summary is generally not more than two percent (2%) of large text content. Consequently, short summaries of sizes 0.67% and 2% of the full text size of the input documents (compression ratio) are considered in this work; the borderline version serves to validate the shorter version.

3.1   Sample Documents

The sample documents extracted from forty-two (42) websites on crops are for the main purpose which describing events during the agriculture life cycle: Durian (14 websites), Banana (6), Corn (6), Okra (5), Pineapple (3), Oil Palm (4), and Rambutan (4).

3.2   Experiment Design

The experiment design is shown in Fig. 1. The experiment goes into five phases, which are:

- Input phase
- Processing phase
- Output phase
- Post-processing phase
- Final Output phase.

The experiments are conducted on three datasets of varying text sizes: Small (14 documents; approx. 15,000 words or about fifty (50) A4 pages of printed standard text), Medium (20 documents; approx. 30,000 words or about one hundred (100) printed A4 pages) and Large (42 documents; approx. 60,000 words or about two hundred (200) printed A4 pages).

In order to produce the LDA summaries, we applied the content of the text documents to Mallet [8], a toolkit that implements LDA. The output consists of weighted topics (*tw*), and the weighted topic words (*ww*) (see Table 1).

Next, we took the entire LDA generated topic words in the Output phase and compute the significant weight of each topic word (*tw* * *ww*) to form a weighted LDA summary. The topic words are ranked, ranging from the most significant to the least significant word. The ranked weighted LDA summary is the source for the LDA Baseline Summary (see Fig. 1).

The post-processing on the weighted LDA summary was carried by first conflating the words using Stanford Lemmatiser and the Paice/Husk Stemmer. Second, the conflated words with same stems are grouped and their significant weights are combined. As a result, the grouped morphological variants and the grouped duplicate words will be assigned new combined weights of significance. Third, the duplicate words are removed, which results in a summary of weighted unique topic words. Finally, the weighted unique words are ranked based on their weights, which is the source for the Unique LDA Summary (as shown in Fig. 1).

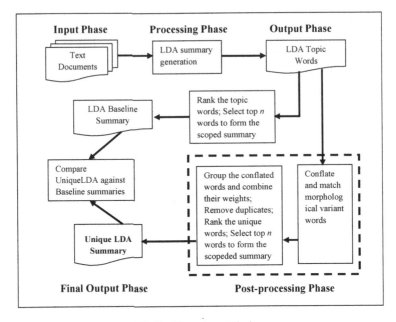

Fig. 1. Experiment design

While the weighty words distinguish the significant words in the ranked LDA summaries, there is a need to restrict the number of words in a scoped summary. We select the top n words according to the predetermined scope (compression ratio) from the ranked summaries. For example, the number of words in the Small dataset (text size) is approximately 15000 words. The size of the summary for 0.67% CR is 100. Therefore we scope our summary to the top 100 unique words in the ranked word list, and so, n for Small dataset is 100. Likewise, for the Medium and Large text sizes, we scope their summaries to 200 and 400 unique weighty words for 0.67% CR, respectively.

Note that a typical LDA Baseline Summary contains many weighty duplicate words, whereas there is no duplicate word in a Unique LDA Summary. The spaces of the removed duplicate words are orderly replaced by morphological variants of the significant words and new weighty topic words in the Unique LDA Summary. These additions are not found in the LDA Baseline Summary.

We compared the final output of the Post-processing phase, i.e. Unique LDA Summary against the final output of the Output phase, i.e. Baseline LDA Summary to find which of these two summaries is of higher content quality. The summaries are compared based on their JSDP values.

4   Experiment Results

We carried out the experiment using three (3) datasets of varying text sizes: Small, Medium and Large, and also considered the scope of the summaries with compression ratios (CR) of 0.67% and 2%. So, in total, we performed six (6) experiments. The result of the Unique LDA Summary (ULDA) is compared against LDA Baseline Summary (LDAB). The results of the short summaries of size 0.67% and 2% are shown in Tables 2 and 3, respectively.

Table 2. Results of experiment for short summaries up to 0.67% CR

Case	Text Size	No. of Docs.	No. of Words in Docs.	No. of Words in Summary	CR %	JSDP	JSDP * CR	Diff (%)
LDAB	Small	14	15,049	100	0.66	0.3716	0.2453	
ULDA						0.3878	0.2559	+4.32
ULDAB				82	0.54	0.4015	0.2168	-11.62
LDAB	Medium	20	30,009	200	0.67	0.3467	0.2323	
ULDA						0.3493	0.2340	+0.73
ULDAB				146	0.49	0.3635	0.1781	-23.33
LDAB	Large	42	60,035	400	0.67	0.3169	0.2123	
ULDA						0.3180	0.2131	+0.38
ULDAB				306	0.51	0.3343	0.1705	-19.69

Table 3. Results of experiment for short summaries up to 2% CR

Case	Text Size	No. of Docs.	No. of Words in Docs.	No. of Words in Summary	CR %	JSDP	JSDP * CR	Diff (%)
LDAB	Small	14	15,049	300	1.99	0.3083	0.6135	
ULDA						0.3257	0.6481	+5.64
ULDAB				274	1.82	0.3312	0.6028	-1.74
LDAB	Medium	20	30,009	600	2.00	0.2709	0.5418	
ULDA						0.2785	0.5570	+2.81
ULDAB				461	1.54	0.3019	0.4649	-14.19
LDAB	Large	42	60,035	1200	2.00	0.2473	0.4946	
ULDA						0.2658	0.5316	+7.48
ULDAB				938	1.56	0.2714	0.4234	-14.40

We also present the results of an extended work in these tables, in which we carried out additional three (3) experiments for each summary size. In here, we considered the removal of the added morphological variants of the significant words and new weighty topic words that are not present in the LDAB from ULDB, which produced the Unique LDA Baseline Summary (ULDAB) for each dataset. As a result of the removal, the CR of ULDAB is comparatively smaller than the pre-determined CR. For that reason, the content quality of the summaries is measured by the value of the product of the divergence value (JSDP) and their corresponding CR. A lower JSDP value signifies a higher quality summary. Accordingly, a lower (JSDP * CR) value signifies a higher quality summary.

For the Small text size, the JSDP*CR value for the Unique LDA Summary (ULDA) is larger than the LDA Baseline Summary (LDAB) at 0.2559 compared to 0.2453 (+4.32%), i.e., an increase of more than 4%. It implies that new weighty words and the morphological variants of the significant words used to fill up the space of the removed duplicate words from the LDAB summary does not improve the content quality of the original summary. See Table 2.

The results of the Medium and Large text sizes are consistent with that of the Small text size for the case of 0.67% CR short summaries where the values of JSDP*CR for ULDB are marginally higher than that of LDAB, i.e., 0.2340 compared to 0.2323 (+0.73%) for Medium text size, and 0.2131 compared to 0.2123 (+0.38%) for Large text size. Even though the trend of the differences between ULDA and LDAB is decreasing as the text size increases, the results show that maintaining only unique words in the LDA summaries cannot effectively improve the content quality of the short summaries with 0.67% CR for text size up to 60000 words.

To validate the findings, we carried out the experiments on borderline short summaries of 2% CR. Once again, the results show that for all data text sizes, i.e., Small, Medium and Large, the JSDP*CR values for ULDB are consistently larger than that of LDAB (see Table 3).

In the case of Small text size, the JSDP*CR value for the ULDB is 0.6481 compared to 0.6135 (+5.64%). For the Medium text size, the JSDP * CR value for ULDB is 0.5570 compared to 0.5418 (+2.8%), and for Large text size, the JSPD*CR value of ULDB is 0.5316 compared to 0.4946 (+7.48%). The results show that there is no improvement in the content quality of short summaries with unique words compared to their corresponding LDAB summaries. Noticeably, there appear to be no trend in the differences between ULDA and LDAB in the case of 2% CR short summaries.

The findings show that the quality of ULDB is weaker than that of LDAB for texts of comparable size and scope. It implies that the addition of new words to fill up the space of the removed duplicate words is not really needed. There are two types of inclusions: new weighty words and the morphological variants of the significant words. In order to further validate the postulation, we performed an extended analysis by removing these additions from ULDA, which in effect is the LDA Baseline Summary without redundant words; we call, Unique LDA Baseline Summary (ULDAB). The results of the experimental analysis are tabulated on the row labelled ULDAB in Tables 2 and 3. Table 2 describes the results of the short summaries whose CR ranges between 0.49% and 0.67% where the CR for the ULDAB summaries has now reduced to around 0.5%; while, Table 3 describes the results of short summaries whose CR ranges between 1.54% and 2% where the CR for the ULDAB summaries has decreased to as low as 1.54%.

In the case of (0.49–0.67%) CR short summaries, the JSDB*CR value of Small text size for ULDAB is smaller than LDAB, i.e., 0.2168 compared to 0.2453 (-11.62%). The result shows that removing the duplicates and maintaining only the unique words in the LDA summary can improved the content quality of the short summaries in the case of Small text size when the CR is factored into the quality measure. The results of the Medium and Large text sizes are consistent with that of the Small text size where the value of JSDP*CR for ULDAB is much smaller, i.e., 0.1781 compared to 0.2340 (-23.33%), and 0.1705 compared to 0.2123 (-19.69%), respectively.

The results for the borderline (1.54–2%) CR short summaries also find the same where the JSDP*CR values for ULDAB are smaller compared to that of LDAB for all three sample text sizes. In the case of Small text size, the JSDP*CR value for the ULDAB is 0.6028 compared to 0.6135 (-1.75%). For the Medium and Large text sizes, the values of JSDP*CR for ULDAB are 0.4649 compared to 0.5418 (-14.19%) and 0.4234 compared to 0.4946 (-14.40%), respectively. The results reinforce the earlier findings that show that if the compression ratios are taken into consideration when evaluating the resulting summaries, then we may say that the removal of duplicates and the morphological variants of the significant words in the LDA summary can improve the content quality of short summaries.

5 Conclusion

The research attempts to address the weakness of LDA with regards to the appearance of redundant topical words in its summary. Since duplicate words do not add values to a content, removing them will create space for other significant words in the summary. However, for reasons given in the introductory section, we retain the morphological variants of the significant words in the summary. Stanford Lemmatiser and Paice/Husk Stemmer are used to conflate and group the similar words.

Consequently, we tested our hypothesis that a short LDA summary of unique words (ULDA) signifies a higher quality summary compared to its corresponding LDA baseline summary (LDAB). The experiment involves datasets of Small, Medium and Large text sizes. Short summaries of size 0.67% and 2% CR were analysed. The product of the divergence value and their corresponding compression ratios (JSDP*CR) was used as the evaluation measure.

The result revealed that the quality of LDA Baseline Summary (LDAB) is better than ULDA when the size of the summaries are fixed, where new weighty words and the morphological variants of the significant words are used to fill up the space of the removed duplicate words. It means the additional effort to maintain only the unique words in the LDA summary does not pay off.

However, if we remove all the duplicate words from a LDA Baseline Summary resulting in a compressed LDA summary of unique words (ULDAB), and if the reduced CR is taken into consideration, then the content quality of the ULDAB is found to be better than the corresponding LDAB summary. Hence, our findings imply that the addition of new words to fill up the space of the removed duplicate words is not required after all.

Future work can validate the findings of this research. First, sampling does play an important part in experimenting. For this larger samples across different subject matter may be considered. Second, the sample documents used are of varying sizes (between

208 and 5493 words each). This may have affected the LDA output. Future research may consider using documents of about the same size, which might produce a more pronounced result. Third, alternative measures of content quality may be used to validate our findings. Finally, future work can also consider medium size summaries of 5% CR.

Acknowledgement. The authors thankfully acknowledge Universiti Teknologi MARA (UiTM) for support of this work, which was funded under the Malaysian Ministry of Higher Education's Fundamental Research Grant Scheme (ref. no FRGS/1/2016/ICT01/UITM/02/3).

References

1. Blei, D., Ng, A., Jordan, M.: Latent Dirichlet allocation. J. Mach. Learn. Res. **3**, 993–1022 (2003)
2. Lee, S., Song, J., Kim, Y.: An empirical comparison of four text mining methods. J. Comput. Inf. **51**(1), 1–10 (2010)
3. Annamalai, M., Mukhlis, S.F.N.: Content quality of clustered latent Dirichlet allocation short summaries. In: Jaafar, A., Mohamad Ali, N., Mohd Noah, S.A., Smeaton, A.F., Bruza, P., Bakar, Z.A., Jamil, N., Sembok, T.M.T. (eds.) AIRS 2014. LNCS, vol. 8870, pp. 494–504. Springer, Cham (2014). https://doi.org/10.1007/978-3-319-12844-3_42
4. Dopichaj, P., Harder, T.: Conflation method and spelling mistakes – a sensitivity analysis in information retrieval. Citeseer (2004)
5. Sirsat, S.R., Chavan, V., Mahalle, H.S.: Strength and accuracy analysis of affix removal stemming algorithms. Int. J. Comput. Sci. Inf. Technol.gies **4**(2), 265–269 (2013)
6. Hobson, S.F.: Text summarization evaluation: correlation human performance on an extrinsic task with automatic intrinsic metrics. Dissertation, University of Maryland, College Park (2007)
7. Kullback, S.: The Kullback-Leibler distance. Am. Stat. **41**(4), 340–341 (1987)
8. McCallum A.K.: MALLET: a machine learning for language toolkit (2002). http://mallet.cs.umass.edu
9. Saranyamol, C., Sindhu, L.: A survey on automatic text summarisation. Int. J. Comput. Sci. Inf. Technol. **5**, 7889–7893 (2014)
10. Hassel, M.: Evaluation of automatic text summarization: a practical implementation. Licentiate Thesis, University of Stockholm, Sweden (2004)
11. Steinberger, J., Jezek, K.: Evaluation measures for text summarization. Comput. Inform. **28**, 1001–1026 (2007)
12. Lloret, E., Palomar, M.: Text summarisation in progress: a literature review. Artif. Intell. Rev. **37**, 1–41 (2011). Springer Science+Business Media B.V.
13. Louis, A., Nenkova, A.: Automatically evaluating content selection in summarization without human models. In: Conference on Empirical Methods in Natural Language Processing, Singapore, pp. 306–314 (2009)
14. Manning, C.D., Surdeanu, M., Bauer, J.F., Jenny, B., Steven, J., McClosky, D.: The Stanford CoreNLP natural language processing toolkit. In: 52nd Annual Meeting of the Association for Computational Linguistics: System Demonstrations, pp. 55–60 (2014)

Author Index

Printed in the United States
By Bookmasters